Lecture Notes in Mathematics

Volume 2282

This series reports on new developments in all areas of mathematics and their applications - quickly, informally and at a high level. Mathematical texts analysing new developments in modelling and numerical simulation are welcome. The type of material considered for publication includes:

1. Research monographs
2. Lectures on a new field or presentations of a new angle in a classical field
3. Summer schools and intensive courses on topics of current research.

Texts which are out of print but still in demand may also be considered if they fall within these categories. The timeliness of a manuscript is sometimes more important than its form, which may be preliminary or tentative.

Titles from this series are indexed by Scopus, Web of Science, Mathematical Reviews, and zbMATH.

More information about this series at http://www.springer.com/series/304

Pavle V. M. Blagojević • Frederick R. Cohen •
Michael C. Crabb • Wolfgang Lück •
Günter M. Ziegler

Equivariant Cohomology of Configuration Spaces Mod 2

The State of the Art

Pavle V. M. Blagojević
Institute of Mathematics, Freie Universität Berlin
Berlin, Germany

Mathematical Institute of Serbian Academy of Sciences and Arts
Belgrade, Serbia

Frederick R. Cohen
Department of Mathematics
University of Rochester
Rochester, NY, USA

Michael C. Crabb
Institute of Mathematics
University of Aberdeen
Aberdeen, UK

Wolfgang Lück
Mathematisches Institut
Universität Bonn
Bonn, Germany

Günter M. Ziegler
Institut für Mathematik
Freie Universität Berlin
Berlin, Germany

ISSN 0075-8434 ISSN 1617-9692 (electronic)
Lecture Notes in Mathematics
ISBN 978-3-030-84137-9 ISBN 978-3-030-84138-6 (eBook)
https://doi.org/10.1007/978-3-030-84138-6

Mathematics Subject Classification: 55R80, 55N25, 57R22, 57R42

This Springer imprint is published by the registered company Springer Nature Switzerland AG.
The registered company address is: Gewerbestrasse 11, 6330 Cham, Switzerland

This book is dedicated to Aleksandra, Helen, Sarah, Torsten, and Vera

Preface

The systematic study of the ordered configuration space

$$\mathrm{F}(M,n) := \{(x_1,\ldots,x_n) \in M^n : x_i \neq x_j \text{ for all } 1 \le i < j \le n\}$$

of all ordered n-tuples of distinct points on a manifold M started in 1962 with the work of Fadell and Neuwirth [47] and Fox and Neuwirth [52], with prehistory going back to the work of Artin [7–9]. Soon after, Arnold, in his seminal work [5] from 1969, gave a description of the integral cohomology ring of the ordered configuration space $\mathrm{F}(\mathbb{R}^2,n)$. From that point on, the topology of the ordered configuration spaces was studied very intensively from many aspects, while finding applications in diverse problems, theories, and even different fields of mathematics and beyond, notably in physics.

Each configuration space $\mathrm{F}(M,n)$ is equipped with a natural free action of the symmetric group on n letters $\mathfrak{S}_n$, given by the permutation of points. The associated orbit space $\mathrm{F}(M,n)/\mathfrak{S}_n$, called the unordered configuration space, is an important and challenging object to study. (The free action of the symmetric group is also an essential ingredient of the little cubes operad structure to be discussed later.)

In his influential 1970 paper [53] using fundamental new ideas, Fuks, gave a description of the cohomology algebra of the unordered configuration space $H^*(\mathrm{F}(\mathbb{R}^2,n)/\mathfrak{S}_n;\mathbb{F}_2)$ as an image of the cohomology $H^*(\mathrm{BO}(n);\mathbb{F}_2)$. In the course of study of infinite and iterated loop spaces, objects of the same homotopy type as the configuration space $\mathrm{F}(\mathbb{R}^d,n)$ were invented by Boardman and Vogt [17] and adapted in a beautiful way by May [76, Sec. 4] for the definition of an important structure that we now call the little cubes operad; see Chap. 7. Frederick Cohen, in his 1976 contribution [33], gave the first descriptions of the cohomology of the unordered configuration space $\mathrm{F}(\mathbb{R}^d,n)/\mathfrak{S}_n$, for n a prime, with trivial coefficients (including the ring structure) and with twisted coefficients (including the relevant module structure) [33, Thm. 5.2 and Thm. 5.3].

The homology of the unordered configuration space for points on a smooth manifold M has been determined in 1989 by Bödigheimer et al. [20] in the case when M is odd-dimensional and coefficients are in an arbitrary field, and in the case when M is even-dimensional and coefficients are in a field of characteristic 2. More precisely, for even-dimensional manifolds, they computed the homology of the unordered configuration space of M with coefficients in the field twisted by the sign representation. These results were given in terms of Cohen's computation for the case $M = \mathbb{R}^n$. Some further results, for an even-dimensional orientable closed manifold and the rationals as the field of coefficients, were obtained by Félix and Thomas [50].

Nguyên Hữu Viêt Hưng, in a series of papers [61–64] from 1981 until 1990, studied the mod 2 cohomology algebra of the symmetric group $\mathfrak{S}_n$ and of the unordered configuration space $\mathrm{F}(\mathbb{R}^d, n)/\mathfrak{S}_n$ in the case when n is a power of two. The key paper in this series, [64], which contained detailed proofs for all results announced in [62], was apparently finished in August of 1982, but after some delays (described in [64, Footnote 1 on p. 286]) was published only in 1990. The central idea was

- To consider a natural embedding

$$\mathrm{Pe}(\mathbb{R}^d, 2^m) \xrightarrow{\mathrm{ecy}_{d,2^m}} \mathrm{F}(\mathbb{R}^d, 2^m)$$

  of the product of spheres $\mathrm{Pe}(\mathbb{R}^d, 2^m) = (S^{d-1})^{2^m-1}$ into the configuration space $\mathrm{F}(\mathbb{R}^d, 2^m)$, which turns out to be equivariant with respect to the action of the Sylow 2-subgroup $\mathcal{S}_{2^m}$ of the symmetric group $\mathfrak{S}_{2^m}$.
- To describe the cohomology ring $H^*(\mathrm{Pe}(\mathbb{R}^d, 2^m)/\mathcal{S}_{2^m}; \mathbb{F}_2)$ of the quotient space $\mathrm{Pe}(\mathbb{R}^d, 2^m)/\mathcal{S}_{2^m}$ using the homeomorphism

$$\mathrm{Pe}(\mathbb{R}^d, 2^{m+1})/\mathcal{S}_{2^{m+1}} \cong \big(\mathrm{Pe}(\mathbb{R}^d, 2^m)/\mathcal{S}_{2^m} \times \mathrm{Pe}(\mathbb{R}^d, 2^m)/\mathcal{S}_{2^m}\big) \times_{\mathbb{Z}_2} S^{d-1}$$

  via an inductive computation.
- To prove that the induced homomorphism in cohomology

$$H^*(\mathrm{F}(\mathbb{R}^d, 2^m)/\mathfrak{S}_{2^m}; \mathbb{F}_2) \xrightarrow{(\mathrm{id}/\mathfrak{S}_{2^m})^*} H^*(\mathrm{F}(\mathbb{R}^d, 2^m)/\mathcal{S}_{2^m}; \mathbb{F}_2) \xrightarrow{(\mathrm{ecy}_{d,2^m}/\mathcal{S}_{2^m})^*} H^*(\mathrm{Pe}(\mathbb{R}^d, 2^m)/\mathcal{S}_{2^m}; \mathbb{F}_2)$$

  is a monomorphism. Here $(\mathrm{id}/\mathfrak{S}_{2^m})^*$ is directly a monomorphism since $\mathcal{S}_{2^m}$ is a Sylow 2-subgroup of $\mathfrak{S}_{2^m}$ and cohomology is considered with $\mathbb{F}_2$ coefficients. Thus, the main difficulty lies in proving that $(\mathrm{ecy}_{d,2^m}/\mathcal{S}_{2^m})^*$ is injective.

In this way, the cohomology ring $H^*(\mathrm{F}(\mathbb{R}^d, 2^m)/\mathfrak{S}_{2^m}; \mathbb{F}_2)$ of the unordered configuration space $\mathrm{F}(\mathbb{R}^d, 2^m)/\mathfrak{S}_{2^m}$ could be seen as a subring of the now-known cohomology ring $H^*(\mathrm{Pe}(\mathbb{R}^d, 2^m)/\mathcal{S}_{2^m}; \mathbb{F}_2)$.

This series of papers, and in particular the paper [64], feature extended and substantial calculations. It turned out to be important and influential. It was quoted, and its main result was used, in quite a number of papers since then, such as

Vassiliev's 1988 and 1998 papers on braid group cohomologies and algorithm complexity [98] and r-neighborly embeddings of manifolds [100], Crabb's 2012 survey on the topological Tverberg theorem and related topics [42], Karasev and Landweber's 2012 paper on higher topological complexity of spheres [68], Karasev and Volovikov's 2013 paper on the waist of the sphere theorem for maps to manifolds [69], Matschke's 2014 paper on a parameterized Borsuk–Ulam–Bourgin–Yang–Volovikov paper [74], as well as Karasev, Hubard, and Aronov's 2014 paper on the "spicy chicken theorem" [67].

None of these papers mentioned the fact that—as we will document in Sect. 4.1 of the present work—Hửng's proof for his main result [62, Thm. 2.3] and [64, Thm. 3.1] is incorrect, as are some of his intermediate and follow-up results. This does not jeopardize the papers listed above, as Hửng's main result, the injectivity of the composition $(\mathrm{ecy}_{d,2^m}/S_{2^m})^* \circ (\mathrm{id}/\mathfrak{S}_{2^m})^*$, holds, as we will demonstrate—by a new, entirely different, homotopy-theoretic proof—in Sect. 4.2 of this book.

In contrast to the above works, the 2016 paper of Blagojević et al. [15]—by three of the present authors—did not only quote Hửng's papers, but it also used some of Hửng's intermediate results in an essential way, specifically the decomposition of the equivariant cohomology claimed in [64, (4.7), page 279]. Our computations in [15] based on this led to results that are not consistent with some of Crabb's computations related to [42]. This led to our discovery of the substantial mistakes in Hửng's paper, including the fact that the decomposition of [64, (4.7), p. 279] is not correct, which also invalidates the main results of [15] and two minor follow-up corollaries given in [14].

Thus, the second main purpose of the present book is to correct our work in [14] and in [15], by presenting alternative arguments, based on the corrected proof for Hửng's theorem, towards estimates for the dimensions of k-regular embeddings and their relatives. The results we get are in some cases weaker than what we had claimed before, in other cases we recreate the previously claimed results in full.

This text is organized as follows. (See below for a summary of notations as well as for definitions and background.)

- In Chap. 2 we describe the S_{2^m}-equivariant embedding

$$\mathrm{ecy}_{d,2^m}\colon \ \mathrm{Pe}(\mathbb{R}^d, 2^m) \longrightarrow \mathrm{F}(\mathbb{R}^d, 2^m)$$

  of the $(d-1)(2^m-1)$-dimensional manifold $\mathrm{Pe}(\mathbb{R}^d, 2^m) \cong (S^{d-1})^{2^m-1}$ into the classical configuration space $\mathrm{F}(\mathbb{R}^d, 2^m)$. Furthermore, we relate the embedding $\mathrm{ecy}_{d,2^m}$ with the structural map of the little cubes operad.
- In Chap. 3 we study the S_{2^m}-equivariant cohomology $H^*_{S_{2^m}}(\mathrm{Pe}(\mathbb{R}^d, 2^m); \mathbb{F}_2)$ using the Serre spectral sequence associated to the fiber bundle

$$X \times X \longrightarrow (X \times X) \times_{\mathbb{Z}_2} \mathrm{E}\mathbb{Z}_2 \longrightarrow \mathrm{B}\mathbb{Z}_2.$$

The highlight of that chapter is the proof of the decomposition of the cohomology given in Theorem 3.11:

$$\begin{aligned} &H^*_{S_{2^m}}(\mathrm{Pe}(\mathbb{R}^d, 2^m); \mathbb{F}_2) \\ &\cong \mathbb{F}_2[V_{m,1}, \ldots, V_{m,m}]/\langle V^d_{m,1}, \ldots, V^d_{m,m}\rangle \oplus \mathrm{I}^*(\mathbb{R}^d, 2^m), \end{aligned}$$

where $\mathrm{I}^*(\mathbb{R}^d, 2^m)$ is an ideal, and $\deg(V_{m,r}) = 2^{r-1}$ for $1 \le r \le m$.

- In Chap. 4 we discuss the claim that the induced homomorphism in cohomology

$$(\mathrm{ecy}_{d,2^m}/S_{2^m})^*\colon H^*(\mathrm{F}(\mathbb{R}^d, 2^m)/S_{2^m}; \mathbb{F}_2) \longrightarrow H^*(\mathrm{Pe}(\mathbb{R}^d, 2^m)/S_{2^m}; \mathbb{F}_2)$$

  is a monomorphism, or equivalently that the homomorphism

$$\begin{aligned} (\mathrm{ecy}_{d,2^m}/S_{2^m})^* \circ (\mathrm{id}/\mathfrak{S}_{2^m})^*\colon H^*(\mathrm{F}(\mathbb{R}^d, 2^m)/\mathfrak{S}_{2^m}; \mathbb{F}_2) \longrightarrow \\ H^*(\mathrm{Pe}(\mathbb{R}^d, 2^m)/S_{2^m}; \mathbb{F}_2) \end{aligned}$$

  is a monomorphism. In Sect. 4.1 we present the proof for injectivity of $(\mathrm{ecy}_{d,2^m}/S_{2^m})^*$ given by Hưng in [64, Thm. 3.1] and document several critical gaps that invalidate this proof. In particular, the failure of decomposition [64, (4.7)] will be illustrated by a counterexample in Claim 4.5. The new inductive proof of the injectivity of $(\mathrm{ecy}_{d,2^m}/S_{2^m})^*$, or $(\mathrm{ecy}_{d,2^m}/S_{2^m})^* \circ (\mathrm{id}/\mathfrak{S}_{2^m})^*$, is given in Sect. 4.2. More precisely, for the inductive step, using the presentation of homology of the configuration space via Araki–Kudo–Dyer–Lashof homology operations, we prove that the structural map of the little cubes operad induces, now in homology, an epimorphism

$$\begin{aligned} (\mu_{d,2^m})_*\colon H_*((C_d(2^{m-1})/\mathfrak{S}_{2^{m-1}} \times C_d(2^{m-1})/\mathfrak{S}_{2^{m-1}}) \times_{\mathbb{Z}_2} C_d(2); \mathbb{F}_2) \longrightarrow \\ H_*(C_d(2^m)/\mathfrak{S}_{2^m}; \mathbb{F}_2); \end{aligned}$$

  see Theorem 4.8.
- Additionally in Sect. 4.3, motivated by the results of Atiyah [10] and Giusti et al. [55] we prove, as an interesting fact, that the homology of the space of all finite subsets of $\mathbb{R}^d$ with addition of a base point and appropriately defined multiplication is a polynomial ring.
- In Chap. 5, based on the results of the previous chapters, we explain the induced gaps in the results given by three of the present authors in [15, Thm. 2.1, Thm. 3.1, Thm. 4.1] and [14, Thm. 5.1, Thm. 6.1] and correct all of them. In particular, corrected lower bounds for the existence of k-regular, ℓ-skew and k-regular-ℓ-skew embeddings of an Euclidean space are given; see Theorems 5.14, 5.18, and 5.22.

– In Chap. 6, using novel techniques for the computation of Stiefel–Whitney classes, we get the Key Lemma 6.6 and from this derive still stronger lower bounds for the existence of k-regular, ℓ-skew, k-regular-ℓ-skew embeddings as well as for their complex analogues. They are summarized in Theorems 6.16 and 6.23.

In Chaps. 5 and 6, we in particular present extensive calculations with characteristic classes of the vector bundles associated with the natural permutation and the standard representation of the symmetric group $\mathfrak{S}_n$ over the unordered configuration space $\mathrm{F}(\mathbb{R}^d, n)/\mathfrak{S}_n$. These are the vector bundles

$$\xi_{\mathbb{R}^d,n}\colon\quad \mathbb{R}^n \longrightarrow \mathrm{F}(\mathbb{R}^d, n) \times_{\mathfrak{S}_n} \mathbb{R}^n \longrightarrow \mathrm{F}(\mathbb{R}^d, n)/\mathfrak{S}_n,$$

and

$$\zeta_{\mathbb{R}^d,n}\colon\quad W_n \longrightarrow \mathrm{F}(\mathbb{R}^d, n) \times_{\mathfrak{S}_k} W_n \longrightarrow \mathrm{F}(\mathbb{R}^d, n)/\mathfrak{S}_n,$$

where

$$W_n = \{(a_1, \ldots, a_n) \in \mathbb{R}^n : a_1 + \cdots + a_n = 0\}$$

denotes the standard representation of $\mathfrak{S}_n$. These vector bundles have been studied intensively over the years. For example, a particular result that can be deduced from [64, Thm. 2.10] about the $(d-1)$st power of the top Stiefel–Whitney class of the vector bundle $\zeta_{\mathbb{R}^d,n}$ was rediscovered, extended, and reproved by many authors. Going back in time, already in 1970 it was known, by the work of Cohen [33, Thm. 8.2], which was published only in 1976, that the Euler class of $\zeta_{\mathbb{R}^d,n}^{\oplus(d-1)}$ does not vanish if n is a prime. At the same time Fuks in [53] showed that the vector bundle $\xi_{\mathbb{R}^2,n}^{\oplus 2}$ is trivial and furthermore that $w_{n-1}(\xi_{\mathbb{R}^2,n}) \neq 0$ if and only if n is a power of 2. In 1978, while working on the existence of k-regular embedding, Cohen and Handel [36, Thm. 3.1] evaluated Stiefel–Whitney classes of the vector bundle $\zeta_{\mathbb{R}^2,n}$. One year later, Chisholm, in his follow-up paper [29, Lem. 3], computed Stiefel–Whitney classes of the vector bundle $\zeta_{\mathbb{R}^d,n}$ in the case when d is a power of 2. Gromov, in his seminal work on the waist of a sphere, sketched an argument that the top Stiefel–Whitney class of $\zeta_{\mathbb{R}^d,n}^{\oplus(d-1)}$ does not vanish in the case when n is a power of 2 [56, Lem. 5.1]. Three different proofs for the fact that the Euler class of $\zeta_{\mathbb{R}^d,n}^{\oplus(d-1)}$ does not vanish if and only if n is a power of a prime were given by three groups of authors: by Karasev et al. in [67, Thm. 1.10], by Blagojević and Ziegler in [16, Thm. 1.2], and by Crabb in [42, Prop. 5.1]. Furthermore, the stable order of the vector bundle $\xi_{\mathbb{R}^d,n}$ was analyzed already in 1978 by Cohen, Mahowald and Milgram in [38, Thm. 1] in the case when $d = 2$. It was completely determined, for all $d \geq 2$, in 1983 by Cohen et al. [35, Thm. 1.1].

We would like to point out that in these lecture notes we treat a number of quite different problems, for which we perform extended and rather complex computations, which use a variety of different tools and methods. In order to make this accessible, and for reasons of completeness, in the following we present many classical proofs with all details, accompanied with all relevant references, rather than just quoting them; see for example Part III. For the same reason we start with the list of notations used in this book, some of which we have already used in the overview.

The research of Pavle V. M. Blagojević was supported by the Serbian Ministry of Education, Science and Technological Development and by the German Science Foundation DFG via the Collaborative Research Center TRR 109 "Discretization in Geometry and Dynamics," Frederick R. Cohen was supported by the University of Rochester, Wolfgang Lück was supported by the ERC Advanced Grant "KL2MG-interactions" (ID 662400), and the work of Günter M. Ziegler was supported by the German Science Foundation DFG via the Collaborative Research Center TRR 109 "Discretization in Geometry and Dynamics," the Berlin Mathematical School, and the Excellence Cluster MATH+. This material is based upon work supported by the National Science Foundation under Grant DMS-1440140 while Pavle V. M. Blagojević and Günter M. Ziegler were in residence at the Mathematical Sciences Research Institute in Berkeley, California, during the fall of 2017.

The authors would like to express their gratitude to Peter Landweber for a great number of excellent comments and recommendations, and to Roman Karasev for useful discussions about different topics related to the content of the book. We also thank Evgeniya Lagoda and Tatiana Levinson for careful reading of the manuscript.

Berlin, Germany and Belgrade, Serbia — Pavle V. M. Blagojević
Rochester, NY, USA — Frederick R. Cohen
Aberdeen, UK — Michael C. Crabb
Bonn, Germany — Wolfgang Lück
Berlin, Germany — Günter M. Ziegler
May 2021

Notation

Groups

- ▷ $C_G(H)$: the centralizer of the subgroup H in the group G
- ▷ $N_G(H)$: the normalizer of the subgroup H in the group G
- ▷ $W_G(H)$: the Weyl group of the subgroup H in the group G is the quotient group $W_G(H) := N_G(H)/H$
- ▷ $\mathfrak{S}_n$: the symmetric group on n letters
- ▷ $\mathfrak{S}_{2^m}$: the symmetric group of the underlying set of the group $\mathbb{Z}_2^{\oplus m}$,
- ▷ $\mathcal{E}_m$: the elementary abelian group, isomorphic to $\mathbb{Z}_2^{\oplus m}$, that is regularly embedded, via translations of $\mathbb{F}_2^{\oplus m}$, into $\mathfrak{S}_{2^m}$
- ▷ $\mathcal{S}_{2^m}$: the Sylow 2-subgroup of $\mathfrak{S}_{2^m}$, isomorphic to $\mathbb{Z}_2^{\wr m} := \mathbb{Z}_2 \wr \mathbb{Z}_2 \wr \cdots \wr \mathbb{Z}_2$ (m times), that contains $\mathcal{E}_m$, and acts freely on $\mathrm{Pe}(\mathbb{R}^d, 2^m)$
- ▷ $\mathrm{GL}_m(\mathbb{F}_2)$: the general linear group of the $\mathbb{F}_2$ vector space $\mathbb{F}_2^{\oplus m}$
- ▷ $\mathrm{L}_m(\mathbb{F}_2)$: the Sylow 2-subgroup of $\mathrm{GL}_m(\mathbb{F}_2)$ of all lower triangular matrices with 1's on the main diagonal
- ▷ $\mathrm{U}_m(\mathbb{F}_2)$: the Sylow 2-subgroup of $\mathrm{GL}_m(\mathbb{F}_2)$ of all upper triangular matrices with 1's on the main diagonal
- ▷ $\mathrm{O}(n)$: the orthogonal group
- ▷ $\mathrm{U}(n)$: the unitary group
- ▷ $\mathcal{B}_n$: Artin's braid group on n strings,
- ▷ $\mathcal{P}_n$: Artin's pure braid group on n strings

Group Representations

- ▷ $\mathbb{R}^n$: the real n-dimensional $\mathfrak{S}_n$-representation given by permutation of the standard base
- ▷ $\mathbb{C}^n$: the complex n-dimensional $\mathfrak{S}_n$-representation given by permutation of the standard base
- ▷ $W_n := \{(a_1, \ldots, a_n) \in \mathbb{R}^n : a_1 + \cdots + a_n = 0\}$: the standard real $(n-1)$-dimensional $\mathfrak{S}_n$-representation and an irreducible $\mathfrak{S}_n$-subrepresentation of $\mathbb{R}^n$
- ▷ $W_n^{\mathbb{C}} := \{(b_1, \ldots, b_n) \in \mathbb{C}^n : b_1 + \cdots + b_n = 0\}$: the standard complex $(n-1)$-dimensional $\mathfrak{S}_n$-representation, and an irreducible $\mathfrak{S}_n$-subrepresentation of $\mathbb{C}^n$

Algebras

- ▷ $\mathfrak{H}_m := \mathbb{F}_2[x_1, \ldots, x_m]^{\mathrm{L}_m(\mathbb{F}_2)}$ the Mùi algebra
- ▷ $\mathfrak{D}_m := \mathbb{F}_2[x_1, \ldots, x_m]^{\mathrm{GL}_m(\mathbb{F}_2)}$ the Dickson algebra

Sets

- ▷ $[n] \ := \{1, 2, \ldots, n\}$
- ▷ $[n]_1 := \{1, 2, \ldots, \frac{n}{2}\}$ for even n
- ▷ $[n]_2 := \{\frac{n}{2} + 1, \frac{n}{2} + 2, \ldots, n\}$ for even n

Categories

- ▷ Top : the category of compactly generated weak Hausdorff spaces with continuous maps as morphisms
- ▷ $\mathrm{Top}_{\mathrm{pt}}$: the category of compactly generated weak Hausdorff spaces with non-degenerate base points and with continuous maps that preserve base points as morphisms
- ▷ $\mathrm{Top}_{\mathrm{cw}}$: the category of CW-complexes with continuous maps as morphisms

Spaces

- ▷ $\mathrm{F}(X, n)$: the ordered configuration space of n pairwise distinct points in X
- ▷ $\mathrm{F}(X, n)/\mathfrak{S}_n$: the unordered configuration space of n pairwise distinct points in X
- ▷ $\mathrm{Pe}(\mathbb{R}^d, 2^m)$: the Ptolemaic epicycles space $(S^{d-1})^{2^m-1}$
- ▷ $\mathrm{Ce}(\mathbb{R}^d, 2^m)$: the little cubes epicycles space
- ▷ $C_d(n)$: the space of ordered n-tuples of interior disjoint little d-cubes
- ▷ $\mathrm{Th}(\xi)$: the Thom space of the vector bundle ξ

Maps

- ▷ $\mathrm{ecy}_{d,2^m}\colon \mathrm{Pe}(\mathbb{R}^d, 2^m) \longrightarrow \mathrm{F}(\mathbb{R}^d, 2^m)$: the $\mathcal{S}_{2^m}$-equivariant Ptolemaic epicycles embedding
- ▷ $\mathrm{cecy}_{d,2^m}\colon \mathrm{Ce}(\mathbb{R}^d, 2^m) \longrightarrow C_d(2^m)$: the $\mathcal{S}_{2^m}$-equivariant little cubes epicycle embedding
- ▷ $\rho_{d,2^m}\colon \mathrm{Pe}(\mathbb{R}^d, 2^m)/\mathcal{S}_{2^m} \longrightarrow \mathrm{F}(\mathbb{R}^d, 2^m)/\mathfrak{S}_{2^m}$: the composition $(\mathrm{id}/\mathfrak{S}_{2^m}) \circ (\mathrm{ecy}_{d,2^m}/\mathcal{S}_{2^m})$
- ▷ $\kappa_{d,2^m}\colon \mathrm{Pe}(\mathbb{R}^d, 2^m) \longrightarrow \mathrm{Pe}(\mathbb{R}^\infty, 2^m)$: the $\mathcal{S}_{2^m}$-equivariant map induced map by the inclusion $\mathbb{R}^d \longrightarrow \mathbb{R}^\infty$, $x \longmapsto (x, 0, 0, \ldots)$ where $x \in \mathbb{R}^d$
- ▷ $\iota_{d,n}\colon \mathrm{F}(\mathbb{R}^d, n) \longrightarrow \mathrm{F}(\mathbb{R}^\infty, n)$: the $\mathfrak{S}_n$-equivariant map induced map by the inclusion $\mathbb{R}^d \longrightarrow \mathbb{R}^\infty$, $x \longmapsto (x, 0, 0, \ldots)$ where $x \in \mathbb{R}^d$
- ▷ $\mathrm{ev}_{d,n}\colon C_d(n) \longrightarrow \mathrm{F}(\mathbb{R}^d, n)$: the evaluation at the centers of cubes map
- ▷ $\beta\colon \mathbb{Z}_2^{\oplus m} \longrightarrow [2^m]$: the bijection given by $(i_1, \ldots, i_m) \longmapsto 1 + \sum_{j=1}^{m} 2^{m-j} i_j$
- ▷ $\alpha\colon \mathbb{N} \longrightarrow \mathbb{N}$: $\alpha(k)$ is the number of 1s in the binary presentation of k
- ▷ $\epsilon\colon \mathbb{N} \longrightarrow \mathbb{N}$: $\epsilon(k)$ is the remainder of k modulo 2
- ▷ $\gamma\colon \mathbb{N} \longrightarrow \mathbb{N}$: $\gamma(k) = \lfloor \log_2 k \rfloor + 1$ is the minimal integer such that $2^{\gamma(k)} \geq k + 1$

Vector Bundles

- ▷ $\xi_{X,k}\colon\ \mathbb{R}^k \longrightarrow \mathrm{F}(X,k)\times_{\mathfrak{S}_k}\mathbb{R}^k \longrightarrow \mathrm{F}(X,k)/\mathfrak{S}_k$,
- ▷ $\zeta_{X,k}\colon\ W_k \longrightarrow \mathrm{F}(X,k)\times_{\mathfrak{S}_k} W_k \longrightarrow \mathrm{F}(X,k)/\mathfrak{S}_k$,
- ▷ $\tau_{X,k}\colon\ \mathbb{R} \longrightarrow \mathrm{F}(X,k)/\mathfrak{S}_k\times\mathbb{R} \longrightarrow \mathrm{F}(X,k)/\mathfrak{S}_k$,
- ▷ $\xi^{\mathbb{C}}_{X,k}\colon\ \mathbb{C}^k \longrightarrow \mathrm{F}(X,k)\times_{\mathfrak{S}_k}\mathbb{C}^k \longrightarrow \mathrm{F}(X,k)/\mathfrak{S}_k$,
- ▷ $\zeta^{\mathbb{C}}_{X,k}\colon\ W^{\mathbb{C}}_k \longrightarrow \mathrm{F}(X,k)\times_{\mathfrak{S}_k} W^{\mathbb{C}}_k \longrightarrow \mathrm{F}(X,k)/\mathfrak{S}_k$,
- ▷ $\tau^{\mathbb{C}}_{X,k}\colon\ \mathbb{C} \longrightarrow \mathrm{F}(X,k)/\mathfrak{S}_k\times\mathbb{C} \longrightarrow \mathrm{F}(X,k)/\mathfrak{S}_k$,
- ▷ $\zeta_{d,k}\colon\ W_k \longrightarrow C_d(k)\times_{\mathfrak{S}_k} W_k \longrightarrow C_d(k)/\mathfrak{S}_k$,
- ▷ $\zeta_{d,k}\colon\ W_k \longrightarrow C_d(k)\times_{\mathfrak{S}_k} W_k \longrightarrow C_d(k)/\mathfrak{S}_k$,
- ▷ $\tau_{d,k}\colon\ \mathbb{R} \longrightarrow C_d(k)/\mathfrak{S}_k\times\mathbb{R} \longrightarrow C_d(k)/\mathfrak{S}_k$,
- ▷ $\lambda_{d,m}\colon\ W_{2^m} \longrightarrow \mathrm{Pe}(\mathbb{R}^d,2^m)\times_{\mathcal{S}_{2^m}} W_{2^m} \longrightarrow \mathrm{Pe}(\mathbb{R}^d,2^m)/\mathcal{S}_{2^m}$,
- ▷ $\tau_{d,m}\colon \mathbb{R} \longrightarrow \mathrm{Pe}(\mathbb{R}^d,2^m)/\mathcal{S}_{2^m}\times\mathbb{R} \longrightarrow \mathrm{Pe}(\mathbb{R}^d,2^m)/\mathcal{S}_{2^m}$,
- ▷ $\gamma_k\colon\ \mathbb{R}^k \longrightarrow \mathrm{EO}(k)\times_{\mathrm{O}(k)}\mathbb{R}^k \longrightarrow \mathrm{BO}(k)$,
- ▷ $\gamma^{\mathbb{C}}_k\colon\ \mathbb{C}^k \longrightarrow \mathrm{EU}(k)\times_{\mathrm{U}(k)}\mathbb{R}^k \longrightarrow \mathrm{BU}(k)$,
- ▷ $\xi_k\colon\ \mathbb{R}^k \longrightarrow \mathrm{E}\mathfrak{S}_k\times_{\mathfrak{S}_k}\mathbb{R}^k \longrightarrow \mathrm{B}\mathfrak{S}_k$,
- ▷ $\eta_{2^m}\colon\ \mathbb{R}^{2^m} \longrightarrow \mathrm{E}\mathcal{S}_{2^m}\times_{\mathcal{S}_{2^m}}\mathbb{R}^{2^m} \longrightarrow \mathrm{B}\mathcal{S}_{2^m}$,
- ▷ $\nu_{2^m}\colon\ \mathbb{R}^{2^m} \longrightarrow \mathrm{E}\mathcal{E}_m\times_{\mathcal{E}_m}\mathbb{R}^{2^m} \longrightarrow \mathrm{B}\mathcal{E}_m$,
- ▷ $\theta_{2^m}\colon\ \mathbb{R}^{2^m} \longrightarrow \mathrm{E}(\mathfrak{S}^2_{2^m-1})\times_{\mathfrak{S}^2_{2^m-1}}\mathbb{R}^{2^m} \longrightarrow \mathrm{B}(\mathfrak{S}^2_{2^m-1})$,
- ▷ $\omega_{2^m}\colon\ \mathbb{R}^{2^m} \longrightarrow \mathrm{E}(\mathcal{S}^2_{2^m-1})\times_{\mathcal{S}^2_{2^m-1}}\mathbb{R}^{2^m} \longrightarrow \mathrm{B}(\mathcal{S}^2_{2^m-1})$.

Contents

1 Snapshots from the History 1
1.1 The Braid Group 2
1.2 The Fundamental Sequence of Fibrations 3
1.3 Artin's Presentation of $\mathcal{B}_n$ and $\pi_1(\mathrm{F}(\mathbb{R}^2, n))$ 5
1.4 The Cohomology Ring $H^*(\mathrm{F}(\mathbb{R}^2, n); \mathbb{Z})$ 7
1.5 The Cohomology of the Braid Group $\mathcal{B}_n$ 9
1.6 The Cohomology Ring $H^*(\mathcal{B}_n; \mathbb{F}_2)$ 10
1.7 Cohomology of Braid Spaces 12
1.8 Homology of Unordered Configuration Spaces 16

Part I Mod 2 Cohomology of Configuration Spaces

2 The Ptolemaic Epicycles Embedding 23

3 The Equivariant Cohomology of Pe($\mathbb{R}^d$, 2^m) 33
3.1 Small Values of m 33
3.2 The Case $m = 2$ 34
3.3 Cohomology of $(X \times X) \times_{\mathbb{Z}_2} S^{d-1}$ and $(X \times X) \times_{\mathbb{Z}_2} \mathrm{E}\mathbb{Z}_2$ 39
3.4 The Induction Step 49
3.5 The Restriction Homomorphisms – Three Aspects 52
3.5.1 A Restriction Homomorphism and the Mùi Invariants 52
3.5.2 A Restriction Homomorphism and the Dickson Invariants 54
3.5.3 Two Lemmas 58

4 Hưng's Injectivity Theorem 63
4.1 Critical Points in Hưng's Proof of His Injectivity Theorem 64
4.2 Proof of the Injectivity Theorem 75
4.2.1 Prerequisites 77
4.2.2 Proof of the Dual Epimorphism Theorem 84
4.3 An Unexpected Corollary 87
4.3.1 Motivation 87
4.3.2 Corollary 89

Part II Applications to the (Non-)Existence of Regular and Skew Embeddings

5 On Highly Regular Embeddings: Revised 95
5.1 k-Regular Embeddings 96
5.2 ℓ-Skew Embeddings 113
5.3 k-Regular-ℓ-Skew Embeddings 128
5.4 Complex Highly Regular Embeddings 132

6 More Bounds for Highly Regular Embeddings 137
6.1 Examples of $\mathcal{S}_{2^m}$-Representations and Associated Vector Bundles 137
6.1.1 Examples of $\mathcal{S}_{2^m}$-Representations 138
6.1.2 Associated Vector Bundles 138
6.2 The Key Lemma and its Consequences 140
6.3 Additional Bounds for the Existence of Highly Regular Embeddings 148
6.4 Additional Bounds for the Existence of Complex Highly Regular Embeddings 157

Part III Technical Tools

7 Operads 163
7.1 Definition and Basic Example 163
7.2 $\mathcal{O}$-Space 166
7.3 Little Cubes Operad 167
7.4 $\mathcal{C}_d$-Spaces, An Example 169
7.5 $\mathcal{C}_d$-Spaces, a Free $\mathcal{C}_d$-Space Over X 170
7.6 Araki–Kudo–Dyer–Lashof Homology Operations 171

8 The Dickson Algebra 173
8.1 Rings of Invariants 173
8.2 The Dickson Invariants as Characteristic Classes 176

9 The Stiefel–Whitney Classes of the Wreath Square of a Vector Bundle 179
9.1 The Wreath Square and the $(d-1)$-Partial Wreath Square of a Vector Bundle 179
9.2 Cohomology of $B(S^2\xi) = S^2B(\xi)$ 181
9.3 The Total Stiefel–Whitney Class of the Wreath Square of a Vector Bundle 183

10 Miscellaneous Calculations 187
10.1 Detecting Group Cohomology 187
10.2 The Image of a Restriction Homomorphism 188
10.3 Weyl Groups of an Elementary Abelian Group 192

10.4 Cohomology of the Real Projective Space with Local Coefficients 194
10.5 Homology of the Real Projective Space with Local Coefficients 197

References 201

Index 207

Chapter 1
Snapshots from the History

In many ways it is much harder to write accurately about the complete history of a subject than to make a contribution to its development. Even with the possibility to take a peek into the past such an endeavor is impossible to complete. For these reasons we give only a brief overview of the study of configuration spaces from the perspective of the contents of this book, choosing both seminal contributions as well as some particular topics to focus on. Our presentation by no means aims to be, or could be, complete.

We begin our story by introducing the object we study; that is, we answer the question: *What is hiding under the name configuration space?*

Let X be a topological space. The **ordered configuration space** of all n-tuples of distinct points on X is the following subspace of the product space X^n:

$$\mathrm{F}(X,n) := \{(x_1,\dots,x_n) \in X^n : x_i \neq x_j \text{ for all } 1 \le i < j \le n\}.$$

The ordered configuration space $\mathrm{F}(X,n)$ is endowed with a natural free (left) action of the symmetric group on n letters $\mathfrak{S}_n$, given by the permutation of points, that is

$$\pi \cdot (x_1,\dots,x_n) = (x_{\pi(1)},\dots,x_{\pi(n)}),$$

where $\pi \in \mathfrak{S}_n$ and $(x_1,\dots,x_n) \in \mathrm{F}(X,n)$. The associated orbit space $\mathrm{F}(X,n)/\mathfrak{S}_n$ is called the **unordered configuration space** of n distinct points on X.

The official history of configuration spaces begins in 1925 with *Theorie der Zöpfe*, the fundamental work of Emil Artin [7], while the prehistory goes back to a work of Adolf Hurwitz [65] from 1891 and a 1930s work of Oscar Zariski [102]. The braids, the oldest of gadgets of man, now transformed into a beautiful algebraic object, became the starting point for the intensive research branching over areas, sciences and decades not ever losing on intensity or its fundamental importance.

P. V. M. Blagojević et al., *Equivariant Cohomology of Configuration Spaces Mod 2*, Lecture Notes in Mathematics 2282, https://doi.org/10.1007/978-3-030-84138-6_1

1.1 The Braid Group

In his seminal paper [7] from 1925 Artin introduced the notion, of what we would call today a **geometric braid**, as a collection of n disjoint arcs $(\beta_1, \ldots, \beta_n)$ connecting two collections of n pairwise distinct points $(x_1, \ldots, x_n)$ and $(x'_1, \ldots, x'_n)$ placed on the planes $z = 0$ and $z = 1$ of the 3-dimensional Euclidean space $\mathbb{R}^3$ respectively. In addition, the arcs $(\beta_1, \ldots, \beta_n)$ need to satisfy the following two properties:

- $\beta_1(0) = x_1, \ldots, \beta_n(0) = x_n$ and $\beta_1(1) = x'_{i_1}, \ldots, \beta_n(1) = x'_{i_n}$ with the index sets $\{i_1, \ldots, i_n\}$ and $\{1, \ldots, n\}$ coinciding, and
- $\beta_i(t)$ belongs to the plane $z = t$ for every $0 \leq t \leq 1$ and every $1 \leq i \leq n$.

Here, an arc is assumed to be a continuous injection of the segment $[0, 1]$ into $\mathbb{R}^3$. Classically the arcs of the braid $\beta_1, \ldots, \beta_n$ are also called strings of the braid. For an illustration of a geometric braid see Fig. 1.1.

Considering braids up to a homotopy (through the space of geometric braids), rather than how they are geometrically presented, Artin was able to define a natural operation on the set $\mathcal{B}_n$ of homotopy classes of all braids between two fixed collections of points. Artin [7] said that two braids are equivalent (homotopic for us) if one braid can be deformed into another without self-intersections—*Zwei solche Zöpfe heißen äquivalent oder kürzer gleich, wenn sie sich ineinander ohne Selbstdurchdringung deformieren lassen*. Concatenation of braids and the shrinking procedure applied to the representatives of homotopy classes of braids nicely fit with the homotopy relation and as such a well defined operation on $\mathcal{B}_n$ was introduced. This operation was associative and with the obvious neutral element—the homotopy class of the trivial braid τ—the collection of pairwise disjoint line segments $([x_1, x'_1], \ldots, [x_n, x'_n])$ as strings. Furthermore, the elementary braids σ_i, for $1 \leq i \leq n-1$, were introduced as braids identical to the trivial braid in all strings except the i-th string crosses "over" $(i + 1)$-th string. Here "over" refers to a side, half-space, of the affine plan spanned by the end points of the strings. An illustration of the plane projections of representatives of the braids τ, σ_i and its inverse σ_i^{-1} are given in Fig. 1.2.

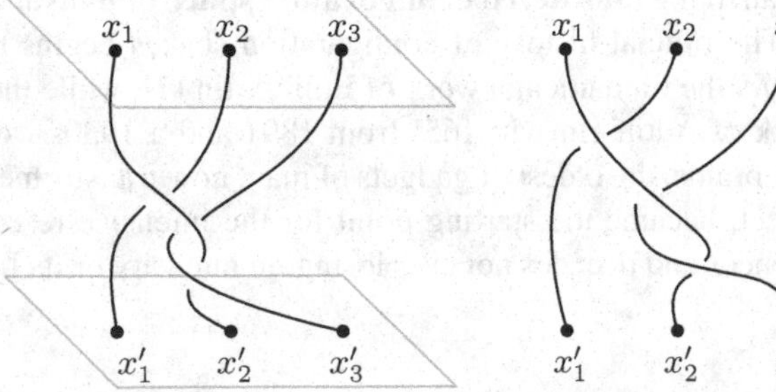

Fig. 1.1 A geometric braid and its projection to the affine plane spanned by the end points of the strings

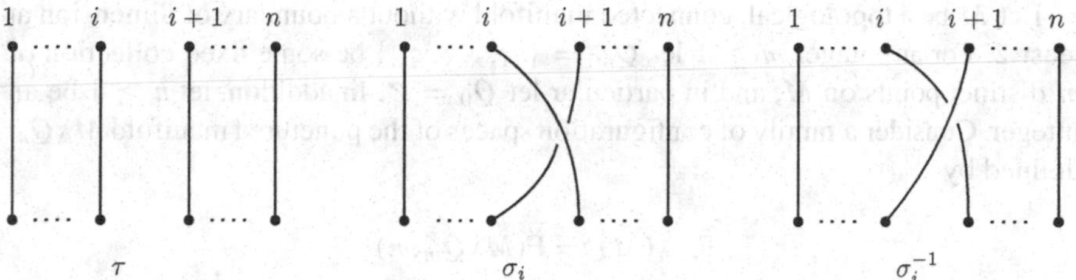

Fig. 1.2 The representatives of the braids τ, σ_i and σ_i^{-1}

The most quoted result of Artin's paper [7, Satz 1] is the presentation of the **braid group** $\mathcal{B}_n$ on n strings of the following form:

$$\mathcal{B}_n = \left\langle \sigma_1, \ldots, \sigma_{n-1} : \begin{array}{ll} \sigma_i\sigma_j = \sigma_j\sigma_i, & \text{for } 1 \leq i < j - 1 \leq n - 2 \\ \sigma_i\sigma_{i+1}\sigma_i = \sigma_{i+1}, \sigma_i\sigma_{i+1} & \text{for } 1 \leq i \leq n - 2 \end{array} \right\rangle,$$

now known as Artin's presentation of the braid group. An alternative and more formal argument for this result was further developed by Artin in his paper [9] from 1947. In particular, a relation of s-isotopy between braids is introduced and it is shown that two braids are s-isotopic if and only if they are homotopic (through the space of geometric braids); see [9, Thm. 8].

Furthermore, Artin in [7, Sec. 4] defines a group homomorphism $\mathfrak{a}_n : \mathcal{B}_n \longrightarrow \mathfrak{S}_n$ which fits into the following short exact sequence of groups:

$$1 \longrightarrow \mathcal{P}_n \longrightarrow \mathcal{B}_n \xrightarrow{\mathfrak{a}_n} \mathfrak{S}_n \longrightarrow 1.$$

The normal subgroup $\mathcal{P}_n := \ker(\mathfrak{a}_n)$ of the braid group $\mathcal{B}_n$ is called the **pure braid group** on n strings. Geometrically, $\mathcal{P}_n$ is the set of all representatives of geometric braids $(\beta_1, \ldots, \beta_n)$ which have the property that $\beta_1(0) = x_1, \ldots, \beta_n(0) = x_n$ and $\beta_1(1) = x'_1, \ldots, \beta_n(1) = x'_n$.

For more aspects in the study of braid groups consult the classical monograph *Braids, links, and mapping class groups* by Joan Birman [12].

1.2 The Fundamental Sequence of Fibrations

The work of Edward Fadell and Lee Neuwirth [47] from 1962 gave birth to the name *configuration space* for the space of all ordered collections of pairwise distinct points on a topological space. The central results of this fundamental work is the construction of the so-called fundamental sequence of fibrations of configuration spaces of a manifold.

Let M be a topological, connected manifold without a boundary of dimension at least 2. For an integer $m \geq 1$ let $Q_m = \{q_1, \ldots, q_m\}$ be some fixed collection of m distinct points on M, and in particular let $Q_0 = \varnothing$. In addition, let $n \geq 1$ be an integer. Consider a family of configuration spaces of the punctured manifold $M \backslash Q_m$ defined by

$$\mathrm{F}_{m,n}(M) := \mathrm{F}(M \backslash Q_m, n)$$

with the corresponding projections

$$p_{m,n,r} \colon \mathrm{F}_{m,n}(M) \longrightarrow \mathrm{F}_{m,r}(M), \quad (x_1, \ldots x_n) \longmapsto (x_1, \ldots, x_r)$$

for $1 \leq r \leq n-1$ and $(x_1, \ldots x_n) \in \mathrm{F}_{m,n}(M)$. Observe that $\mathrm{F}(M, n) = \mathrm{F}_{0,n}(M)$. Fadell and Neuwirth, in [47, Thm. 1 and Thm. 3], for $n \geq 2$ showed that $p_{m,n,1}$ is a locally trivial fibration with fibre $\mathrm{F}_{m+r,n-r}(M)$. In addition, they proved that the fibration $p_{m,n,1}$ admits a (continuous) cross-section.

Using the sequence of fibrations

$$\begin{array}{ccccccccc}
\mathrm{F}_{n-1,1}(M) & \longrightarrow & \mathrm{F}_{n-2,2}(M) & \longrightarrow \cdots \longrightarrow & \mathrm{F}_{2,n-2}(M) & \longrightarrow & \mathrm{F}_{1,n-1}(M) & \longrightarrow & \mathrm{F}_{0,n}(M) \\
 & & \downarrow & & \downarrow & & \downarrow & & \downarrow \\
 & & \mathrm{F}_{n-2,1}(M) & & \mathrm{F}_{2,1}(M) & & \mathrm{F}_{1,1}(M) & & \mathrm{F}_{0,1}(M),
\end{array}$$

and associated long exact sequences in homotopy, in combination with the existence of the corresponding cross-sections, they obtained various descriptions of homotopy groups of ordered configuration spaces. For example, they showed that for $d \geq 2$ and $i \geq 2$ there exist isomorphisms

$$\pi_i\big(\mathrm{F}(\mathbb{R}^d, n)\big) \cong \bigoplus_{k=1}^{n-1} \pi_i\big(\underbrace{S^{d-1} \vee \cdots \vee S^{d-1}}_{k}\big),$$

and

$$\pi_i\big(\mathrm{F}(S^{2d-1}, n)\big) \cong \pi_i\big(S^{2d-1}\big) \oplus \bigoplus_{k=1}^{n-2} \pi_i\big(\underbrace{S^{2d-2} \vee \cdots \vee S^{2d-2}}_{k}\big).$$

In particular, they obtained that $\mathrm{F}(\mathbb{R}^2, n)$ is an Eilenberg–Mac Lane space $\mathrm{K}(\mathcal{P}_n, 1)$. Hence, the (co)homology of the pure braid group $\mathcal{P}_n$ coincides with the corresponding (co)homology of the configuration space $\mathrm{F}(\mathbb{R}^2, n)$. Consult [47, Cor. 2.1, Cor. 5.1].

Furthermore, almost at the same time, Fadell in [46, Thm. 1 and Thm. 2] gave additional descriptions of homotopy groups of configuration spaces for $d \geq 4$:

$$\pi_i\big(\mathrm{F}(S^d, n)\big) \cong \pi_i\big(V_2(\mathbb{R}^{d+1})\big) \oplus \bigoplus_{k=1}^{n-2} \pi_i\big(\underbrace{S^{d-1} \vee \cdots \vee S^{d-1}}_{k}\big),$$

and

$$\pi_i\big(\mathrm{F}(\mathbb{R}\mathrm{P}^{d-1}, n)\big) \cong \pi_i\big(V_2(\mathbb{R}^{d+1})\big) \oplus \bigoplus_{k=1}^{n-2} \pi_i\big(\underbrace{S^{d-1} \vee \cdots \vee S^{d-1}}_{2k+1}\big).$$

Here $V_2(\mathbb{R}^{d+1})$ denotes the Stiefel manifold of orthonormal 2-frames in $\mathbb{R}^{d+1}$.

For a detailed exposition of these results and much more about topology of configuration spaces consult the book *Geometry and Topology of Configuration Spaces* by Fadell and Husseini [49].

1.3 Artin's Presentation of $\mathcal{B}_n$ and $\pi_1(\mathrm{F}(\mathbb{R}^2, n))$

The work of Fadell and Neuwirth [47] we discussed in the previous section intertwines in an essential way with the work of Ralph Fox with Lee Neuwirth [52] from the same year. As explained at the beginning of their papers they aimed at "a straightforward derivation" of Artin's presentation of the braid group $\mathcal{B}_n$.

Since the braid group $\mathcal{B}_n$ can also be seen as the fundamental group of the unordered configuration space of n distinct points in the plane, as already pointed out by Artin, Fox and Neuwirth aimed to use classical knowledge for presentation of fundamental groups to obtain exactly Artin's presentation as a presentation of $\pi_1(\mathrm{F}(\mathbb{R}^2, n)/\mathfrak{S}_n)$.

More precisely, using the lexicographic ordering on the coordinates of $\mathbb{R}^2$ the power set $(\mathbb{R}^2)^n$ can be stratified by a family of convex cones. For example, for $n = 7$ the stratum defined by the symbol

$$\theta = (3 < 5 = 1 < 6 \veebar 4 \veebar 2 = 7)$$

is the convex cone of $(\mathbb{R}^2)^7$ given by

$$\Big\{(p_1, \ldots, p_7) = (x_1, y_1, \ldots, x_7, y_7) \in (\mathbb{R}^2)^7 : \begin{array}{l} x_3 < x_5 = x_1 < x_6 = x_4 = x_2 = x_7 \\ \quad\quad\; y_5 = y_1 \quad\;\; y_6 < y_4 < y_2 = y_7 \end{array}\Big\}.$$

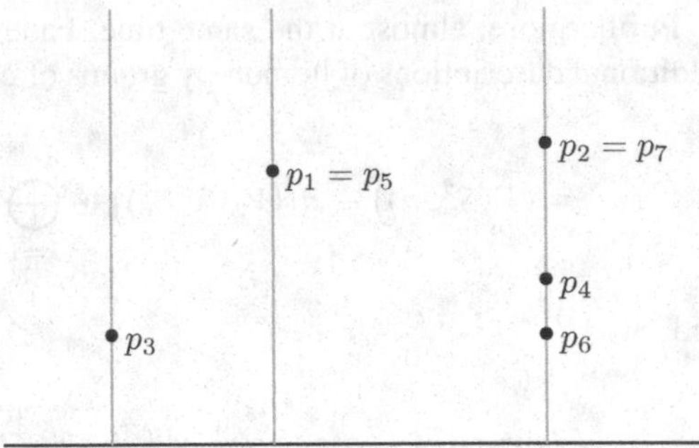

Fig. 1.3 A configuration that corresponds to a point in the stratum $(3 < 5 = 1 < 6 \underset{=}{\vee} 4 \underset{=}{\vee} 2 = 7)$

For an illustration of a point inside θ see Fig. 1.3. The action of the symmetric group $\mathfrak{S}_n$ on $(\mathbb{R}^2)^n$ induces a stratum-wise action—translations given by the action send strata to strata homeomorphically. Furthermore, such a stratification induces a regular $\mathfrak{S}_n$-invariant cell complex structure on the one-point compactification S^{2n} of the ambient $(\mathbb{R}^2)^n$.

Consider the $\mathfrak{S}_n$-CW-subcomplex Δ of S^{2n} given by all the cells whose symbols have at least one sign of equality "=". In particular such a cell is induced by the stratum θ. Then the configuration space $\mathrm{F}(\mathbb{R}^2, n)$ is the complement $S^{2n} \backslash \Delta$ of the $(2n-2)$-dimensional $\mathfrak{S}_n$-CW-subcomplex Δ inside the $(2n)$-dimensional $\mathfrak{S}_n$-CW-complex S^{2n}. Furthermore, the unordered configuration space can also be seen as the complement

$$\mathrm{F}(\mathbb{R}^2, n)/\mathfrak{S}_n = (S^{2n}/\mathfrak{S}_n) \setminus (\Delta/\mathfrak{S}_n)$$

of the $(2n-2)$-dimensional CW-subcomplex $\Delta/\mathfrak{S}_n$ inside the $(2n)$-dimensional CW-complex $S^{2n}/\mathfrak{S}_n$. In particular, the CW-complex $S^{2n}/\mathfrak{S}_n$ has one maximal $(2n)$-cell given by the symbol (representative) $(1 < 2 < \cdots < n)$.

In such a situation, a presentation of the fundamental group of a complement of a cell complex, in our case $\pi_1(\mathrm{F}(\mathbb{R}^2, n)/\mathfrak{S}_n) = \pi_1((S^{2n}/\mathfrak{S}_n) \setminus (\Delta/\mathfrak{S}_n))$, can be obtained from:

- the specified generators contained in the complement indexed by all $(2n-1)$-cells in the boundary of the $(2n)$-cell, and
- the relations corresponding to $(2n-2)$-cells contained in the complement induced from theirs coboundaries in a specific way.

Knowing this technical gadget Fox and Neuwirth "only" needed to list generators and identify the corresponding relations, as the "recipe" suggested. They did this in a beautiful and geometrically clear way in [52, Sec. 7] showing that, in the end, the braid group $\mathcal{B}_n \cong \pi_1(\mathrm{F}(\mathbb{R}^2, n)/\mathfrak{S}_n)$ can be given via Artin's presentation.

Furthermore, Fadell and Neuwirth noticed that the unordered configuration space $\mathrm{F}(\mathbb{R}^2, n)/\mathfrak{S}_n$ is an Eilenberg–Mac Lane space $\mathrm{K}(\mathcal{B}_n, 1)$. This implies, in particular, that the braid group $\mathcal{B}_n$ has no elements of finite order [52, Cor. 1].

In parallel, Fadell and James Van Buskirk in [48], based on the work of Wei-Liang Chow [30], offered an argument of different flavour for the fact that $\mathcal{B}_n \cong \pi_1(\mathrm{F}(\mathbb{R}^2, n)/\mathfrak{S}_n)$ can be described via Artin's presentation.

The idea of Fox and Neuwirth to stratify the ambient $(\mathbb{R}^2)^n$ of the configuration space $\mathrm{F}(\mathbb{R}^2, n)$ into cones motivated Anders Björner and Günter M. Ziegler [13] to use stratifications in construction of cell complex models for general complements of subspace arrangements; it was essential in the work of Pavle Blagojević and Ziegler [16].

1.4 The Cohomology Ring $H^*(\mathrm{F}(\mathbb{R}^2, n); \mathbb{Z})$

The next chapter in our story about configuration spaces is dedicated to the seminal work of Vladimir Igorovich Arnol'd [5] from 1969. He showed that the integral cohomology ring of the configuration space $\mathrm{F}(\mathbb{R}^2, n)$ can be presented as a quotient of the exterior algebra (over the ring of integers) as follows:

$$H^*(\mathrm{F}(\mathbb{R}^2, n); \mathbb{Z}) \cong \frac{\Lambda\big(\omega_{i,j} : 1 \le i < j \le n\big)}{(\omega_{i,j}\omega_{j,k} + \omega_{j,k}\omega_{i,k} + \omega_{i,k}\omega_{j,k} : 1 \le i < j < k \le n)} =: A(n).$$

In particular, he described an additive basis of the cohomology $H^*(\mathrm{F}(\mathbb{R}^2, n); \mathbb{Z})$ and showed that the Poincaré polynomial of $\mathrm{F}(\mathbb{R}^2, n); \mathbb{Z})$ is given by

$$p_{\mathrm{F}(\mathbb{R}^2,n)}(t) = (1+t)(2+t)\cdots(1+(n-1)t).$$

See for example [5, Thm. 1, Cor. 1, Cor. 3].

The presented proof of these results was based on a masterful use of the Serre spectral sequence. According to Arnol'd, Dmitry Fuks made a substantial contribution to this proof. More precisely, Arnol'd considered the projection

$$p\colon \mathrm{F}(\mathbb{R}^2, n) \longrightarrow \mathrm{F}(\mathbb{R}^2, n-1), \qquad (z_1, \dots, z_n) \longmapsto (z_1, \dots, z_{n-1})$$

and showed that it is a fiber bundle which additionally admits a continuous cross-section. Furthermore, in [5, Lem. 1] he showed that the fundamental group of the base space—the pure braid group $\pi_1(\mathrm{F}(\mathbb{R}^2, n-1)) \cong \mathcal{P}_{n-1}$—acts trivially on the cohomology of the fiber

$$\mathbb{R}^2 \backslash \{z_1, \dots, z_{n-1}\} \simeq \underbrace{S^1 \vee \cdots \vee S^1}_{n-1}.$$

Here the point $(z_1, \ldots, z_{n-1}) \in \mathrm{F}(\mathbb{R}^2, n-1)$ is assumed to be fixed. With these ingredients the E_2-term of the Serre spectral sequence associated to the fiber bundle p is:

$$\begin{aligned} E_2^{r,s} &= H^r\big(\mathrm{F}(\mathbb{R}^2, n-1); \mathcal{H}^s(\mathbb{R}^2\backslash\{z_1, \ldots, z_{n-1}\}; \mathbb{Z})\big) \\ &\cong H^r\big(\mathrm{F}(\mathbb{R}^2, n-1); \mathbb{Z}\big) \otimes H^s(\mathbb{R}^2\backslash\{z_1, \ldots, z_{n-1}\}; \mathbb{Z}\big). \end{aligned}$$

This spectral sequence converges to the cohomology of the base space of the fiber bundle—in this case $H^*(\mathrm{F}(\mathbb{R}^2, n); \mathbb{Z})$. In the description of the E_2-term only the triviality of the action of $\pi_1(\mathrm{F}(\mathbb{R}^2, n-1))$ on the cohomology of the fiber is used. Now the existence of the cross-section of the fiber bundle p implies that the only possible non-zero differential ∂_2 has to vanish. Consequently, the spectral sequence is completely determined and collapses at the E_2-term, that is $E_2^{r,s} \cong E_\infty^{r,s}$ for all integers r and s. Since the cohomology groups of the fiber $H^*(\mathbb{R}^2\backslash\{z_1, \ldots, z_{n-1}\}; \mathbb{Z})$ are free abelian groups (free $\mathbb{Z}$-modules) in every dimension using induction on n we see that all entries of the E_∞-term are free abelian groups. Hence, the spectral sequence has no extension problem and the additive structure of the cohomology of the configuration space $H^*(\mathrm{F}(\mathbb{R}^2, n); \mathbb{Z})$ is completely determined. For example, one can write

$$H^*(\mathrm{F}(\mathbb{R}^2, n); \mathbb{Z}) \cong H^*(S^1 \times (S^1 \vee S^1) \times \cdots \times \underbrace{(S^1 \vee \cdots \vee S^1)}_{n-1}; \mathbb{Z}).$$

What about the ring structure on the cohomology $H^*(\mathrm{F}(\mathbb{R}^2, n); \mathbb{Z})$? To answer this question Arnol'd brought into play a beautiful new idea which was to be used over and over again for years to come. He defined the map $\varphi\colon A(n) \longrightarrow H^*(\mathrm{F}(\mathbb{R}^2, n); \mathbb{R})$, between the quotient of the exterior algebra and the de Rham cohomology of the configuration space $\mathrm{F}(\mathbb{R}^2, n)$, by sending the algebra generator $\omega_{i,j}$, $1 \le i < j \le n$, to the cohomology class of the logarithmic differential form

$$w_{i,j} := \frac{1}{2\pi i} \cdot \frac{dz_i - dz_j}{z_i - z_j} = \frac{d\log(z_i - z_j)}{2\pi i}.$$

The map turns out to be well defined and, in particular, a ring homomorphism. Now, since the cohomology classes of $w_{i,j}$, for all $1 \le i < j \le n$, are integral, the ring homomorphism φ factors as follows:

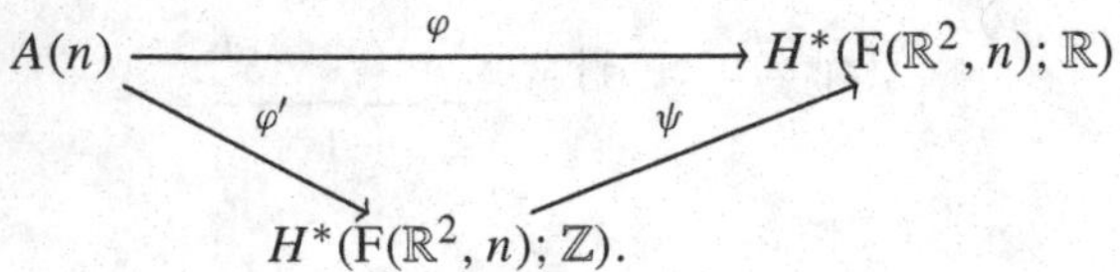

Here, the ring homomorphism ψ is induced by the coefficient inclusion $\mathbb{Z} \longrightarrow \mathbb{R}$. Finally, combining the knowledge of additive structures of $A(n)$ and the cohomology $H^*(\mathrm{F}(\mathbb{R}^2, n); \mathbb{Z})$, Arnol'd proved that the ring homomorphism φ' is actually a ring isomorphism.

1.5 The Cohomology of the Braid Group $\mathcal{B}_n$

A year later, in 1970 Arnol'd published yet another breakthrough paper [6] in which he studied the cohomology, now of the unordered configuration space $\mathrm{F}(\mathbb{R}^2, n)/\mathfrak{S}_n$.

In order to argue the importance of understanding the topology of the unordered configuration space he gave a list of various important incarnations of the unordered configuration space, like the space of all monic polynomials of degree n in the polynomial ring $\mathbb{C}[z]$ (algebraic functions) without multiple roots, the space of all hyperelliptic curves of degree n (see [3, 97]), and the set of regular values of the mapping Σ^{1n} (see [4]).

Using the fact that the unordered configuration space $\mathrm{F}(\mathbb{R}^2, n)/\mathfrak{S}_n$ is a $\mathrm{K}(\mathcal{B}_n, 1)$-space he utilized the isomorphism $H^i(\mathcal{B}_n; \mathbb{Z}) \cong H^i(\mathrm{F}(\mathbb{R}^2, n)/\mathfrak{S}_n; \mathbb{Z})$ (with trivial integer coefficients), to obtain the following fundamental facts about the additive cohomology structure of the braid group.

Theorem 1.1 (Finiteness, Repetition and Stability Theorem) *Let $n \geq 1$ be an integer.*

(1) *The cohomology groups of the braid group $\mathcal{B}_n$ are all finite, except for $H^0(\mathcal{B}_n; \mathbb{Z}) \cong H^1(\mathcal{B}_n; \mathbb{Z}) \cong \mathbb{Z}$. Furthermore, $H^i(\mathcal{B}_n; \mathbb{Z}) = 0$ for all $i \geq n$.*

(2) *For all integers $n \geq 1$ and $i \geq 0$ there is an isomorphism*

$$H^i(\mathcal{B}_{2n+1}; \mathbb{Z}) \cong H^i(\mathcal{B}_{2n}; \mathbb{Z}).$$

(3) *For all integers $n \geq 1$ and $i \geq 0$ with the property that $n \geq 2i - 2$ there is an isomorphism*

$$H^i(\mathcal{B}_n; \mathbb{Z}) \cong H^i(\mathcal{B}_{2i-2}; \mathbb{Z}).$$

Furthermore, by direct computations Arnol'd completed the following table of cohomologies for the first ten braid groups $\mathcal{B}_2, \ldots, \mathcal{B}_{11}$.

$i =$	0	1	2	3	4	5	6	7	8	9
$H^i(\mathcal{B}_2; \mathbb{Z}) \cong H^i(\mathcal{B}_3; \mathbb{Z}) \cong$	$\mathbb{Z}$	$\mathbb{Z}$	0	0	0	0	0	0	0	0
$H^i(\mathcal{B}_4; \mathbb{Z}) \cong H^i(\mathcal{B}_5; \mathbb{Z}) \cong$	$\mathbb{Z}$	$\mathbb{Z}$	0	$\mathbb{Z}_2$	0	0	0	0	0	0
$H^i(\mathcal{B}_6; \mathbb{Z}) \cong H^i(\mathcal{B}_7; \mathbb{Z}) \cong$	$\mathbb{Z}$	$\mathbb{Z}$	0	$\mathbb{Z}_2$	$\mathbb{Z}_2$	$\mathbb{Z}_3$	0	0	0	0
$H^i(\mathcal{B}_8; \mathbb{Z}) \cong H^i(\mathcal{B}_9; \mathbb{Z}) \cong$	$\mathbb{Z}$	$\mathbb{Z}$	0	$\mathbb{Z}_2$	$\mathbb{Z}_2$	$\mathbb{Z}_6$	$\mathbb{Z}_3$	$\mathbb{Z}_2$	0	0
$H^i(\mathcal{B}_{10}; \mathbb{Z}) \cong H^i(\mathcal{B}_{11}; \mathbb{Z}) \cong$	$\mathbb{Z}$	$\mathbb{Z}$	0	$\mathbb{Z}_2$	$\mathbb{Z}_2$	$\mathbb{Z}_6$	$\mathbb{Z}_6$ or $\mathbb{Z}_3$	$\mathbb{Z}_2$ or 0	$\mathbb{Z}_2$	$\mathbb{Z}_5$

1.6 The Cohomology Ring $H^*(\mathcal{B}_n;\mathbb{F}_2)$

In parallel with the work of Arnol'd, a flood of new ideas was presented by Dmitry Borisovich Fuks in his seminal paper [53], which aimed to describe the cohomology ring of the braid group $H^*(\mathcal{B}_n;\mathbb{F}_2)$.

The approach Fuks used differed from the one applied by Arnol'd. The initial idea was to consider the sequence of group embeddings

$$\mathcal{B}_n \xrightarrow{\mathfrak{a}_n} \mathfrak{S}_n \longrightarrow \mathrm{O}(n) \longrightarrow \mathrm{U}(n). \tag{1.1}$$

The sequence of homomorphisms (1.1) induces the corresponding sequence of continuous maps between the classifying spaces:

$$\mathrm{F}(\mathbb{R}^2,n)/\mathfrak{S}_n \cong \mathrm{B}\mathcal{B}_n \xrightarrow{\mathrm{B}\mathfrak{a}_n} \mathrm{B}\mathfrak{S}_n \longrightarrow \mathrm{BO}(n) \longrightarrow \mathrm{BU}(n). \tag{1.2}$$

The classifying space $\mathrm{B}\mathcal{B}_n$ can be substituted with $\mathrm{F}(\mathbb{R}^2,n)/\mathfrak{S}_n$, because $\mathrm{F}(\mathbb{R}^2,n)/\mathfrak{S}_n$ is a $\mathrm{K}(\mathrm{B}\mathcal{B}_n,1)$-space. The sequence of continuous maps (1.2) induces the following sequence of morphisms of (pull-back) real vector bundles $\xi_{\mathbb{R}^2,n} \longrightarrow \xi_n \longrightarrow \gamma_n$, that is,

$$\begin{array}{ccccc}
\mathrm{F}(\mathbb{R}^2,n)\times_{\mathfrak{S}_n}\mathbb{R}^n & \longrightarrow & \mathrm{E}\mathfrak{S}_n\times_{\mathfrak{S}_n}\mathbb{R}^n & \longrightarrow & \mathrm{EO}(n)\times_{\mathrm{O}(n)}\mathbb{R}^n \\
\downarrow & & \downarrow & & \downarrow \\
\mathrm{F}(\mathbb{R}^2,n)/\mathfrak{S}_n & \longrightarrow & \mathrm{B}\mathfrak{S}_n & \longrightarrow & \mathrm{BO}(n).
\end{array} \tag{1.3}$$

In addition, the following morphisms between complex vector bundles $\xi^C_{\mathbb{R}^2,n} \longrightarrow \gamma^C_n$ can also be induced, that is

$$\begin{array}{ccc}
\mathrm{F}(\mathbb{R}^2,n)\times_{\mathfrak{S}_n}\mathbb{C}^n & \longrightarrow & \mathrm{EU}(n)\times_{\mathrm{U}(n)}\mathbb{C}^n \\
\downarrow & & \downarrow \\
\mathrm{F}(\mathbb{R}^2,n)/\mathfrak{S}_n & \longrightarrow & \mathrm{BU}(n).
\end{array} \tag{1.4}$$

Now, the basic results of Fuks' paper [53] can be stated as follows.

Theorem 1.2 *Let $n \geq 1$ be an integer.*

(1) *The homomorphism in cohomology*

$$H^*(\mathrm{BO}(n);\mathbb{F}_2) \longrightarrow H^*(\mathrm{F}(\mathbb{R}^2,n)/\mathfrak{S}_n;\mathbb{F}_2),$$

induced by the bundle morphism (1.3), *is an epimorphism. In other words, the cohomology (algebra) ring* $H^*(\mathcal{B}_n; \mathbb{F}_2) \cong H^*(\mathrm{F}(\mathbb{R}^2, n)/\mathfrak{S}_n; \mathbb{F}_2)$ *is generated by the Stiefel–Whitney classes of the vector bundle* $\xi_{\mathbb{R}^2,n}$*, that is*

$$H^*(\mathcal{B}_n; \mathbb{F}_2) \cong H^*(\mathrm{F}(\mathbb{R}^2, n)/\mathfrak{S}_n; \mathbb{F}_2) \cong \mathbb{F}_2[w_1(\xi_{\mathbb{R}^2,n}), \ldots, w_{n-1}(\xi_{\mathbb{R}^2,n})]/I_n, \tag{1.5}$$

where

$$I_n = \ker\big(H^*(\mathrm{BO}(n); \mathbb{F}_2) \longrightarrow H^*(\mathrm{F}(\mathbb{R}^2, n)/\mathfrak{S}_n; \mathbb{F}_2)\big).$$

(2) *The homomorphism in cohomology*

$$H^*(\mathrm{BU}(n); \mathbb{F}_2) \longrightarrow H^*(\mathrm{F}(\mathbb{R}^2, n)/\mathfrak{S}_n; \mathbb{F}_2), \tag{1.6}$$

is the zero homomorphism in all positive degrees.

Note that the bundle $\xi_{\mathbb{R}^2,n}$ has a non-vanishing cross section and consequently $w_n(\xi_{\mathbb{R}^2,n}) = 0$ does not appear in (1.5). Furthermore, the zero homomorphism (1.6) factors through the non-zero homomorphism $H^*(\mathrm{BU}(n); \mathbb{F}_2) \longrightarrow H^*(\mathrm{B}\mathfrak{S}_n; \mathbb{F}_2)$.

How did Fuks obtain these results? First, he realised that the lexicographic stratification of $(\mathbb{R}^2)^n$, introduced by Fox and Neuwirth in [52], can be used to obtain a non-regular cell complex model for the one-point compactification $\widehat{\mathrm{F}(\mathbb{R}^2, n)/\mathfrak{S}_n}$ of the unordered configuration space $\mathrm{F}(\mathbb{R}^2, n)/\mathfrak{S}_n$. In particular, the cell complex model has one 0-cell, the infinity point, and for every integer partition $(n_1, \ldots, n_k)$ of the integer $n = n_1+\cdots+n_k$ a cell $e(n_1, \ldots, n_k)$ of dimension $n+k$. Furthermore, he computed the boundary operator for all associated generators in the cellular chain complex, that is

$$\partial e(n_1, \ldots, n_k) = \sum_{i=1}^{k} \binom{n_i + n_{i+1}}{n_i} e(n_1, \ldots, n_{i-1}, n_i + n_{i+1}, \ldots, n_k).$$

(Here, with the usual abuse of notation, $e(n_1, \ldots, n_k)$ also denotes the corresponding generator in the chain group $C_{n+k}(\widehat{\mathrm{F}(\mathbb{R}^2, n)/\mathfrak{S}_n}; \mathbb{F}_2)$.) Then the Poincaré duality isomorphism for relative homology manifolds, as in [84, Thm 70.2]:

$$H^*(\mathrm{F}(\mathbb{R}^2, n)/\mathfrak{S}_n; \mathbb{F}_2) \cong \widetilde{H}_{2n-*}(\widehat{\mathrm{F}(\mathbb{R}^2, n)/\mathfrak{S}_n}; \mathbb{F}_2)$$

allowed Fuks to go back and forth between the cohomology $H^*(\mathrm{F}(\mathbb{R}^2, n)/\mathfrak{S}_n; \mathbb{F}_2)$ and the homology $H_*(\widehat{\mathrm{F}(\mathbb{R}^2, n)/\mathfrak{S}_n}; \mathbb{F}_2)$ of the explicitly given CW-complex for $\widehat{\mathrm{F}(\mathbb{R}^2, n)/\mathfrak{S}_n}$. In this way he was able, for example, to show the following results.

Theorem 1.3 *Let $n \geq 1$ and $k \geq 0$ be integers.*

(1) *The dimension of the $\mathbb{F}_2$ vector space*

$$H^k(\mathcal{B}_n; \mathbb{F}_2) \cong H^k(\mathrm{F}(\mathbb{R}^2, n)/\mathfrak{S}_n; \mathbb{F}_2)$$

is equal to the number of representations of the integer n as a sum of $n-k$ powers of 2, that is the number of sets $\{i_1, \ldots, i_{n-k}\}$ of non-negative integers such that $n = 2^{i_1} + \cdots + 2^{i_{n-k}}$.

(2) *The group*

$$H^{n-1}(\mathcal{B}_n; \mathbb{F}_2) \cong H^{n-1}(\mathrm{F}(\mathbb{R}^2, n)/\mathfrak{S}_n; \mathbb{F}_2)$$

does not vanish if and only if n is power of two.

For every integer $n \geq 1$ there is the natural inclusion homomorphism of groups $\varphi_n : \mathcal{B}_n \longrightarrow \mathcal{B}_{n+1}$ induced by extending a collection of n strings with a trivial $(n+1)$st string. The approach Fuks employed allowed him also to prove the following stability results.

Theorem 1.4 *Let $n \geq 1$ be an integer.*

(1) *The homomorphism $\varphi_n^* : H^*(\mathcal{B}_{n+1}; \mathbb{F}_2) \longrightarrow H^*(\mathcal{B}_n; \mathbb{F}_2)$ is an epimorphism.*
(2) *If n is even, then the homomorphism φ_n^* is an isomorphism.*

For an additional presentation of results by Fuks consult the famous book of Vassiliev *Complements of Discriminants of Smooth Maps: Topology and Applications*, [99, Ch. I].

1.7 Cohomology of Braid Spaces

In 1973, in a paper of Frederick Cohen [31] titled *Cohomology of braid spaces*, came an announcement, a teaser, for the landmark Springer Lecture Notes in Mathematics volume 533, *The Homology of Iterated Loop Spaces*, written by Cohen et al. [37]. Cohen presented two theorems [31, Thm. 1 and Thm. 2] and outlined a proof, with details appearing in [33], as an auxiliary tool on the road towards the homology of C_d-spaces. Both the results of [31] and [33] and even more the proof methods played a key role in various applications over the years.

What was announced in [31] and then proved in [33], or in other words what is the cohomology of braid spaces? In this article under the name of a braid space was hidden an unordered configuration space $\mathrm{F}(M, n)/\mathfrak{S}_n$ of a manifold M of dimension at least 2. Motivated by the fundamental work of May [75, 76] related to the study of iterated loop spaces and corresponding homology operations, Cohen computed specific cohomologies of the unordered configuration space $\mathrm{F}(\mathbb{R}^d, n)/\mathfrak{S}_n$ in the case when $n = p$ is a prime.

To be more precise let us fix an odd prime p, and integers $d \geq 2$ and $q \geq 0$. Furthermore, let $\mathcal{F}_p(q)$ denote the $\mathfrak{S}_p$-module defined on the ground vector space $\mathbb{F}_p$ by $\pi \cdot x = (-1)^{q \cdot \operatorname{sgn}(\pi)} x$. The cohomology Cohen considered is the cohomology of the cochain complex

$$\hom_{\mathfrak{S}_p}\big(C_* \operatorname{F}(\mathbb{R}^d, p); \mathcal{F}_p(q)\big) \tag{1.7}$$

of all $\mathfrak{S}_p$-equivariant cochains in the cochain complex $\hom\big(C_* \operatorname{F}(\mathbb{R}^d, p); \mathcal{F}_p(q)\big)$. Here $C_* \operatorname{F}(\mathbb{R}^d, p)$ denotes the singular chain complex of the ordered configuration space $\operatorname{F}(\mathbb{R}^d, p)$ with the natural structure of a free $\mathfrak{S}_p$-module, or free $\mathbb{Z}[\mathfrak{S}_p]$-module, inherited from the free $\mathfrak{S}_p$-action on $\operatorname{F}(\mathbb{R}^d, p)$.

In the case when q is even and since the action of $\mathfrak{S}_p$ on $\operatorname{F}(\mathbb{R}^d, p)$ is proper the cohomology of the cochain complex (1.7) coincides with the (usual) cohomology

$$H^*(\operatorname{F}(\mathbb{R}^d, p)/\mathfrak{S}_p; \mathbb{F}_p)$$

of the unordered configuration space $\operatorname{F}(\mathbb{R}^d, p)/\mathfrak{S}_p$ with (trivial) coefficients in the field $\mathbb{F}_p$. Furthermore, in this case the cohomology has a structure of a ring. In the case when $d \geq 2$ the cohomology of the cochain complex (1.7) coincides with the cohomology

$$H^*(\operatorname{F}(\mathbb{R}^d, p)/\mathfrak{S}_p; \mathcal{F}_p(q))$$

of the unordered configuration space $\operatorname{F}(\mathbb{R}^d, p)/\mathfrak{S}_p$ with local (twisted) coefficients in the $\mathfrak{S}_p$-module $\mathcal{F}_p(q)$. For simplicity, by an abuse of notation we denote the cohomology of the cochain complex (1.7) always by $H^*(\operatorname{F}(\mathbb{R}^d, p)/\mathfrak{S}_p; \mathcal{F}_p(q))$. Note that in the case when q is odd the cohomology $H^*(\operatorname{F}(\mathbb{R}^d, p)/\mathfrak{S}_p; \mathcal{F}_p(q))$ does not have a structure of a ring, only an additional structure of $\mathbb{F}_p$-module.

Before we state the main results announced in [31], and proved in all the details in [33], we need to set the stage.

For integers $n \geq 2$ and $d \geq 1$ let $\iota_{d,n} \colon \operatorname{F}(\mathbb{R}^d, n) \longrightarrow \operatorname{F}(\mathbb{R}^\infty, n)$ denote the $\mathfrak{S}_n$-equivariant continuous map induced by the inclusion $\mathbb{R}^d \longrightarrow \mathbb{R}^\infty$, $x \longmapsto (x, 0, 0, \dots)$ where $x \in \mathbb{R}^d$. It induces the following morphism of fibrations:

$$\begin{array}{ccc} \operatorname{F}(\mathbb{R}^d, n) & \xrightarrow{\ \iota_{d,n}\ } & \operatorname{F}(\mathbb{R}^\infty, n) \\ \big\downarrow & & \big\downarrow \\ \operatorname{F}(\mathbb{R}^d, n)/\mathfrak{S}_n & \xrightarrow{\ \iota_{d,n}/\mathfrak{S}_n\ } & \operatorname{F}(\mathbb{R}^\infty, n)/\mathfrak{S}_n . \end{array}$$

Since the configuration space $\operatorname{F}(\mathbb{R}^\infty, n)$ is contractible and equipped with a free $\mathfrak{S}_n$-action the orbit space $\operatorname{F}(\mathbb{R}^\infty, n)/\mathfrak{S}_n$ is a model for $B\mathfrak{S}_n$. In particular, $H^*(\mathfrak{S}_n) \cong H^*(\operatorname{F}(\mathbb{R}^\infty, n)/\mathfrak{S}_n)$ with any appropriately defined coefficients.

Let A and B be connected $\mathbb{Z}_{\geq 0}$-graded $\mathbb{F}_p$-algebras, where *connected* refers to $A_0 \cong B_0 \cong \mathbb{F}_p$. The $\sqcap$-product of A and B is the connected graded $\mathbb{F}_p$-algebra $A \sqcap B$ given for an integer $n \geq 0$ by

$$(A \sqcap B)_n := \begin{cases} \mathbb{F}_p, & n = 0, \\ A_n \times B_n, & n \geq 1. \end{cases}$$

The product structure on $A \sqcap B$ is specified by $A_s \cdot B_r = 0$ for all $s \geq 1$ and $r \geq 1$, and by the requirement that both projection maps $A \sqcap B \longrightarrow A$ and $A \sqcap B \longrightarrow B$ are algebra homomorphisms. Consult also [33, pp. 245–246].

Now we can present the following result of Cohen [31, Thm. 1] and [33, Thm. 5.2].

Theorem 1.5 *Let $d \geq 2$ be an integer, p an odd prime and q an even integer. Then*

$$H^*(\mathrm{F}(\mathbb{R}^d, p)/\mathfrak{S}_p; \mathcal{F}_p(q)) \cong A_d \sqcap \mathrm{im}(\iota_{d,n}/\mathfrak{S}_n)^*$$

as a connected $\mathbb{F}_p$-algebra. Here the $\mathbb{F}_p$-algebra $\mathrm{im}(\iota_{d,n}/\mathfrak{S}_n)^*$ is given by

$$\begin{aligned} \mathrm{im}(\iota_{d,n}/\mathfrak{S}_n)^* &\cong H^*(\mathfrak{S}_p; \mathcal{F}_p(q))/\ker(\iota_{d,n}/\mathfrak{S}_n)^* \\ &\cong H^*(\mathfrak{S}_p; \mathcal{F}_p(q))/H^{\geq (d-1)(p-1)+1}(\mathfrak{S}_p; \mathcal{F}_p(q)), \end{aligned}$$

and the graded $\mathbb{F}_p$-algebra A_d by

$$A_d = \begin{cases} \Lambda(a), & d \text{ even}, \\ \mathbb{F}_p, & d \text{ odd}. \end{cases}$$

The element a is of degree $d-1$, $\Lambda(a)$ is the exterior algebra generated by a, and $\mathbb{F}_p$ denotes the trivial connected $\mathbb{F}_p$-algebra.

Implicitly, we said that $\ker(\iota_{d,n}/\mathfrak{S}_n)^*$ is the ideal of $H^*(\mathfrak{S}_p; \mathcal{F}_p(q))$ consisting of all elements of the ring of degree $\geq (d-1)(p-1)+1$.

It is important to mention that the cohomology ring of the symmetric group $\mathfrak{S}_p$ with trivial $\mathbb{F}_p$ coefficients was already known at that time [75, p. 158]. Concretely, for q even

$$H^*(\mathfrak{S}_p; \mathbb{F}_p) \cong H^*(\mathfrak{S}_p; \mathcal{F}_p(q)) \cong \Lambda(b) \otimes \mathbb{F}_p[\beta b],$$

where b is an element of degree $2(p-1)-1$, $\Lambda(b)$ is the exterior algebra generated by b, βb is the Bockstein of b, and $\mathbb{F}_p[\beta b]$ the polynomial algebra generated by βb.

Next we give the following result of Cohen [31, Thm. 2] and [33, Thm. 5.3].

Theorem 1.6 *Let $d \geq 2$ be an integer, p an odd prime and q an odd integer. Then*

$$H^*(\mathrm{F}(\mathbb{R}^d, p)/\mathfrak{S}_p; \mathcal{F}_p(q)) \cong M_d \oplus \mathrm{im}(\iota_{d,n}/\mathfrak{S}_n)^*$$

as an $\mathbb{F}_p$-vector space, or as an $H^*(\mathfrak{S}_p;\mathbb{F}_p)$-module, or as an $H^*(\mathrm{F}(\mathbb{R}^d,p)/\mathfrak{S}_p;\mathbb{F}_p)$-module. Here the $\mathbb{F}_p$-vector space, or $H^*(\mathfrak{S}_p;\mathbb{F}_p)$-module, or $H^*(\mathrm{F}(\mathbb{R}^d,p)/\mathfrak{S}_p;\mathbb{F}_p)$-module, $\mathrm{im}(\iota_{d,n}/\mathfrak{S}_n)^*$ is given by

$$\begin{aligned}\mathrm{im}(\iota_{d,n}/\mathfrak{S}_n)^* &\cong H^*(\mathfrak{S}_p;\mathcal{F}_p(q))/\ker(\iota_{d,n}/\mathfrak{S}_n)^* \\ &\cong H^*(\mathfrak{S}_p;\mathcal{F}_p(q))/H^{\geq(d-1)(p-1)+1}(\mathfrak{S}_p;\mathcal{F}_p(q)),\end{aligned}$$

and the $\mathbb{F}_p$-vector space, or the $H^*(\mathfrak{S}_p;\mathbb{F}_p)$-module, or $H^*(\mathrm{F}(\mathbb{R}^d,p)/\mathfrak{S}_p;\mathbb{F}_p)$-module, M_d is determined by

$$M_d = \begin{cases} 0, & d \text{ even}, \\ \mathbb{F}_p = \langle\lambda\rangle, & d \text{ odd and } \deg(\lambda) = \frac{1}{2}(d-1)(p-1). \end{cases}$$

The $H^*(\mathfrak{S}_p;\mathbb{F}_p)$-module structure on M_d is trivial, which means that the generator λ is annihilated by all elements of positive degree of the ring $H^*(\mathfrak{S}_p;\mathbb{F}_p)$.

Again, implicitly we have that $\ker(\iota_{d,n}/\mathfrak{S}_n)^*$ is the $H^*(\mathfrak{S}_p;\mathbb{F}_p)$-submodule of $H^*(\mathfrak{S}_p;\mathcal{F}_p(q))$ generated by all elements of degree $\geq (d-1)(p-1)+1$.

The methods used in the proofs of Theorems 1.5 and 1.6 are at least as important as the results themselves. The proofs are presented on more than 50 pages in [33, Sec. 5–11]. One of the technical highlights is the so-called *Vanishing Theorem* [33, Thm. 8.2], which was also announced in [31, Thm. 4]. It gives a description of the cohomology Serre spectral sequences, with coefficients in the appropriately interpreted $\mathfrak{S}_p$-module associated with fiber bundles

$$\mathrm{F}(\mathbb{R}^d,p) \longrightarrow \mathrm{F}(\mathbb{R}^d,p)\times_{\mathfrak{S}_p}\mathrm{E}\mathfrak{S}_p \longrightarrow \mathrm{B}\mathfrak{S}_p \tag{1.8}$$

and

$$\mathrm{F}(\mathbb{R}^d,p) \longrightarrow \mathrm{F}(\mathbb{R}^d,p)\times_{\mathbb{Z}_p}\mathrm{E}\mathfrak{S}_p \longrightarrow \mathrm{B}\mathbb{Z}_p. \tag{1.9}$$

Here the cyclic group $\mathbb{Z}_p$ is the Sylow p-subgroup of the symmetric group $\mathfrak{S}_p$ generated by the cyclic shift, or in other words by the p-cycle $(12\ldots p)$. Note that $\mathrm{E}\mathfrak{S}_p$ is also a model for $\mathrm{E}\mathbb{Z}_p$. An illustration of the $E_2 \cong E_{(d-1)(p-1)+1}$-term of the Serre spectral sequence with trivial $\mathbb{F}_p$ coefficients associated to the fiber bundle (1.9) shaped by the Vanishing theorem is given in Fig. 1.4.

In this section, so far, we touched only the achievements of [33] announced in [31], but as the title *The homology of C_{n+1}-spaces* indicates, the paper has much more to offer; see also the announcement [32]. In the context of the previously presented results we only point out that the homology of the unordered configuration space $\mathrm{F}(\mathbb{R}^d,n)/\mathfrak{S}_n$, with arbitrary number of particles $n\geq 1$, and coefficients in the field $\mathbb{F}_p$, can be recovered from [33, Thm. 3.1]. In addition, a recipe for the

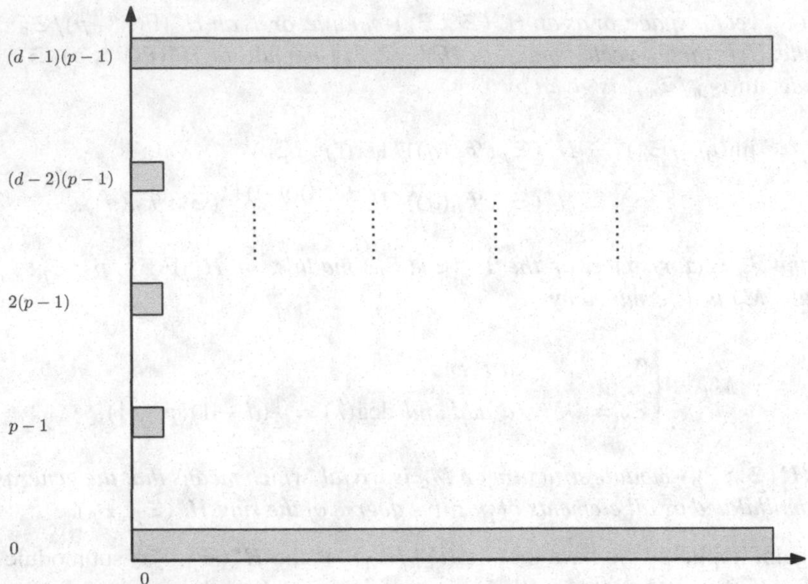

Fig. 1.4 The shape of E_2-term of the Serre spectral sequence associated to the fiber bundle (1.9)

coalgebra structure on $\mathbb{F}_p$-homology of $\mathrm{F}(\mathbb{R}^d, n)/\mathfrak{S}_n$ was given by showing that $H_*(\mathrm{F}(\mathbb{R}^d, n)/\mathfrak{S}_n; \mathbb{F}_p)$ injects into the coalgebra $H_*(\Omega^d S^d; \mathbb{F}_p)$.

Now, it was natural to ask: *How much further can these computations be extended? For which classes of spaces are similar formulas true?*

1.8 Homology of Unordered Configuration Spaces

The breakthroughs made in the 1970s first by May [75–77], and then by F. Cohen [31–33], Dusa McDuff [78, 79], Victor Snaith [94], Graeme Segal [92], Cohen et al. in [38, Thm. 1], and Cohen et al. [39, 40], opened a pathway for applications of homotopy methods in the study of topology of configuration spaces.

The next decade brought more excitement of different flavor with the work of Cohen et al. [35], Cohen et al. [41], Caruso et al. [28], followed by the results of Bödigheimer et al. [20] and Bödigheimer and Cohen [19].

The highlight of the 1970s and 1980s in the study of configuration spaces, from the perspective of this book, are the results given in the paper [20]. For this reason, we give a (simplified) presentation of how Bödigheimer, Cohen and Taylor computed homologies of unordered configuration spaces of manifolds.

In this section we consider configuration spaces of smooth, compact connected manifolds M of (fixed) dimension $d \geq 2$. Furthermore, by $\mathbb{F}$ we denote the field $\mathbb{F}_p$ with prime number p of elements, or a field of characteristic zero. Let $n \geq 1$ be an integer, and assume that in the case when $\mathbb{F}$ is not the field with two elements $\mathbb{F}_2$ the sum $d+n$ is odd.

The objective is to compute the homology of the unordered configuration spaces of the manifold M with coefficients in the field $\mathbb{F}$. The main idea is to describe the graded vector space $H_*(\mathrm{F}(M,k)/\mathfrak{S}_k;\mathbb{F})$ as a part of the homology of a much larger space, namely the quotient space

$$C(M;X) := \Big(\coprod_{k\geq 1} \mathrm{F}(M,k)\times_{\mathfrak{S}_k} X^k\Big)/\approx,$$

where X is a CW-complex with the base point $\mathrm{pt}\in X$, and the equivalence relation $\approx$ is generated by $(m_1,\ldots,m_k;x_1,\ldots,x_k)\approx(m_1,\ldots,m_{k-1};x_1,\ldots,x_{k-1})$ if $x_k=\mathrm{pt}$. The computation is done in several steps.

In the first step, based on a result from [33] and proceeding by an induction on the number of handles in a handle decomposition of M, the homology of $C(M;S^n)$ is described in terms of homologies of iterated loop spaces of spheres as follows; see [20, Thm. A].

Theorem 1.7 *There is an isomorphism of graded vector spaces*

$$\theta\colon H_*(C(M;S^n);\mathbb{F}) \longrightarrow \bigotimes_{i=0}^{d} H_*(\Omega^{d-i}S^{d+n};\mathbb{F})^{\otimes \dim(H_i(M;\mathbb{F}))}.$$

It is important to point out that, as an artefact of the proof, the isomorphism θ depends on the choice of a handle decomposition of M. On the other hand, the isomorphism is natural for embeddings which preserve the handle decomposition.

The next step is a more delicate one. We can say that it gives us a "filtration refinement" of the isomorphism θ. Indeed, the space $C(M;S^n)$ can be naturally filtered by the number of points in a configuration.
More precisely, let

$$\mathrm{F}_kC(M;S^n) := \mathrm{im}\Big(\coprod_{0\leq m\leq k} \mathrm{F}(M,k)\times_{\mathfrak{S}_m}(S^n)^m$$
$$\longrightarrow \coprod_{m\geq 0}\mathrm{F}(M,k)\times_{\mathfrak{S}_m}(S^n)^m \longrightarrow \Big(\coprod_{m\geq 0}\mathrm{F}(M,k)\times_{\mathfrak{S}_m}(S^n)^m\Big)/\approx\Big),$$

with the first map being the obvious inclusion and the second map the identification map. In this way we have the filtration of $C(M;S^n)$:

$$\varnothing = \mathrm{F}_{-1}C(M;S^n)\subseteq \mathrm{F}_0C(M;S^n)\subseteq \mathrm{F}_1C(M;S^n)\subseteq\cdots$$
$$\subseteq \mathrm{F}_{k-1}C(M;S^n)\subseteq \mathrm{F}_kC(M;S^n)\subseteq\cdots,$$

where each consecutive pair of spaces $(\mathrm{F}_kC(M;S^n), \mathrm{F}_{k-1}C(M;S^n))$ is an NDR-pair; consult [76, Prop. 2.6]. According to the work of Segal [92], Cohen [34] and Bödigheimer [18], the filtration stably splits. In particular,

$$\widetilde{H}_*(C(M;S^n);\mathbb{F}) \cong \bigoplus \widetilde{H}_*(\mathrm{F}_kC(M;S^n)/\mathrm{F}_{k-1}C(M;S^n);\mathbb{F}).$$

On the other hand $C(H_*, M, S^n) := \bigotimes_{i=0}^{d} H_*(\Omega^{d-i}S^{d+n};\mathbb{F})^{\otimes \dim(H_i(M;\mathbb{F}))}$ is an algebra with each generator equipped with a weight—as described in the language of Araki–Kudo–Dyer–Lashof homology operations much earlier by Shôrô Araki and Tatsuji Kudo [70] in the case $\mathbb{F} = \mathbb{F}_2$, by Eldon Dyer and Richard Lashof [45] and Cohen [33] for $\mathbb{F} = \mathbb{F}_p$, and by May [77] using the framework of E_∞-operads. The weight assignment induces the so-called product filtration on the algebra $C(H_*, M, S^n)$:

$$\begin{aligned} 0 = \mathcal{F}_0C(H_*, M, S^n) &\subseteq \mathcal{F}_1C(H_*, M; S^n) \subseteq \cdots \\ &\subseteq \mathcal{F}_{k-1}C(H_*, M, S^n) \subseteq \mathcal{F}_kC(H_*, M, S^n) \subseteq \cdots . \end{aligned}$$

It turns out that these two filtrations agree under the isomorphism θ of Theorem 1.7. In other words, the following theorem holds [20, Thm. B].

Theorem 1.8 *There are isomorphisms of graded vector spaces*

$$\theta_k\colon H_*(\mathrm{F}_kC(M;S^n);\mathbb{F}) \longrightarrow \mathcal{F}_kC(H_*, M, S^n),$$

such that for every $k \geq 0$ the following diagram commutes:

$$\begin{array}{ccc} H_*(\mathrm{F}_kC(M;S^n);\mathbb{F}) & \xrightarrow{\theta_k} & \mathcal{F}_kC(H_*, M, S^n) \\ \downarrow & & \downarrow \\ H_*(C(M;S^n);\mathbb{F}) & \xrightarrow{\theta} & C(H_*, M, S^n). \end{array}$$

The left vertical homomorphism in the diagram is induced by the inclusion of spaces $\mathrm{F}_kC(M;S^n) \subseteq C(M;S^n)$, while the right vertical homomorphism is the inclusion homomorphism $\mathcal{F}_kC(H_*, M, S^n) \subseteq C(H_*, M, S^n)$.

In the final step we consider the successive quotients:

$$\begin{aligned} \mathrm{D}_kC(M;S^n) &:= \mathrm{F}_kC(M;S^n)/\mathrm{F}_{k-1}C(M;S^n), \\ \mathcal{D}_kC(M;S^n) &:= \mathcal{F}_kC(H_*, M, S^n)/\mathcal{F}_{k-1}C(H_*, M, S^n). \end{aligned}$$

From Theorem 1.8 it follows directly that the family of isomorphisms θ_k induce the sequence of isomorphisms

$$\overline{\theta}_k\colon H_*(\mathrm{D}_kC(M;S^n);\mathbb{F}) \longrightarrow \mathcal{D}_kC(H_*, M, S^n).$$

Next, consider the vector bundle $\xi_{M,k}$ given by

$$\mathbb{R}^k \longrightarrow \mathrm{F}(M,k) \times_{\mathfrak{S}_k} \mathbb{R}^k \longrightarrow \mathrm{F}(M,k)/\mathfrak{S}_k.$$

It is not hard to see that the space $\mathrm{D}_k C(M; S^n)$ is the Thom space [80, Sec, 18] of the Whitney power vector bundle $\xi_{M,k}^{\oplus n}$. Consequently, applying the Thom isomorphism theorem [80, Cor. 10.7 and Lem. 18.2] to the previous result, a description of the homology of the unordered configuration space of the manifold M can be obtained; see [20, Thm. C].

Theorem 1.9 *There is an isomorphism of graded vector spaces*

$$H_{*-kn}(\mathrm{F}(M,k)/\mathfrak{S}_k; \mathbb{F}) \cong \mathcal{D}_k C(H_*, M, S^n).$$

In the case when $M = \mathbb{R}^d$ the last result is used in Sect. 4.2, more precisely in the proof of Theorem 4.1. For relevant details see Corollary 4.14.

For n odd and d even a similar result was deduced by a slight modification of the coefficients. Instead of coefficients in the field $\mathbb{F}$ with trivial $\mathfrak{S}_k$-module structure one considers the local coefficient system given on the field $\mathbb{F}$ by $\pi \cdot a = (-1)^{\mathrm{sgn}(\pi)} a$ for $\pi \in \mathfrak{S}_k$ and $a \in \mathbb{F}$.

The isomorphism of Theorem 1.9, in combination with understanding of Araki–Kudo–Dyer–Lashof homology operations, allows one to do explicit computations of the homology of unordered configuration spaces of manifolds. For example, Bödigheimer, Cohen and Taylor illustrated such computations in [20, Sec. 5] and in particular computed the dimensions of the vector spaces $H_*(\mathrm{F}(S^2,k)/\mathfrak{S}_k; \mathbb{F}_2)$ for $k \leq 10$:

$k =$	1	2	3	4	5	6	7	8	9	10
$\dim(H_0(\mathrm{F}(S^2,k)/\mathfrak{S}_k; \mathbb{F})) =$	1	1	1	1	1	1	1	1	1	1
$\dim(H_1(\mathrm{F}(S^2,k)/\mathfrak{S}_k; \mathbb{F})) =$	0	1	1	1	1	1	1	1	1	1
$\dim(H_2(\mathrm{F}(S^2,k)/\mathfrak{S}_k; \mathbb{F})) =$	1	1	1	2	2	2	2	2	2	2
$\dim(H_3(\mathrm{F}(S^2,k)/\mathfrak{S}_k; \mathbb{F})) =$			1	2	2	3	3	3	3	3
$\dim(H_4(\mathrm{F}(S^2,k)/\mathfrak{S}_k; \mathbb{F})) =$					1	2	2	3	3	3
$\dim(H_5(\mathrm{F}(S^2,k)/\mathfrak{S}_k; \mathbb{F})) =$					1	1	2	3	3	4
$\dim(H_6(\mathrm{F}(S^2,k)/\mathfrak{S}_k; \mathbb{F})) =$							1	2	3	4
$\dim(H_7(\mathrm{F}(S^2,k)/\mathfrak{S}_k; \mathbb{F})) =$								1	2	3
$\dim(H_8(\mathrm{F}(S^2,k)/\mathfrak{S}_k; \mathbb{F})) =$									1	2
$\dim(H_9(\mathrm{F}(S^2,k)/\mathfrak{S}_k; \mathbb{F})) =$									1	1

Finally, in the case when the sum $n + d$ is even and the coefficients are taken in the field of rational numbers, Bödigheimer and Cohen in [19] computed the cohomology $H^*(C(M_g; S^{2n}); \mathbb{Q})$ for $n \geq 1$, as a $\mathbb{Q}$ vector space. Here M_g denotes the open manifold obtained by deleting a point from an orientable surface of genus

g. In this way they demonstrated the importance of the parity assumption on the sum $n + d$ for the results of [20]. More precisely, they showed in [19, Thm. A] that

$$H^*(C(M_g; S^{2n}); \mathbb{Q}) \cong \mathbb{Q}[v, u_1, \dots, u_{2g}] \otimes H_*(\Lambda(w, z_1, \dots, z_{2g}), \partial),$$

where $\deg(v) = 2n$, $\deg(u_1) = \cdots = \deg(u_{2g}) = 4n + 2$, $\deg(w) = 4n + 1$, $\deg(z_1) = \cdots = \deg(z_{2g}) = 2n + 1$, and the differential ∂ on $\Lambda(w, z_1, \dots, z_{2g})$ is given by $\partial w = 2(z_1 z_2 + \cdots + z_{2g-1} z_{2g})$, and $\partial z_1 = \cdots = \partial z_{2g} = 0$.

Part I
Mod 2 Cohomology of Configuration Spaces

Chapter 2
The Ptolemaic Epicycles Embedding

In this chapter we follow the work of Hưng [62, Sec. 2], [64, Sec. 1 and Sec. 2]. Using the analogy with the structural map of the little cubes operad, we introduce and study an embedding of a product of spheres into the ordered configuration space of a Euclidean space.

Let M be a topological space. The **ordered configuration space** of n pairwise distinct points on the space M is the following space:

$$\mathrm{F}(M, n) := \{(x_1, \dots, x_n) \in X^n : x_i \neq x_j \text{ for all } 1 \leq i < j \leq n\},$$

equipped with the subspace topology. The symmetric group $\mathfrak{S}_n$ acts freely on the configuration space by

$$\pi \cdot (x_1, \dots, x_n) := (x_{\pi(1)}, \dots, x_{\pi(n)}),$$

where $\pi \in \mathfrak{S}_n$ and $(x_1, \dots, x_n) \in \mathrm{F}(M, n)$.

Now let $m \geq 0$ be an integer and $n = 2^m$. We define a bijection $\beta \colon \mathbb{Z}_2^{\oplus m} \longrightarrow [2^m]$ as follows:

$$(i_1, \dots, i_m) \longrightarrow 1 + \sum_{j=1}^{m} 2^{m-j} i_j.$$

In particular, we have:

$$\begin{aligned} (0, 0, \dots, 0, 0) &\longmapsto 1, & (0, 1, \dots, 1, 1) &\longmapsto 2^{m-1}, \\ (1, 0, \dots, 0, 0) &\longmapsto 2^{m-1} + 1, & (1, 1, \dots, 1, 1) &\longmapsto 2^m. \end{aligned}$$

The symmetric group $\mathfrak{S}_{2^m}$ for us is the group of permutations of the set $\mathbb{Z}_2^{\oplus m} = [2^m]$, where the last equality (set identification) is given via the bijection β.

P. V. M. Blagojević et al., *Equivariant Cohomology of Configuration Spaces Mod 2*, Lecture Notes in Mathematics 2282, https://doi.org/10.1007/978-3-030-84138-6_2

Definition 2.1 Let $d \geq 2$ be an integer or $d = \infty$, and let $m \geq 0$ be also an integer. Furthermore, let S^{d-1} denote the unit sphere in $\mathbb{R}^d$ with the base point $* := (1, 0, \ldots, 0) \in S^{d-1}$. Fix a real number $0 < \varepsilon < \frac{1}{3}$.

- The space $\mathrm{Pe}(\mathbb{R}^d, 2^m)$, a product of spheres,
- its embedding into the configuration space

$$\mathrm{ecy}_{d,2^m}\colon \mathrm{Pe}(\mathbb{R}^d, 2^m) \longrightarrow \mathrm{F}(\mathbb{R}^d, 2^m),$$

which is called the **Ptolemaic epicycles embedding**, and
- the group $\mathcal{S}_{2^m}$ that acts on the space $\mathrm{Pe}(\mathbb{R}^d, 2^m)$,

are defined inductively as follows.

(1) If $m = 0$ then $\mathrm{Pe}(\mathbb{R}^d, 1) := \{\mathrm{pt}\}$ is a point, and

$$\mathrm{ecy}_{d,1}\colon \mathrm{Pe}(\mathbb{R}^d, 1) \longrightarrow \mathrm{F}(\mathbb{R}^d, 1), \qquad \mathrm{pt} \longmapsto 0 \in (\mathbb{R}^d)^1.$$

The group $\mathcal{S}_1 := 1$ acts on both spaces $\mathrm{Pe}(\mathbb{R}^d, 1)$ and $\mathrm{F}(\mathbb{R}^d, 1)$ trivially. Since $\mathcal{S}_1 = 1 = \mathfrak{S}_1$ is the trivial group, the map $\mathrm{ecy}_{d,1}$ is an $\mathcal{S}_1$-equivariant map.

(2) If $m = 1$ then $\mathrm{Pe}(\mathbb{R}^d, 2) := S^{d-1} = (\mathrm{Pe}(\mathbb{R}^d, 1) \times \mathrm{Pe}(\mathbb{R}^d, 1)) \times S^{d-1}$ is a $(d-1)$-sphere, and

$$\mathrm{ecy}_{d,2}\colon \mathrm{Pe}(\mathbb{R}^d, 2) \longrightarrow \mathrm{F}(\mathbb{R}^d, 2), \qquad x \longmapsto (x, -x).$$

The group $\mathcal{S}_2 := (\mathcal{S}_1 \times \mathcal{S}_1) \rtimes \mathbb{Z}_2 = \mathcal{S}_1 \wr \mathbb{Z}_2 \cong \mathbb{Z}_2$ acts on $\mathrm{Pe}(\mathbb{R}^d, 2)$ antipodally. The groups $\mathcal{S}_2$ and $\mathfrak{S}_2$ are isomorphic via the unique isomorphism $\iota_1\colon \mathcal{S}_2 \longrightarrow \mathfrak{S}_2$. Hence, $\mathrm{F}(\mathbb{R}^d, 2)$ is a $\mathcal{S}_2$-space where the $\mathcal{S}_2$-action on $\mathrm{F}(\mathbb{R}^d, 2)$ is induced via the isomorphism ι_1. Consequently, $\mathrm{ecy}_{d,2}$ is an $\mathcal{S}_2$-equivariant map.

(3) Let us now assume that for $m = k$ we have defined

- the space

$$\mathrm{Pe}(\mathbb{R}^d, 2^k) = (\mathrm{Pe}(\mathbb{R}^d, 2^{k-1}) \times \mathrm{Pe}(\mathbb{R}^d, 2^{k-1})) \times S^{d-1} = (S^{d-1})^{2^k-1},$$

- the embedding of the spaces

$$\mathrm{ecy}_{d,2^k}\colon \mathrm{Pe}(\mathbb{R}^d, 2^k) \longrightarrow \mathrm{F}(\mathbb{R}^d, 2^k),$$

- the group embedding

$$\iota_k\colon \mathcal{S}_{2^k} \longrightarrow \mathfrak{S}_{2^k}$$

such that $\iota_k(\mathcal{S}_{2^k})$ is a Sylow 2-subgroup of $\mathfrak{S}_{2^k}$, and
- the action of the group $\mathcal{S}_{2^k}$ on $\mathrm{Pe}(\mathbb{R}^d, 2^k)$ in such a way that $\mathrm{ecy}_{d,2^k}$ is an $\mathcal{S}_{2^k}$-equivariant map, assuming that the action of $\mathcal{S}_{2^k}$ on the configuration space $\mathrm{F}(\mathbb{R}^d, 2^k)$ is given via ι_k.

For convenience we denote the coordinate functions of the embedding $\mathrm{ecy}_{d,2^k}$ by

$$\mathrm{ecy}_{d,2^k}(y) = (\mathrm{ecy}^1_{d,2^k}(y), \dots, \mathrm{ecy}^{2^k}_{d,2^k}(y)) \in \mathrm{F}(\mathbb{R}^d, 2^k),$$

where $y \in \mathrm{Pe}(\mathbb{R}^d, 2^k)$. That is, $\mathrm{ecy}^i_{d,2^k} \colon \mathrm{Pe}(\mathbb{R}^d, 2^k) \longrightarrow \mathbb{R}^d$ for $1 \le i \le 2^k$.

(4) Let $m = k + 1$, then we define

- the space

$$\begin{aligned}\mathrm{Pe}(\mathbb{R}^d, 2^{k+1}) &:= (\mathrm{Pe}(\mathbb{R}^d, 2^k) \times \mathrm{Pe}(\mathbb{R}^d, 2^k)) \times S^{d-1}\\ &= (S^{d-1})^{2^{k+1}-1},\end{aligned}$$

- the group

$$\begin{aligned}\mathcal{S}_{2^{k+1}} &:= (\mathcal{S}_{2^k} \times \mathcal{S}_{2^k}) \rtimes \mathbb{Z}_2\\ &= \mathcal{S}_{2^k} \wr \mathbb{Z}_2 = \mathbb{Z}_2 \wr \cdots \wr \mathbb{Z}_2 \qquad (k+1 \text{ times}),\end{aligned}$$

- the action of the group $\mathcal{S}_{2^{k+1}}$ on $\mathrm{Pe}(\mathbb{R}^d, 2^{k+1})$ by

$$\begin{aligned}(h_1, h_2) \cdot (y_1, y_2, x) &:= (h_1 \cdot y_1, h_2 \cdot y_2, x),\\ (h_1, h_2, \omega) \cdot (y_1, y_2, x) &:= (h_2 \cdot y_2, h_1 \cdot y_1, -x),\end{aligned}$$

  where $(h_1, h_2) \in \mathcal{S}_{2^k} \times \mathcal{S}_{2^k} \subseteq (\mathcal{S}_{2^k} \times \mathcal{S}_{2^k}) \rtimes \mathbb{Z}_2$, ω is the generator of $\mathbb{Z}_2 \subseteq (\mathcal{S}_{2^k} \times \mathcal{S}_{2^k}) \rtimes \mathbb{Z}_2$, and $(y_1, y_2, x) \in (\mathrm{Pe}(\mathbb{R}^d, 2^k) \times \mathrm{Pe}(\mathbb{R}^d, 2^k)) \times S^{d-1}$,
- the Ptolemaic epicycles embedding

$$\mathrm{ecy}_{d,2^{k+1}} \colon \mathrm{Pe}(\mathbb{R}^d, 2^{k+1}) \longrightarrow \mathrm{F}(\mathbb{R}^d, 2^{k+1})$$

  by

$$\begin{aligned}\mathrm{ecy}_{d,2^{k+1}}(y_1, y_2, x) := (&x + \varepsilon\, \mathrm{ecy}^1_{d,2^k}(y_1), \dots, x + \varepsilon\, \mathrm{ecy}^{2^k}_{d,2^k}(y_1),\\ &-x + \varepsilon\, \mathrm{ecy}^1_{d,2^k}(y_2), \dots, -x + \varepsilon\, \mathrm{ecy}^{2^k}_{d,2^k}(y_2)),\end{aligned}$$

- the embedding $\iota_{k+1} \colon \mathcal{S}_{2^{k+1}} \longrightarrow \mathfrak{S}_{2^{k+1}}$ that is defined by

$$\iota_{k+1}(h_1, h_2)(i) := \begin{cases}\iota_k(h_1)(i), & 1 \le i \le 2^k,\\ \iota_k(h_2)(i - 2^k) + 2^k, & 2^k + 1 \le i \le 2^{k+1},\end{cases}$$

$$\iota_{k+1}(\omega)(i) := \begin{cases} i + 2^k, & 1 \le i \le 2^k,\\ i - 2^k, & 2^k + 1 \le i \le 2^{k+1}.\end{cases}$$

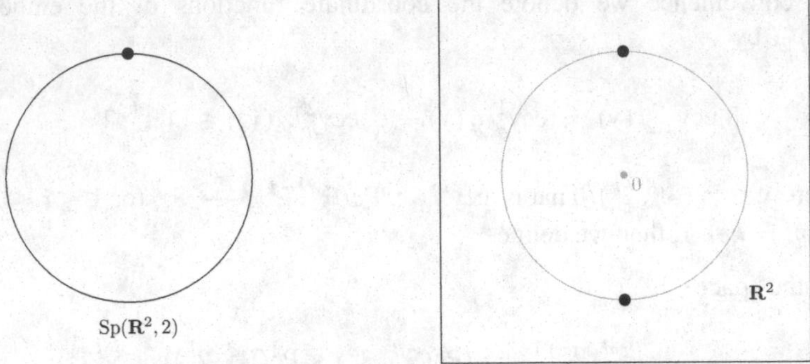

Fig. 2.1 An illustration of the embedding $\mathrm{ecy}_{2,2}\colon \mathrm{Pe}(\mathbb{R}^2, 2) \longrightarrow \mathrm{F}(\mathbb{R}^2, 2)$

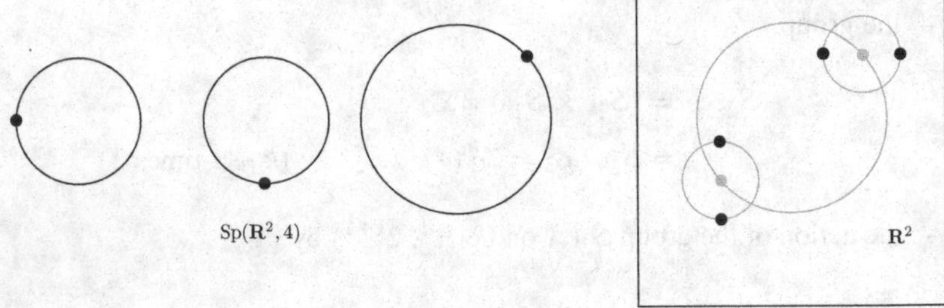

Fig. 2.2 An illustration of the embedding $\mathrm{ecy}_{2,4}\colon \mathrm{Pe}(\mathbb{R}^2, 4) \longrightarrow \mathrm{F}(\mathbb{R}^2, 4)$

This, in particular, means that the subgroup $\mathcal{S}_{2^k} \times 1$ permutes elements of $[2^{k+1}]_1$ while keeping elements of $[2^{k+1}]_2$ fixed, the subgroup $1 \times \mathcal{S}_{2^k}$ on the other hand permutes elements of $[2^{k+1}]_2$ and fixes elements of $[n]_1$. The subgroup generated by ω interchanges the blocks $[2^{k+1}]_1$ and $[2^{k+1}]_2$. In addition $\iota_{k+1}(\mathcal{S}_{2^{k+1}})$ is a Sylow 2-subgroup of $\mathfrak{S}_{2^{k+1}}$. For an illustration of the embedding $\mathrm{ecy}_{2,2}\colon \mathrm{Pe}(\mathbb{R}^2, 2) \longrightarrow \mathrm{F}(\mathbb{R}^2, 2)$ see Fig. 2.1.

Then $\mathrm{ecy}_{d,2^{k+1}}$ is an $\mathcal{S}_{2^{k+1}}$-equivariant map if the action of $\mathcal{S}_{2^{k+1}}$ on $\mathrm{F}(\mathbb{R}^d, 2^{k+1})$ is given by

$$(h_1, h_2) \cdot (z_1, \ldots, z_{2^{k+1}}) := (h_1 \cdot (z_1, \ldots, z_{2^k}), h_2 \cdot (z_{2^k+1}, \ldots, z_{2^{k+1}})),$$

$$(h_1, h_2, \omega) \cdot (z_1, \ldots, z_{2^{k+1}}) := (h_2 \cdot (z_{2^k+1}, \ldots, z_{2^{k+1}}), h_1 \cdot (z_1, \ldots, z_{2^k})),$$

for $(h_1, h_2) \in \mathcal{S}_{2^k} \times \mathcal{S}_{2^k}$, ω the generator of $\mathbb{Z}_2$, and $(z_1, \ldots, z_{2^{k+1}}) \in \mathrm{F}(\mathbb{R}^d, 2^{k+1})$. In other words, the action of $\mathcal{S}_{2^{k+1}}$ on $\mathrm{F}(\mathbb{R}^d, 2^{k+1})$ is given via the embedding ι_{k+1}. For an illustration of the embedding $\mathrm{ecy}_{2,4}$ see Fig. 2.2.

Remark 2.2 An analogous construction can be given using the little cubes operad $C_d(2)$ in the place of the sphere S^{d-1}. (For more details on little cubes operad see for example [76] or consult Sect. 7.3.) Indeed, let $d \geq 2$ be an integer or $d = \infty$, and let $m \geq 0$ be also an integer. We define the space $\mathrm{Ce}(\mathbb{R}^d, 2^m)$ of **little cubes epicycles space** and the corresponding $\mathcal{S}_{2^m}$-equivariant map $\mathrm{cecy}_{d,2^m}\colon \mathrm{Ce}(\mathbb{R}^d, 2^m) \longrightarrow \mathcal{C}_d(2^m)$ as follows.

(1) If $m = 0$, then we set $\mathrm{Ce}(\mathbb{R}^d, 1) := \mathcal{C}_d(1)$, and $\mathrm{cecy}_{d,1}\colon \mathrm{Ce}(\mathbb{R}^d, 1) \longrightarrow \mathcal{C}_d(1)$ is the identity map. The group $\mathcal{S}_1 = 1$ acts on both spaces $\mathrm{Ce}(\mathbb{R}^d, 1)$ and $\mathcal{C}_d(1)$ trivially, and so $\mathrm{cecy}_{d,1}$ is an $\mathcal{S}_1$-equivariant map.

(2) If $m = 1$, then we set $\mathrm{Ce}(\mathbb{R}^d, 2) := (\mathrm{Ce}(\mathbb{R}^d, 1) \times \mathrm{Ce}(\mathbb{R}^d, 1)) \times \mathcal{C}_d(2)$, and the map

$$\mathrm{cecy}_{d,2}\colon \mathrm{Ce}(\mathbb{R}^d, 2) \longrightarrow \mathcal{C}_d(2)$$

is the composition map

$$\begin{array}{ccc} (\mathrm{Ce}(\mathbb{R}^d, 1) \times \mathrm{Ce}(\mathbb{R}^d, 1)) \times \mathcal{C}_d(2) & \xrightarrow{(\mathrm{cecy}_{d,1} \times \mathrm{cecy}_{d,1}) \times \mathrm{id}} & (\mathcal{C}_d(1) \times \mathcal{C}_d(1)) \times \mathcal{C}_d(2) \\ & \searrow & \downarrow \mu \\ & & \mathcal{C}_d(2), \end{array}$$

where μ denotes the structural map of the little cubes operad, as defined in Sect. 7.3. The group $\mathcal{S}_2 = (\mathcal{S}_1 \times \mathcal{S}_1) \rtimes \mathfrak{S}_2 \xhookrightarrow{\iota_1} \mathfrak{S}_2$ coincides with the group $\mathcal{S}_{1,1;2}$ defined in Lemma 7.2 and acts on $\mathrm{Ce}(\mathbb{R}^d, 2)$ as follows:

$$(h_1, h_2) \cdot (y_1, y_2, x) := (h_1 \cdot y_1, h_2 \cdot y_2, x),$$
$$(h_1, h_2, \omega) \cdot (y_1, y_2, x) := (h_2 \cdot y_2, h_1 \cdot y_1, -x),$$

where $(h_1, h_2) \in \mathcal{S}_1 \times \mathcal{S}_1 \subseteq (\mathcal{S}_1 \times \mathcal{S}_1) \rtimes \mathfrak{S}_2$, ω is the generator of $\mathfrak{S}_2 \subseteq (\mathcal{S}_1 \times \mathcal{S}_1) \rtimes \mathfrak{S}_2$, and $(y_1, y_2, x) \in (\mathrm{Ce}(\mathbb{R}^d, 1) \times \mathrm{Ce}(\mathbb{R}^d, 1)) \times \mathcal{C}_d(2)$. The assumed action of $\mathcal{S}_2 = \mathcal{S}_{1,1;2}$ on $(\mathcal{C}_d(1) \times \mathcal{C}_d(1)) \times \mathcal{C}_d(2)$ is described in Sect. 7.3. Finally the action of $\mathcal{S}_2$ on $\mathcal{C}_d(2)$ is given via embedding $\iota_1 : \mathcal{S}_2 \longrightarrow \mathfrak{S}_2$. With these actions both maps $(\mathrm{cecy}_{d,1} \times \mathrm{cecy}_{d,1}) \times \mathrm{id}$ and μ are $\mathcal{S}_2$-equivariant, and consequently the composition map $\mathrm{cecy}_{d,2}$ is an $\mathcal{S}_2$-equivariant. It is important to notice that $\mathcal{S}_1 = 1$ and so $\mathcal{S}_2 \cong \mathbb{Z}_2 \cong \mathfrak{S}_2$.

(3) Let us assume that for $m = k$ we have defined the space

$$\mathrm{Ce}(\mathbb{R}^d, 2^k) = (\mathrm{Ce}(\mathbb{R}^d, 2^{k-1}) \times \mathrm{Ce}(\mathbb{R}^d, 2^{k-1})) \times \mathcal{C}_d(2),$$

the embedding $\mathrm{cecy}_{d,2^k}\colon \mathrm{Ce}(\mathbb{R}^d, 2^k) \longrightarrow \mathcal{C}_2(2^k)$, the embedding ι_k of the Sylow 2-subgroup $\mathcal{S}_{2^k}$ into $\mathfrak{S}_{2^k}$, and the action of the group $\mathcal{S}_{2^k}$ on $\mathrm{Ce}(\mathbb{R}^d, 2^k)$ in such a way that $\mathrm{cecy}_{d,2^k}$ is an $\mathcal{S}_{2^k}$-equivariant map.

For $m = k + 1$ we define the space

$$\mathrm{Ce}(\mathbb{R}^d, 2^{k+1}) := (\mathrm{Ce}(\mathbb{R}^d, 2^k) \times \mathrm{Ce}(\mathbb{R}^d, 2^k)) \times C_d(2).$$

The action of the group $\mathcal{S}_{2^{k+1}}$ on $\mathrm{Ce}(\mathbb{R}^d, 2^{k+1})$ is given by

$$(h_1, h_2) \cdot (y_1, y_2, x) := (h_1 \cdot y_1, h_2 \cdot y_2, x),$$
$$(h_1, h_2, \omega) \cdot (y_1, y_2, x) := (h_2 \cdot y_2, h_1 \cdot y_1, -x),$$

where $(h_1, h_2) \in \mathcal{S}_{2^k} \times \mathcal{S}_{2^k} \subseteq (\mathcal{S}_{2^k} \times \mathcal{S}_{2^k}) \rtimes \mathfrak{S}_2$, ω is the generator of $\mathfrak{S}_2 \subseteq (\mathcal{S}_{2^k} \times \mathcal{S}_{2^k}) \rtimes \mathfrak{S}_2$, and $(y_1, y_2, x) \in (\mathrm{Ce}(\mathbb{R}^d, 2^k) \times \mathrm{Ce}(\mathbb{R}^d, 2^k)) \times C_d(2)$. The map

$$\mathrm{cecy}_{d,2^{k+1}} \colon \mathrm{Ce}(\mathbb{R}^d, 2^{k+1}) \longrightarrow C_d(2^{k+1})$$

is defined to be the following composition

$$\begin{array}{ccc} (\mathrm{Ce}(\mathbb{R}^d, 2^k) \times \mathrm{Ce}(\mathbb{R}^d, 2^k)) \times C_d(2) & \xrightarrow{(\mathrm{cecy}_{d,2^k} \times \mathrm{cecy}_{d,2^k}) \times \mathrm{id}} & (C_d(2^k) \times C_d(2^k)) \times C_d(2) \\ & \searrow^{\mathrm{cecy}_{d,2^{k+1}}} & \downarrow \mu \\ & & C_d(2^{k+1}), \end{array}$$

where μ denotes the structural map of the little cubes operad. Under assumed actions, by direct inspection, we get that the two maps $(\mathrm{cecy}_{d,2^k} \times \mathrm{cecy}_{d,2^k}) \times \mathrm{id}$ and μ are both $\mathcal{S}_{2^{k+1}}$-equivariant. Consequently the composition map $\mathrm{cecy}_{d,2^{k+1}}$ is also $\mathcal{S}_{2^{k+1}}$-equivariant.

Example 2.3 Let $m = 2$. Then the bijection $\beta \colon \mathbb{Z}_2^{\oplus 2} \longrightarrow [4]$ is given by

$$(0,0) \longmapsto 1, \quad (0,1) \longmapsto 2, \quad (1,0) \longmapsto 3, \quad (1,1) \longmapsto 4.$$

The group $\mathcal{S}_2 = (\mathcal{S}_1 \times \mathcal{S}_1) \rtimes \mathbb{Z}_2 = \langle \varepsilon_1, \varepsilon_2 \rangle \rtimes \langle \omega \rangle \cong (\mathbb{Z}_2 \times \mathbb{Z}_2) \rtimes \mathbb{Z}_2$ embeds via ι_2 into the symmetric group $\mathfrak{S}_4$ by sending generators to the following permutations

$$\varepsilon_1 \longmapsto \begin{pmatrix} 1234 \\ 2134 \end{pmatrix}, \qquad \varepsilon_2 \longmapsto \begin{pmatrix} 1234 \\ 1243 \end{pmatrix}, \qquad \omega \longmapsto \begin{pmatrix} 1234 \\ 3412 \end{pmatrix}.$$

Let $n \geq 1$ be an integer. We consider the following vector subspace of $\mathbb{R}^n$:

$$W_n := \{(a_1, \ldots, a_n) \in \mathbb{R}^n : a_1 + \cdots + a_n = 0\}.$$

Then the subspace $\{(x_1, \ldots, x_n) \in (\mathbb{R}^d)^n : x_1 + \cdots + x_n = 0\}$ of $(\mathbb{R}^d)^n$ can be identified with the direct sum $W_n^{\oplus d}$. The map $\mathrm{ecy}_{d,n}$ that we have defined has the following property.

Proposition 2.4 *Let $d \geq 2$ be an integer or $d = \infty$, and let $m \geq 0$ be an integer. Then* $\mathrm{im}(\mathrm{ecy}_{d,2^m}) \subseteq W_{2^m}^{\oplus d}$*, that is, for every* $(y_1, y_2, x) \in \mathrm{Pe}(\mathbb{R}^d, 2^m)$

$$\mathrm{ecy}^1_{d,2^m}(y_1, y_2, x) + \cdots + \mathrm{ecy}^{2^m}_{d,2^m}(y_1, y_2, x) = 0.$$

Proof We use induction on the integer $m \geq 0$. For $m = 0$ we have that

$$\mathrm{im}(\mathrm{ecy}_{d,1}) = \{0\} = W_1^{\oplus d}.$$

Assume that for $m = k > 0$ we have $\mathrm{im}(\mathrm{ecy}_{d,2^k}) \subseteq W_{2^k}^{\oplus d}$. Then for $m = k+1$ and $(y_1, y_2, x) \in \mathrm{Pe}(\mathbb{R}^d, 2^{k+1}) = (\mathrm{Pe}(\mathbb{R}^d, 2^k) \times \mathrm{Pe}(\mathbb{R}^d, 2^k)) \times S^{d-1}$, using the induction hypothesis, we get

$$\begin{aligned}
\sum_{j=1}^{2^{k+1}} \mathrm{ecy}^j_{d,2^{k+1}}(y_1, y_2, x) &= (x + \varepsilon\, \mathrm{ecy}^1_{d,2^k}(y_1)) + \cdots + (x + \varepsilon\, \mathrm{ecy}^{2^k}_{d,2^k}(y_1)) \\
&\quad + (-x + \varepsilon\, \mathrm{ecy}^1_{d,2^k}(y_2)) + \cdots + (-x + \varepsilon\, \mathrm{ecy}^{2^k}_{d,2^k}(y_2)) \\
&= \varepsilon\Big(\sum_{j=1}^{2^k} \mathrm{ecy}^j_{d,2^k}(y_1)\Big) + \varepsilon\Big(\sum_{j=1}^{2^k} \mathrm{ecy}^j_{d,2^k}(y_2)\Big) \\
&= 0.
\end{aligned}$$

Consequently $\mathrm{im}(\mathrm{ecy}_{d,2^{k+1}}) \subseteq W_{2^{k+1}}^{\oplus d}$, and the induction is completed. □

Next we verify that the Ptolemaic epicycles embedding $\mathrm{ecy}_{d,2^m}$ is indeed an embedding, see [64, Lem. 1.6].

Proposition 2.5 *Let $d \geq 2$ be an integer or $d = \infty$, and let $m \geq 0$ be an integer. For any fixed real number $0 < \varepsilon < \frac{1}{3}$ the map*

$$\mathrm{ecy}_{d,2^m} \colon \mathrm{Pe}(\mathbb{R}^d, 2^m) \longrightarrow \mathrm{F}(\mathbb{R}^d, 2^m)$$

is an embedding.

Proof The continuity of the map $\mathrm{ecy}_{d,2^m}$ follows directly from Definition 2.1. Thus, we only need to show that $\mathrm{ecy}_{d,2^m}$ is an injective map. For that we use induction on integer $m \geq 0$.

The map $\mathrm{ecy}_{d,1}\colon \mathrm{Pe}(\mathbb{R}^d, 1) \longrightarrow \mathrm{F}(\mathbb{R}^d, 1)$ is evidently injective, since $\mathrm{Pe}(\mathbb{R}^d, 1) = \mathrm{pt}$. Assume that for $m = k \geq 0$ the map $\mathrm{ecy}_{d,2^k}$ is injective. Now, for $m = k + 1$, suppose that

$$\mathrm{ecy}_{d,2^{k+1}}(y_1, y_2, x) = \mathrm{ecy}_{d,2^{k+1}}(y_1', y_2', x'), \tag{2.1}$$

where $(y_1, y_2, x), (y_1', y_2', x') \in \mathrm{Pe}(\mathbb{R}^d, 2^{k+1})$. Consequently, the sums of first 2^k coordinates must coincide

$$\sum_{j=1}^{2^k} \mathrm{ecy}^j_{d,2^{k+1}}(y_1, y_2, x) = \sum_{j=1}^{2^k} \mathrm{ecy}^j_{d,2^{k+1}}(y_1', y_2', x').$$

From the definition of the map $\mathrm{ecy}_{d,2^{k+1}}$ it follows that

$$2^k x + \varepsilon\Big(\sum_{j=1}^{2^k} \mathrm{ecy}^j_{d,2^k}(y_1)\Big) = 2^k x' + \varepsilon\Big(\sum_{j=1}^{2^k} \mathrm{ecy}^j_{d,2^k}(y_1')\Big).$$

From Proposition 2.4 we get

$$\sum_{j=1}^{2^k} \mathrm{ecy}^j_{d,2^k}(y_1) = 0 \qquad \text{and} \qquad \sum_{j=1}^{2^k} \mathrm{ecy}^j_{d,2^k}(y_1') = 0,$$

implying that $x = x'$. Furthermore, using Definition 2.1 and (2.1) we have that

$$\mathrm{ecy}_{d,2^k}(y_1) = \mathrm{ecy}_{d,2^k}(y_1') \qquad \text{and} \qquad \mathrm{ecy}_{d,2^k}(y_2) = \mathrm{ecy}_{d,2^k}(y_2').$$

Finally the induction hypothesis implies that $y_1 = y_1'$ and $y_2 = y_2'$ concluding the proof of the proposition. □

In the case when $d = \infty$ the space $\mathrm{Pe}(\mathbb{R}^d, 2^m)$ is a contractible space with a free $\mathcal{S}_{2^m}$-action, and therefore is a model for $\mathrm{E}\mathcal{S}_{2^m}$. In particular, we can observe that

$$\mathrm{Pe}(\mathbb{R}^\infty, 2^m) \cong \mathrm{colim}_{d\to\infty} \mathrm{Pe}(\mathbb{R}^d, 2^m),$$

where the colimit is defined via the inclusions $\mathbb{R}^d \longrightarrow \mathbb{R}^{d+1}$, $x \longmapsto (x, 0)$, which induce the corresponding inclusion maps $\mathrm{Pe}(\mathbb{R}^d, 2^m) \longrightarrow \mathrm{Pe}(\mathbb{R}^{d+1}, 2^m)$. Furthermore, the induced $\mathcal{S}_{2^m}$-equivariant map $\mathrm{Pe}(\mathbb{R}^d, 2^m) \longrightarrow \mathrm{colim}_{d\to\infty} \mathrm{Pe}(\mathbb{R}^d, 2^m)$ is given by the inclusion maps $\mathbb{R}^d \longrightarrow \mathbb{R}^\infty$, $x \longmapsto (x, 0, 0, \ldots)$, and is denoted by

$$\kappa_{d,2^m}\colon \mathrm{Pe}(\mathbb{R}^d, 2^m) \longrightarrow \mathrm{Pe}(\mathbb{R}^\infty, 2^m). \tag{2.2}$$

In summary, we have the following.

Proposition 2.6 *Let $m \geq 0$ be an integer. Then* $\mathrm{Pe}(\mathbb{R}^\infty, 2^m)$ *is a free contractible* $\mathcal{S}_{2^m}$*-CW complex.*

Before we continue towards the study of the equivariant cohomology of the space $\mathrm{Pe}(\mathbb{R}^d, 2^m)$, we specify an elementary abelian subgroup $\mathcal{E}_m$ of the Sylow 2-subgroup $\mathcal{S}_{2^m}$ of the symmetric group $\mathfrak{S}_{2^m}$.

Definition 2.7 Let $m \geq 0$ be an integer. The subgroup $\mathcal{E}_m$ of the group $\mathcal{S}_{2^m}$ is defined inductively as follows.

(1) If $m = 0$, then we set $\mathcal{E}_0 := \mathcal{S}_1 = 1$.
(2) If $m = 1$, then we set $\mathcal{E}_1 := \mathcal{S}_2 \cong \mathbb{Z}_2$.
(3) Let us assume that for $m = k \geq 1$ we have defined the subgroup $\mathcal{E}_k$ of $\mathcal{S}_{2^k}$.
(4) If $m = k + 1$ and $\delta \colon \mathcal{S}_{2^k} \longrightarrow \mathcal{S}_{2^k} \times \mathcal{S}_{2^k}$ denotes the diagonal monomorphism given by $s \mapsto (s, s)$, then we set

$$\mathcal{E}_{k+1} := \delta(\mathcal{E}_k) \times \mathbb{Z}_2 \cong \mathcal{E}_k \times \mathbb{Z}_2 \cong \mathbb{Z}_2^{\oplus k+1} \subseteq (\mathcal{S}_{2^k} \times \mathcal{S}_{2^k}) \rtimes \mathbb{Z}_2 = \mathcal{S}_{2^{k+1}}.$$

Furthermore, let $C_1, \dots, C_m$ be cyclic groups isomorphic to $\mathbb{Z}_2$ with the property that

$$\mathcal{E}_i = (C_1 \times \cdots \times C_{i-1}) \times C_i = \delta(\mathcal{E}_{i-1}) \times C_i.$$

Here we make slight abuse of notation identifying $\mathcal{E}_{i-1}$ with a subgroup of $\mathcal{E}_i$. Having this decomposition fixed we see that $\mathcal{S}_{2^m} = C_1 \wr \cdots \wr C_m$. In summary, we have inclusions of the groups

$$\mathcal{E}_m \subseteq \mathcal{S}_{2^m} \subseteq \mathfrak{S}_{2^m}, \tag{2.3}$$

where $\mathcal{E}_m$ is the subgroup of $\mathfrak{S}_{2^m}$ given by translations (seen as permutations) of $\mathbb{Z}_2^{\oplus m}$, the so called regular embedded subgroup [2, Ex. III.2.7], and the inclusion $\mathcal{S}_{2^m} \subseteq \mathfrak{S}_{2^m}$ is defined via the monomorphism ι_m.

The cohomology of the elementary abelian group $\mathcal{E}_m$ with $\mathbb{F}_2$ coefficients is well known. In particular, we fix the following presentation

$$H^*(\mathcal{E}_m; \mathbb{F}_2) = \mathbb{F}_2[y_1, \dots, y_m],$$

where $\deg(y_k) = 1$ for $1 \leq k \leq m$, in such a way that the exact sequence of groups

$$1 \longrightarrow \mathcal{E}_{m-1} \longrightarrow \mathcal{E}_{m-1} \times C_m \longrightarrow C_m \longrightarrow 1$$

induces the following sequence of algebras

$$0 \longleftarrow \mathbb{F}_2[y_1, \dots, y_{m-1}] \longleftarrow \mathbb{F}_2[y_1, \dots, y_{m-1}, y_m] \longleftarrow \mathbb{F}_2[y_m] \longleftarrow 0,$$

which is exact in each positive degree.

Chapter 3
The Equivariant Cohomology of $\mathrm{Pe}(\mathbb{R}^d, 2^m)$

Let $d \geq 2$ be an integer or $d = \infty$, and let $m \geq 0$ be an integer. In this chapter we study the equivariant cohomology of the space $\mathrm{Pe}(\mathbb{R}^d, 2^m)$ with respect to the already defined free action of the group $\mathcal{S}_{2^m}$, that is,

$$\begin{aligned} H^*_{\mathcal{S}_{2^m}}(\mathrm{Pe}(\mathbb{R}^d, 2^m); \mathbb{F}_2) &\cong H^*(\mathrm{Pe}(\mathbb{R}^d, 2^m)/\mathcal{S}_{2^m}; \mathbb{F}_2) \\ &\cong H^*(\mathrm{E}\mathcal{S}_{2^m} \times_{\mathcal{S}_{2^m}} \mathrm{Pe}(\mathbb{R}^d, 2^m); \mathbb{F}_2). \end{aligned}$$

In particular, if $d = \infty$, then according to Proposition 2.6 we have that

$$H^*_{\mathcal{S}_{2^m}}(\mathrm{Pe}(\mathbb{R}^\infty, 2^m); \mathbb{F}_2) \cong H^*(\mathrm{Pe}(\mathbb{R}^\infty, 2^m)/\mathcal{S}_{2^m}; \mathbb{F}_2) \cong H^*(\mathcal{S}_{2^m}; \mathbb{F}_2).$$

Since the space $\mathrm{Pe}(\mathbb{R}^d, 2^m)$ is defined inductively, our computation utilizes this feature, and is an alternative to the calculation presented in [64, Sec. 2]. The methods we use are applicable in a more general setting.

3.1 Small Values of m

Let us first consider the case when $m = 0$. In this case, according to Definition 2.1, we have that

$$\mathrm{Pe}(\mathbb{R}^d, 1) = \mathrm{pt} \qquad \text{and} \qquad \mathcal{S}_1 = 1.$$

Consequently

$$H^r_{\mathcal{S}_1}(\mathrm{Pe}(\mathbb{R}^d, 1); \mathbb{F}_2) \cong H^r(\mathrm{Pe}(\mathbb{R}^d, 1)/\mathcal{S}_1; \mathbb{F}_2) \cong H^r(\mathrm{pt}; \mathbb{F}_2) \cong \begin{cases} \mathbb{F}_2, & r = 0, \\ 0, & r \neq 0. \end{cases}$$

P. V. M. Blagojević et al., *Equivariant Cohomology of Configuration Spaces Mod 2*, Lecture Notes in Mathematics 2282, https://doi.org/10.1007/978-3-030-84138-6_3

We specify a particular additive basis for the cohomology $H^*(\mathrm{Pe}(\mathbb{R}^d, 1)/\mathcal{S}_1; \mathbb{F}_2)$ by

$$\mathcal{B}(\mathbb{R}^d, 1) := \{1\},$$

where 1 is the generator of $H^0(\mathrm{Pe}(\mathbb{R}^d, 1)/\mathcal{S}_1; \mathbb{F}_2) \cong \mathbb{F}_2$. In addition we partition the set $\mathcal{B}(\mathbb{R}^d, 1)$ into two subsets

$$\mathcal{B}_a(\mathbb{R}^d, 1) := \mathcal{B}(\mathbb{R}^d, 1) \quad \text{and} \quad \mathcal{B}_i(\mathbb{R}^d, 1) := \mathcal{B}(\mathbb{R}^d, 1) \backslash \mathcal{B}_a(\mathbb{R}^d, 1).$$

In particular, $\mathcal{B}_i(\mathbb{R}^d, 1) = \varnothing$.

Next, let $m = 1$. Then from Definition 2.1 we get

$$\mathrm{Pe}(\mathbb{R}^d, 2) = S^{d-1} \quad \text{and} \quad \mathcal{S}_2 = \mathbb{Z}_2.$$

Therefore

$$\begin{aligned} H^r_{\mathcal{S}_2}(\mathrm{Pe}(\mathbb{R}^d, 2); \mathbb{F}_2) &\cong H^r(\mathrm{Pe}(\mathbb{R}^d, 2)/\mathbb{Z}_2; \mathbb{F}_2) \cong H^r(\mathbb{RP}^{d-1}; \mathbb{F}_2) \\ &\cong \begin{cases} \mathbb{F}_2, & 0 \le r \le d-1, \\ 0, & \text{otherwise.} \end{cases} \end{aligned}$$

Again we specify an additive basis for the cohomology $H^*(\mathrm{Pe}(\mathbb{R}^d, 2)/\mathcal{S}_2; \mathbb{F}_2)$ as follows:

$$\mathcal{B}(\mathbb{R}^d, 2) := \{1, e, e^2, \ldots, e^{d-1}\},$$

where e^i is the generator of $H^i(\mathbb{RP}^{d-1}; \mathbb{F}_2) \cong \mathbb{F}_2$. The partition of the basis $\mathcal{B}(\mathbb{R}^d, 2)$ we use is defined by:

$$\mathcal{B}_a(\mathbb{R}^d, 2) := \{(x \otimes x) \otimes_{\mathbb{Z}_2} e^i : x \in \mathcal{B}_a(\mathbb{R}^d, 1),\ 0 \le i \le d-1\} = \{1, e, e^2, \ldots, e^{d-1}\},$$

and

$$\mathcal{B}_i(\mathbb{R}^d, 2) := \mathcal{B}(\mathbb{R}^d, 2) \backslash \mathcal{B}_a(\mathbb{R}^d, 2) = \varnothing.$$

3.2 The Case $m = 2$

The first interesting case is when $m = 2$. Again from Definition 2.1 we have that

$$\mathrm{Pe}(\mathbb{R}^d, 2^m) = \mathrm{Pe}(\mathbb{R}^d, 4) = (\mathrm{Pe}(\mathbb{R}^d, 2) \times \mathrm{Pe}(\mathbb{R}^d, 2)) \times S^{d-1} = (S^{d-1} \times S^{d-1}) \times S^{d-1},$$

and

$$\mathcal{S}_{2^m} = \mathcal{S}_4 = (\mathbb{Z}_2 \times \mathbb{Z}_2) \rtimes \mathbb{Z}_2 = \mathbb{Z}_2 \wr \mathbb{Z}_2 = \big(\langle \varepsilon_1 \rangle \times \langle \varepsilon_2 \rangle\big) \rtimes \langle \omega \rangle.$$

(The group $\mathcal{S}_4$ is isomorphic to the dihedral group D_8.) The free action of the group $\mathcal{S}_4$ on $\mathrm{Pe}(\mathbb{R}^d, 4)$ we introduced is given by

$$\begin{aligned} \varepsilon_1 \cdot (y_1, y_2, x) &= (-y_1, y_2, x), \\ \varepsilon_2 \cdot (y_1, y_2, x) &= (y_1, -y_2, x), \\ \omega \cdot (y_1, y_2, x) &= (y_2, y_1, -x), \end{aligned}$$

for $(y_1, y_2, x) \in (S^{d-1} \times S^{d-1}) \times S^{d-1}$. Thus

$$\begin{aligned} \mathrm{Pe}(\mathbb{R}^d, 4)/\mathcal{S}_4 &= \big((S^{d-1} \times S^{d-1}) \times S^{d-1}\big)/\mathcal{S}_4 \\ &\cong \big((S^{d-1}/\mathbb{Z}_2 \times S^{d-1}/\mathbb{Z}_2) \times S^{d-1}\big)/\mathbb{Z}_2 \\ &\cong \big((\mathrm{RP}^{d-1} \times \mathrm{RP}^{d-1}) \times S^{d-1}\big)/\mathbb{Z}_2 =: (\mathrm{RP}^{d-1} \times \mathrm{RP}^{d-1}) \times_{\mathbb{Z}_2} S^{d-1}. \end{aligned}$$

The action of $\mathbb{Z}_2$ on $(\mathrm{RP}^{d-1} \times \mathrm{RP}^{d-1}) \times S^{d-1}$ we have is given by

$$(u_1, u_2, x) \longmapsto (u_2, u_1, -x)$$

for $(u_1, u_2, x) \in (\mathrm{RP}^{d-1} \times \mathrm{RP}^{d-1}) \times S^{d-1}$. Since this $\mathbb{Z}_2$-action is free then the corresponding quotient space $(\mathrm{RP}^{d-1} \times \mathrm{RP}^{d-1}) \times_{\mathbb{Z}_2} S^{d-1}$ is the total space of the fiber bundle

$$\mathrm{RP}^{d-1} \times \mathrm{RP}^{d-1} \longrightarrow (\mathrm{RP}^{d-1} \times \mathrm{RP}^{d-1}) \times_{\mathbb{Z}_2} S^{d-1} \xrightarrow{\ p\ } \mathrm{RP}^{d-1}, \tag{3.1}$$

where the map p is induced by the ($\mathbb{Z}_2$-equivariant) projection on the second factor

$$(\mathrm{RP}^{d-1} \times \mathrm{RP}^{d-1}) \times S^{d-1} \longrightarrow S^{d-1}.$$

Furthermore, the $\mathcal{S}_4$-equivariant inclusion $\kappa_{d,4} \colon \mathrm{Pe}(\mathbb{R}^d, 4) \longrightarrow \mathrm{Pe}(\mathbb{R}^\infty, 4)$, introduced in (2.2), induces the following morphism of fiber bundles

$$\begin{array}{ccc} (\mathrm{RP}^{d-1} \times \mathrm{RP}^{d-1}) \times_{\mathbb{Z}_2} S^{d-1} & \xrightarrow{\ \kappa_{d,4}/\mathcal{S}_4\ } & (\mathrm{RP}^{\infty} \times \mathrm{RP}^{\infty}) \times_{\mathbb{Z}_2} S^{\infty} \\ \big\downarrow & & \big| \\ \mathrm{RP}^{d-1} & \longrightarrow & \mathrm{RP}^{\infty}. \end{array} \tag{3.2}$$

For every integer $d \geq 2$ or $d = \infty$ the fiber bundle (3.1) induces a Serre spectral sequence that converges to $H^*((\mathbb{RP}^{d-1} \times \mathbb{RP}^{d-1}) \times_{\mathbb{Z}_2} S^{d-1}; \mathbb{F}_2)$. The E_2-term of this spectral sequence is of the form

$$E_2^{r,s}(d) = H^r(\mathbb{RP}^{d-1}; \mathcal{H}^s(\mathbb{RP}^{d-1} \times \mathbb{RP}^{d-1}; \mathbb{F}_2)). \tag{3.3}$$

Here $\mathcal{H}^*(\mathbb{RP}^{d-1} \times \mathbb{RP}^{d-1}; \mathbb{F}_2)$ denotes a local coefficient system determined by the action of the fundamental group of the base $\pi_1(\mathbb{RP}^{d-1})$ on the cohomology of the fiber $H^*(\mathbb{RP}^{d-1} \times \mathbb{RP}^{d-1}; \mathbb{F}_2)$. The cohomology ring of the fiber, via the Künneth formula [23, Thm. VI.3.2], can be presented in the following way

$$\begin{aligned} H^*(\mathbb{RP}^{d-1} \times \mathbb{RP}^{d-1}; \mathbb{F}_2) &\cong H^*(\mathbb{RP}^{d-1}; \mathbb{F}_2) \otimes H^*(\mathbb{RP}^{d-1}; \mathbb{F}_2) \\ &\cong \mathbb{F}_2[e_1]/\langle e_1^d \rangle \otimes \mathbb{F}_2[e_2]/\langle e_2^d \rangle \\ &\cong \mathbb{F}_2[e_1, e_2]/\langle e_1^d, e_2^d \rangle, \end{aligned}$$

where $\deg(e_1) = \deg(e_2) = 1$. Here $\langle e_1^d, e_2^d \rangle$ denotes the ideal in $\mathbb{F}_2[e_1, e_2]$ generated by the polynomials e_1^d and e_2^d. The fundamental group $\pi_1(\mathbb{RP}^{d-1}) = \langle t \rangle$ is a cyclic group. Indeed, $\pi_1(\mathbb{RP}^1) \cong \mathbb{Z}$ and $\pi_1(\mathbb{RP}^{d-1}) \cong \mathbb{Z}_2$ for $d \geq 3$. The action of $\pi_1(\mathbb{RP}^{d-1})$ on the cohomology ring $H^*(\mathbb{RP}^{d-1} \times \mathbb{RP}^{d-1}; \mathbb{F}_2)$ is given by $t \cdot e_1 = e_2$.

In the case when $d = \infty$ the spectral sequence (3.3) becomes

$$E_2^{r,s}(\infty) = H^r(\mathbb{RP}^\infty; \mathcal{H}^s(\mathbb{RP}^\infty \times \mathbb{RP}^\infty; \mathbb{F}_2)) \cong H^r(\mathbb{Z}_2; H^s(\mathbb{Z}_2 \times \mathbb{Z}_2; \mathbb{F}_2)). \tag{3.4}$$

Now, the cohomology ring of the fiber, via the Künneth formula [23, Thm. VI.3.2], can be presented in the following way

$$\begin{aligned} H^*(\mathbb{RP}^\infty \times \mathbb{RP}^\infty; \mathbb{F}_2) &\cong H^*(\mathbb{RP}^\infty; \mathbb{F}_2) \otimes H^*(\mathbb{RP}^\infty; \mathbb{F}_2) \\ &\cong \mathbb{F}_2[e_1] \otimes \mathbb{F}_2[e_2] \cong \mathbb{F}_2[e_1, e_2], \end{aligned}$$

where $\deg(e_1) = \deg(e_2) = 1$. Notice an abuse of notation occurring in naming of the generators of the cohomology of the fibers in spectral sequences (3.3) and (3.4). The fundamental group $\pi_1(\mathbb{RP}^\infty) = \langle t \rangle \cong \mathbb{Z}_2$ acts on the cohomology ring of the fiber $H^*(\mathbb{RP}^\infty \times \mathbb{RP}^\infty; \mathbb{F}_2)$ by $t \cdot e_1 = e_2$.

Next, we consider the morphism between the spectral sequences (3.3) and (3.4) induced by the morphism of fiber bundles (3.2) on the level of E_2-terms:

$$\begin{array}{ccc} E_2^{r,s}(d) & \xleftarrow{\;E_2^{r,s}(\kappa_{d,4}/S_4)\;} & E_2^{r,s}(\infty) \\ \| & & \| \\ H^r(\mathbb{RP}^{d-1}; \mathcal{H}^s(\mathbb{RP}^{d-1} \times \mathbb{RP}^{d-1}; \mathbb{F}_2)) & \xleftarrow{\;E_2^{r,s}(\kappa_{d,4}/S_4)\;} & H^r(\mathbb{RP}^\infty; \mathcal{H}^s(\mathbb{RP}^\infty \times \mathbb{RP}^\infty; \mathbb{F}_2)). \end{array}$$

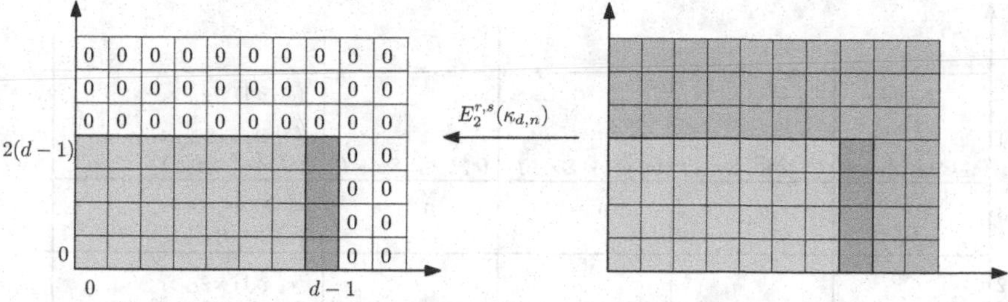

Fig. 3.1 The morphism $E_2^{r,s}(\kappa_{d,n})$

Then we have that

- $E_2^{r,s}(\kappa_{d,4}/S_4)$ is an isomorphism for $(r, s) \in \{0, \ldots, d-2\} \times \{0, \ldots, d-1\}$,
- $E_2^{r,s}(\kappa_{d,4}/S_4)$ is an epimorphism for $(r, s) \in \{0, \ldots, d-2\} \times \{0, \ldots, 2(d-1)\}$,
- $E_2^{r,s}(\kappa_{d,4}/S_4) \neq 0$ is a monomorphism for $(r, s) \in \{d-1\} \times \{1, \ldots, d-1\}$,
- $E_2^{r,s}(\kappa_{d,4}/S_4) = 0$ for $(r, s) \notin \{0, \ldots, d-1\} \times \{0, \ldots, 2(d-1)\}$, because then all $E_2^{r,s}(d)$ vanish.

(For an illustration of the morphism $E_2^{r,s}(\kappa_{d,n})$ see Fig. 3.1.) Thus, if we prove that the spectral sequence (3.4) collapses at the E_2-term the same will be true for the spectral sequence (3.3). Indeed, in the next section we prove Theorem 3.4 which guaranties that the spectral sequence (3.4) collapses at the E_2-term. Consequently, the spectral sequence (3.3) also collapses at the E_2-term because all the differentials emanating from positions $(d-1, s) \in \{d-1\} \times \mathbb{Z}$ are zero.

Now, the calculation of the cohomology of the real projective space with local coefficients $H^*(\mathbb{RP}^{d-1}; \mathcal{M})$, presented in Sect. 10.4 of the appendix, in combination with Lemma 3.2 from the next section, gives the complete description of the E_2-terms of the both spectral sequences.

- For the spectral sequence (3.3) we have

$$E_2^{r,s}(d) = H^r(\mathbb{RP}^{d-1}; \mathcal{H}^s(\mathbb{RP}^{d-1} \times \mathbb{RP}^{d-1}; \mathbb{F}_2))$$
$$= \begin{cases} H^0(\mathbb{RP}^{d-1}; \mathbb{F}_2) \oplus \mathbb{F}_2^{\oplus q(s)}, & r = 0, s \text{ even}, 0 \le s \le 2d-2, \\ \mathbb{F}_2^{\oplus q(s)}, & r = 0, s \text{ odd}, 1 \le s \le 2d-3, \\ H^{d-1}(\mathbb{RP}^{d-1}; \mathbb{F}_2) \oplus \mathbb{F}_2^{\oplus q(s)}, & r = d-1, s \text{ even}, 0 \le s \le 2d-2, \\ \mathbb{F}_2^{\oplus q(s)}, & r = d-1, s \text{ odd}, 1 \le s \le 2d-3, \\ H^r(\mathbb{RP}^{d-1}; \mathbb{F}_2), & 1 \le r \le d-2, s \text{ even}, 0 \le s \le 2d-2, \\ 0, & \text{otherwise}, \end{cases}$$

where $q(s) := |\{(i, j) \in \mathbb{Z} \times \mathbb{Z} : 0 \le i < j \le d-1, i + j = s\}|$.

4	$e_1^4 \otimes 1$ $e_1^3 \otimes e_2$ $(e_1^2 \otimes e_2^2) \otimes_{\mathbf{Z}/2} 1$	$(e_1^2 \otimes e_2^2) \otimes_{\mathbf{Z}/2} f$	$(e_1^2 \otimes e_2^2) \otimes_{\mathbf{Z}/2} f^2$	$(e_1^2 \otimes e_2^2) \otimes_{\mathbf{Z}/2} f^3$	$(e_1^4 \otimes 1) \otimes_{\mathbf{Z}/2} z_4$ $(e_1^3 \otimes e_2) \otimes_{\mathbf{Z}/2} z_4$ $(e_1^2 \otimes e_2^2) \otimes_{\mathbf{Z}/2} f^4$
3	$e_1^2 \otimes e_2$ $e_2^3 \otimes 1$	0	0	0	$(e_1^2 \otimes e_2) \otimes_{\mathbf{Z}/2} z_4$ $(e_1^3 \otimes 1) \otimes_{\mathbf{Z}/2} z_4$
2	$e_2^2 \otimes 1$ $(e_1 \otimes e_2) \otimes_{\mathbf{Z}/2} 1$	$(e_1 \otimes e_2) \otimes_{\mathbf{Z}/2} f$	$(e_1 \otimes e_2) \otimes_{\mathbf{Z}/2} f^2$	$(e_1 \otimes e_2) \otimes_{\mathbf{Z}/2} f^3$	$(e_1^2 \otimes 1) \otimes_{\mathbf{Z}/2} z_4$ $(e_1 \otimes e_2) \otimes_{\mathbf{Z}/2} f^4$
1	$e_1 \otimes 1$	0	0	0	$(e_1 \otimes 1) \otimes_{\mathbf{Z}/2} z_4$
0	$(1 \otimes 1) \otimes_{\mathbf{Z}/2} 1$	$(1 \otimes 1) \otimes_{\mathbf{Z}/2} f$	$(1 \otimes 1) \otimes_{\mathbf{Z}/2} f^2$	$(1 \otimes 1) \otimes_{\mathbf{Z}/2} f^3$	$(1 \otimes 1) \otimes_{\mathbf{Z}/2} f^4$
	0	1	2	3	4

Fig. 3.2 $E_2 = E_\infty$-term of the Serre spectral sequence (3.3) for the fibration (3.1) when $d = 5$. In the picture, for example, by $e_1 \otimes 1 \in E_2^{0,1}$ we denote the invariant element $e_1 \otimes 1 + 1 \otimes e_1$

– For the spectral sequence (3.4) we have

$$
\begin{aligned}
E_2^{r,s}(\infty) &= H^r(\mathrm{B}\mathbb{Z}_2; \mathcal{H}^s(\mathbb{R}\mathrm{P}^\infty \times \mathbb{R}\mathrm{P}^\infty; \mathbb{F}_2)) \\
&\cong H^r(\mathbb{Z}_2; H^s(\mathbb{R}\mathrm{P}^\infty \times \mathbb{R}\mathrm{P}^\infty; \mathbb{F}_2)) \\
&= \begin{cases} H^0(\mathbb{Z}_2; \mathbb{F}_2) \oplus \mathbb{F}_2^{\oplus q(s)}, & r = 0, s \text{ is even}, s \geq 0, \\ \mathbb{F}_2^{\oplus q(s)}, & r = 0, s \text{ is odd}, s \geq 1, \\ H^r(\mathbb{Z}_2; \mathbb{F}_2), & r \geq 1, s \text{ is even}, s \geq 0, \\ 0, & \text{otherwise}. \end{cases}
\end{aligned}
$$

In Fig. 3.2 the E_2-term of the Serre spectral sequence (3.3) associated with the fibration (3.1) in the case $d = 5$ is presented.

The additive basis $\mathcal{B}(\mathbb{R}^d, 4)$ for the cohomology $H^*(\mathrm{Pe}(\mathbb{R}^d, 4)/S_4; \mathbb{F}_2)$ is specified using already defined basis $\mathcal{B}(\mathbb{R}^d, 2)$ in the following way:

(a) $(e^i \otimes e^i) \otimes_{\mathbb{Z}_2} f^k = (e_1^i \otimes e_2^i) \otimes_{\mathbb{Z}_2} f^k \in \mathcal{B}(\mathbb{R}^d, 4)$ for $e^i \in \mathcal{B}(\mathbb{R}^d, 2)$, $0 \leq k \leq d-1$,

(b) $(e^i \otimes e^j) \otimes_{\mathbb{Z}_2} 1 = (e_1^i \otimes e_2^j) \otimes_{\mathbb{Z}_2} 1 \in \mathcal{B}(\mathbb{R}^d, 4)$ for $e^i, e^j \in \mathcal{B}(\mathbb{R}^d, 2)$, $0 \leq j < i \leq d-1$,

(c) $(e^i \otimes e^j) \otimes_{\mathbb{Z}_2} z_{d-1} = (e_1^i \otimes e_2^j) \otimes_{\mathbb{Z}_2} z_{d-1} \in \mathcal{B}(\mathbb{R}^d, 4)$ when $d < \infty$ for $e^i, e^j \in \mathcal{B}(\mathbb{R}^d, 2)$, $0 \leq j < i \leq d-1$.

Here $f \in H^1(\mathbb{R}\mathrm{P}^{d-1}; \mathbb{F}_2)$ denotes the multiplicative generator of the cohomology of the base space of the fibration (3.1). Moreover, in the definition of the basis $\mathcal{B}(\mathbb{R}^d, 4)$ we used two notations. One of them is analogues to the notation used in Sect. 10.4 for the generator $(e^i \otimes e^j) \otimes_{\mathbb{Z}_2} z_{d-1}$ of the cohomology group $H^{d-1}(\mathbb{R}\mathrm{P}^{d-1}; \mathcal{M}_{i,j})$ where the local coefficient system $\mathcal{M}_{i,j} \cong \mathbb{F}_2 \oplus \mathbb{F}_2$ is given by cohomology classes $e^i \otimes e^j$ and $e^j \otimes e^i$.

Now, a partition of the basis $\mathcal{B}(\mathbb{R}^d, 4)$ is defined by

$$\begin{aligned}\mathcal{B}_a(\mathbb{R}^d, 4) &:= \big\{(x \otimes x) \otimes_{\mathbb{Z}_2} f^k : x \in \mathcal{B}_a(\mathbb{R}^d, 2),\ 0 \le k \le d-1\big\} \\ &= \big\{(e^i \otimes e^i) \otimes_{\mathbb{Z}_2} f^k : e^i \in \mathcal{B}(\mathbb{R}^d, 2),\ 0 \le k \le d-1\big\},\end{aligned}$$

and in addition

$$\begin{aligned}\mathcal{B}_i(\mathbb{R}^d, 4) &:= \mathcal{B}(\mathbb{R}^d, 4) \backslash \mathcal{B}_a(\mathbb{R}^d, 4) \\ &= \big\{(e^i \otimes e^j) \otimes_{\mathbb{Z}_2} 1 : e^i, e^j \in \mathcal{B}(\mathbb{R}^d, 2),\ 0 \le j < i \le d-1\big\} \\ &\cup \big\{(e^i \otimes e^j) \otimes_{\mathbb{Z}_2} z_{d-1} : d < \infty,\ 0 \le j < i \le d-1\big\}.\end{aligned}$$

Furthermore, denote by

$$\mathrm{A}^*(\mathbb{R}^d, 4) := \langle \mathcal{B}_a(\mathbb{R}^d, 4) \rangle \qquad \text{and} \qquad \mathrm{I}^*(\mathbb{R}^d, 4) := \langle \mathcal{B}_i(\mathbb{R}^d, 4) \rangle.$$

Then there is an additive decomposition of the cohomology of $\mathrm{Pe}(\mathbb{R}^d, 4)/\mathcal{S}_4$ as follows:

$$H^*(\mathrm{Pe}(\mathbb{R}^d, 4)/\mathcal{S}_4; \mathbb{F}_2) \cong \mathrm{A}^*(\mathbb{R}^d, 4) \oplus \mathrm{I}^*(\mathbb{R}^d, 4),$$

where $\mathrm{A}^*(\mathbb{R}^d, 4)$ turns out to be a subalgebra of $H^*(\mathrm{Pe}(\mathbb{R}^d, 4)/\mathcal{S}_4; \mathbb{F}_2)$ while $\mathrm{I}^*(\mathbb{R}^d, 4)$ is an ideal. More precisely, there is an isomorphism of algebras

$$\mathrm{A}^*(\mathbb{R}^d, 4) \cong \langle (1 \otimes 1) \otimes_{\mathbb{Z}_2} f, (e_1 \otimes e_2) \otimes_{\mathbb{Z}_2} 1 \rangle \cong \mathbb{F}_2[V_{2,1}, V_{2,2}]/\langle V_{2,1}^d, V_{2,2}^d \rangle,$$

where $V_{2,1} = (1 \otimes 1) \otimes_{\mathbb{Z}_2} f$ and $V_{2,2} = (e \otimes e) \otimes_{\mathbb{Z}_2} 1 = (e_1 \otimes e_2) \otimes_{\mathbb{Z}_2} 1$. Thus, we have proved that the cohomology $H^*(\mathrm{Pe}(\mathbb{R}^d, 4)/\mathcal{S}_4; \mathbb{F}_2)$ can be decomposed into the sum of a subalgebra and an ideal as follows

$$H^*(\mathrm{Pe}(\mathbb{R}^d, 4)/\mathcal{S}_4; \mathbb{F}_2) \cong \mathbb{F}_2[V_{2,1}, V_{2,2}]/\langle V_{2,1}^d, V_{2,2}^d \rangle \oplus \mathrm{I}^*(\mathbb{R}^d, 4), \tag{3.5}$$

where $\deg(V_{2,1}) = 2^{1-1} = 1$ and $\deg(V_{2,2}) = 2^{2-1} = 2$. The part of the Serre spectral sequence in the case $d = 5$ induced by the subalgebra $\mathrm{A}^*(\mathbb{R}^5, 4)$ is illustrated in Fig. 3.3.

3.3 Cohomology of $(X \times X) \times_{\mathbb{Z}_2} S^{d-1}$ and $(X \times X) \times_{\mathbb{Z}_2} \mathrm{E}\mathbb{Z}_2$

Let X be a CW-complex (not necessarily finite dimensional). Consider an action of the group $\mathbb{Z}_2 = \langle t \rangle$ on the product $X \times X$ given by $t \cdot (x_1, x_2) := (x_2, x_1)$. Then the product spaces $(X \times X) \times S^{d-1}$ and $(X \times X) \times \mathrm{E}\mathbb{Z}_2$ are equipped with the diagonal $\mathbb{Z}_2$-action where the action on $\mathrm{E}\mathbb{Z}_2$ comes with the space definition and the

	0	1	2	3	4
4	$(e_1^2 \otimes e_2^2) \otimes_{\mathbb{Z}/2} 1$	$(e_1^2 \otimes e_2^2) \otimes_{\mathbb{Z}/2} f$	$(e_1^2 \otimes e_2^2) \otimes_{\mathbb{Z}/2} f^2$	$(e_1^2 \otimes e_2^2) \otimes_{\mathbb{Z}/2} f^3$	$(e_1^2 \otimes e_2^2) \otimes_{\mathbb{Z}/2} f^4$
3	0	0	0	0	0
2	$(e_1 \otimes e_2) \otimes_{\mathbb{Z}/2} 1$	$(e_1 \otimes e_2) \otimes_{\mathbb{Z}/2} f$	$(e_1 \otimes e_2) \otimes_{\mathbb{Z}/2} f^2$	$(e_1 \otimes e_2) \otimes_{\mathbb{Z}/2} f^3$	$(e_1 \otimes e_2) \otimes_{\mathbb{Z}/2} f^4$
1	0	0	0	0	0
0	$(1 \otimes 1) \otimes_{\mathbb{Z}/2} 1$	$(1 \otimes 1) \otimes_{\mathbb{Z}/2} f$	$(1 \otimes 1) \otimes_{\mathbb{Z}/2} f^2$	$(1 \otimes 1) \otimes_{\mathbb{Z}/2} f^3$	$(1 \otimes 1) \otimes_{\mathbb{Z}/2} f^4$

Fig. 3.3 The algebra $\mathrm{A}^*(\mathbb{R}^5, 4)$

action on S^{d-1} is assumed to be antipodal. In this section, using the classical work of Minoru Nakaoka [87] and following the fundamental book of Alejandro Adem and James Milgram [2, Sec. IV.1], we describe the cohomology with $\mathbb{F}_2$-coefficients of the following quotient spaces

$$(X \times X) \times_{\mathbb{Z}_2} S^{d-1} := \big((X \times X) \times S^{d-1}\big)/\mathbb{Z}_2,$$
$$(X \times X) \times_{\mathbb{Z}_2} \mathrm{E}\mathbb{Z}_2 := \big((X \times X) \times \mathrm{E}\mathbb{Z}_2\big)/\mathbb{Z}_2.$$

For this we will use Serre spectral sequences of the following fibrations.

The spaces $(X \times X) \times_{\mathbb{Z}_2} S^{d-1}$ and $(X \times X) \times_{\mathbb{Z}_2} \mathrm{E}\mathbb{Z}_2$ are the total spaces of the following fiber bundles

$$X \times X \longrightarrow (X \times X) \times_{\mathbb{Z}_2} S^{d-1} \longrightarrow \mathbb{R}\mathrm{P}^{d-1}, \tag{3.6}$$

where we use that $S^{d-1}/\mathbb{Z}_2 \cong \mathbb{R}\mathrm{P}^{d-1}$, and

$$X \times X \longrightarrow (X \times X) \times_{\mathbb{Z}_2} \mathrm{E}\mathbb{Z}_2 \longrightarrow \mathbb{R}\mathrm{P}^{\infty} \cong \mathrm{B}\mathbb{Z}_2. \tag{3.7}$$

If for a model of $\mathrm{E}\mathbb{Z}_2$ we take S^∞ with antipodal action the natural inclusion map $S^{d-1} \longrightarrow S^\infty$ induces the following map between corresponding quotient spaces

$$(X \times X) \times_{\mathbb{Z}_2} S^{d-1} \xrightarrow{\;i_d\;} (X \times X) \times_{\mathbb{Z}_2} S^\infty \cong (X \times X) \times_{\mathbb{Z}_2} \mathrm{E}\mathbb{Z}_2. \tag{3.8}$$

This map induces a morphism of the fiber bundles (3.6) and (3.7):

$$\begin{array}{ccc} (X \times X) \times_{\mathbb{Z}_2} S^{d-1} & \xrightarrow{\ i_d\ } & (X \times X) \times_{\mathbb{Z}_2} S^{\infty} \cong (X \times X) \times_{\mathbb{Z}_2} \mathrm{E}\mathbb{Z}_2 \\ \downarrow & & \downarrow \\ \mathbb{R}\mathrm{P}^{d-1} & \longrightarrow & \mathbb{R}\mathrm{P}^{\infty} \cong \mathrm{B}\mathbb{Z}_2. \end{array} \tag{3.9}$$

The cohomology of the total spaces of both fibrations (3.6) and (3.7) can be computed using Serre spectral sequences. In the case of the fibration (3.6) the E_2-term of the associated Serre spectral sequence is of the form

$$E_2^{r,s}(d) = H^r(\mathbb{R}\mathrm{P}^{d-1}; \mathcal{H}^s(X \times X; \mathbb{F}_2)). \tag{3.10}$$

The local coefficient system $\mathcal{H}^*(X \times X; \mathbb{F}_2)$ is determined by the action of the fundamental group $\pi_1(\mathbb{R}\mathrm{P}^{d-1})$ of the base space on the cohomology of the fiber $H^*(X \times X; \mathbb{F}_2)$. Recall that $\pi_1(\mathbb{R}\mathrm{P}^{d-1})$ is a cyclic group. In particular, the Künneth formula [23, Thm. VI.3.2] gives a presentation of the cohomology of the fiber

$$H^*(X \times X; \mathbb{F}_2) \cong H^*(X; \mathbb{F}_2) \otimes H^*(X; \mathbb{F}_2),$$

and the action of $\pi_1(\mathbb{R}\mathrm{P}^{d-1})$ is given by the cyclic shift of factors in the tensor product.

The Serre spectral sequence associated to the fibration (3.7) has the E_2-term

$$\begin{aligned} E_2^{r,s}(\infty) = H^r(\mathbb{R}\mathrm{P}^{\infty}; \mathcal{H}^s(X \times X; \mathbb{F}_2)) &\cong H^r(\mathrm{B}\mathbb{Z}_2; \mathcal{H}^s(X \times X; \mathbb{F}_2)) \\ &\cong H^r(\mathbb{Z}_2; H^s(X \times X; \mathbb{F}_2)). \end{aligned} \tag{3.11}$$

The fundamental group of the base $\pi_1(\mathrm{B}\mathbb{Z}_2) \cong \mathbb{Z}_2$ acts on the cohomology of the fiber and defines the local coefficient system $\mathcal{H}^s(X \times X; \mathbb{F}_2)$, or defines the $\mathbb{Z}_2$-module structure on $H^r(\mathbb{Z}_2; H^s(X \times X; \mathbb{F}_2))$.

The morphism (3.9) between the fibrations (3.6) and (3.7) induces a morphism between the spectral sequences (3.10) and (3.11):

$$E_2^{r,s}(d) = H^r(\mathbb{R}\mathrm{P}^{d-1}; \mathcal{H}^s(X \times X; \mathbb{F}_2)) \xleftarrow{E_2^{r,s}(i_d)} E_2^{r,s}(\infty) = H^r(\mathbb{R}\mathrm{P}^{\infty}; \mathcal{H}^s(X \times X; \mathbb{F}_2)).$$

The homomorphisms $E_2^{r,s}(i_d)$ are isomorphisms whenever $E_2^{r,s}(d) \neq 0$ and $(r, s) \notin \{d-1\} \times \mathbb{Z}$, or $E_2^{r,s}(d) = E_2^{r,s}(\infty) = 0$. In the case when $(r, s) \in \{d-1\} \times \mathbb{Z}$ we have that

$$E_2^{d-1,s}(d) \cong E_2^{d-1,s}(\infty) \oplus \mathbb{F}_2^{b(d,X)},$$

where for s odd

$$b(d, X) = \sum_{0\le i<j,\, i+j=s} \big(\mathrm{rank}(H^i(X;\mathbb{F}_2)) \cdot \mathrm{rank}(H^j(X;\mathbb{F}_2))\big),$$

and for s even

$$b(d, X) = \sum_{0\le i<j,\, i+j=s} \big(\mathrm{rank}(H^i(X;\mathbb{F}_2)) \cdot \mathrm{rank}(H^j(X;\mathbb{F}_2))\big)$$
$$+\frac{1}{2}\big(\mathrm{rank}(H^{\frac{s}{2}}(X;\mathbb{F}_2))\big)^2 - \mathrm{rank}(H^{\frac{s}{2}}(X;\mathbb{F}_2)).$$

Consult also Sect. 10.4. In particular, all the homomorphisms $E_2^{r,s}(i_d)$ are monomorphism, and for $(r, s) \notin \{d-1\} \times \mathbb{Z}$ they are non-zero epimorphisms.

To make any further progress in computation of the spectral sequences (3.10) and (3.11) we need to understand the action of the fundamental groups of the base spaces on the cohomology of the fibers. For that we use the following lemma, which is a particular case of [2, Lem. IV.1.4].

Lemma 3.1 *Let $V = \bigoplus_{n\ge 0} V_n$ be a graded $\mathbb{F}_2$ vector space, and let $\mathcal{B} := \{v_i : i \in I\}$ be a basis of V where the index set I is equipped with a linear order. The tensor product vector space $V \otimes V$ as a $\mathbb{F}_2[\mathbb{Z}_2]$-module, where the action of $\mathbb{Z}_2$ is given by the cyclic shift, is a direct sum of free and trivial $\mathbb{F}_2[\mathbb{Z}_2]$-modules. The trivial modules are generated by the elements of the form $v_i \otimes v_i$ where v_i is an element of the basis $\mathcal{B}$, while the free modules are generated by the elements of the form $v_i \otimes v_j$ where $i < j$ and v_i, v_j belong to $\mathcal{B}$.*

Proof A basis of the vector space $V \otimes V$ is given by all vectors of the form $v_i \otimes v_j$ where $v_i, v_j \in \mathcal{B}$. The $\mathbb{Z}_2 = \langle t\rangle$-action on $V \otimes V$ preserves this basis. Since $t \cdot (v_i \otimes v_j) = v_j \otimes v_i$ we have that each element $v_i \otimes v_i$ generates a copy of the trivial $\mathbb{F}_2[\mathbb{Z}_2]$-module and each element $v_i \otimes v_j$, $i < j$, generates a copy of the free $\mathbb{F}_2[\mathbb{Z}_2]$-module. Thus, $V \otimes V$ is a direct sum of free and trivial $\mathbb{F}_2[\mathbb{Z}_2]$-modules. □

Since the homomorphisms $E_2^{r,s}(i_d)$ are isomorphisms in the case when $(r, s) \in \{0, \ldots, d-2\} \times \mathbb{Z}$, monomorphism for $(r, s) \in \{d-1\} \times \mathbb{Z}$, and otherwise zero homomorphisms, we study first the spectral sequence (3.11). If we prove that $E_2^{r,s}(\infty) = E_\infty^{r,s}(\infty)$ it would imply that all its differentials vanish and consequently the same would hold for the spectral sequence (3.10) implying that $E_2^{r,s}(d) = E_\infty^{r,s}(d)$. In the next step we describe the E_2-term of the spectral sequence (3.11), see also [2, Cor. IV.1.6].

Lemma 3.2 *Let $\mathcal{B} := \{v_i : i \in I\}$ be a basis of the $\mathbb{F}_2$ vector space $H^*(X;\mathbb{F}_2)$ where the index set I is equipped with a linear order. The E_2-term of the spectral sequence* (3.11) *can be presented as follows:*

$$E_2^{r,s}(\infty) = H^r(\mathrm{B}\mathbb{Z}_2; \mathcal{H}^s(X \times X; \mathbb{F}_2)) \cong H^r(\mathbb{Z}_2; H^s(X \times X; \mathbb{F}_2))$$

$$\cong \begin{cases} H^s(X \times X; \mathbb{F}_2)^{\mathbb{Z}_2}, & r = 0, \\ H^{\frac{s}{2}}(X; \mathbb{F}_2), & r > 0, \ s \text{ even}, \\ 0, & \text{otherwise.} \end{cases}$$

Moreover, $E_2^{*,*}(\infty)$ as a $H^*(\mathbb{Z}_2; \mathbb{F}_2)$-module, ignoring the grading, decomposes into the direct sum

$$\bigoplus_{i \in I} H^*(\mathbb{Z}_2; \mathbb{F}_2) \oplus \bigoplus_{i<j \in I} \mathbb{F}_2$$

where the action of $H^*(\mathbb{Z}_2; \mathbb{F}_2)$ on each summand of the first sum is given by the cup product, and on the each summand of the second sum is trivial.

Proof This is a direct consequence of Lemma 3.1, and the facts that:

- $H^0(G; M) = M^G$ for any group G and any G-module M, and
- $H^i(G; F) = 0$ for $i \geq 1$ when F is a projective (free) G-module. □

Now, for the spectral sequence (3.10), using the calculation presented in Sect. 10.4, we get the following presentation of the corresponding E_2-term.

Lemma 3.3 *The E_2-term of the spectral sequence* (3.10) *can be presented as follows*

$$E_2^{r,s}(d) = H^r(\mathbb{R}\mathrm{P}^{d-1}; \mathcal{H}^s(X \times X; \mathbb{F}_2))$$

$$\cong \begin{cases} H^s(X \times X; \mathbb{F}_2)^{\mathbb{Z}_2}, & r \in \{0, d-1\}, \\ H^{\frac{s}{2}}(X; \mathbb{F}_2), & 1 \leq r \leq d-2, \ s \text{ even}, \\ 0, & \text{otherwise.} \end{cases}$$

The fact that the E_2-term of the spectral sequence (3.11) collapses dates back to work of Paul Smith [93], Norman Steenrod [95] and Nakaoka [87], for a result in more general setup consult more recent work of Ian Leary [71, Thm. 2.1]. The following theorem is a special case of [2, Thm. IV.1.7].

Theorem 3.4 *Let X be a CW-complex. The Serre spectral sequence of the fibration*

$$X \times X \longrightarrow (X \times X) \times_{\mathbb{Z}_2} \mathrm{E}\mathbb{Z}_2 \longrightarrow \mathrm{B}\mathbb{Z}_2. \tag{3.12}$$

collapses at the E_2-term, that is, $E_2^{r,s}(\infty) \cong E_\infty^{r,s}(\infty)$ for all $(r, s) \in \mathbb{Z} \times \mathbb{Z}$.

Proof Let $\mathcal{B} := \{v_i : i \in I\}$ be a basis of the $\mathbb{F}_2$ vector space $H^*(X; \mathbb{F}_2)$ where the index set I is equipped with a linear order. The E_2-term (3.11) of the Serre spectral sequence of the fibration (3.12) is calculated in Lemma 3.2. Further on, the $\mathbb{F}_2[\mathbb{Z}_2]$-module structure on the cohomology of the fiber—the coefficient system—is described in Lemma 3.1. Since the differentials of these spectral sequences are $H^*(\mathbb{Z}_2; \mathbb{F}_2)$-module maps we concentrate on the generators of the $H^*(\mathbb{Z}_2; \mathbb{F}_2)$-module structure of the rows of the spectral sequences. In this situation it means that we consider elements of the zero column and prove that they survive to the E_∞-term. Consequently the proof of the theorem proceeds in two steps.

(A) Let v_i and v_j be two different cohomology classes from the basis $\mathcal{B}$. In the first step we prove that all the elements (of the form $v_i \otimes v_j + v_j \otimes v_i$) in $E_2^{0,s}$ associated to the invariants of free $\mathbb{F}_2[\mathbb{Z}_2]$-modules (generated by $v_i \otimes v_j$) in the decomposition of $H^*(X \times X; \mathbb{F}_2)$ survive to the E_∞-term. Since $\mathrm{E}\mathbb{Z}_2$ is a contractible and free $\mathbb{Z}_2$-space we have that $(X \times X) \times \mathrm{E}\mathbb{Z}_2 \simeq (X \times X)$ is a free $\mathbb{Z}_2$-space and the quotient map

$$\pi\colon (X \times X) \times \mathrm{E}\mathbb{Z}_2 \longrightarrow (X \times X) \times_{\mathbb{Z}_2} \mathrm{E}\mathbb{Z}_2$$

is a covering map. Denote by $p\colon (X \times X) \times \mathrm{E}\mathbb{Z}_2 \longrightarrow X \times X$ the projection. Since it is a homotopy equivalence it induces an isomorphism in cohomology. Furthermore, there is a transfer homomorphism

$$\mathrm{tr}\colon H^*((X \times X) \times \mathrm{E}\mathbb{Z}_2; \mathbb{F}_2) \longrightarrow H^*((X \times X) \times_{\mathbb{Z}_2} \mathrm{E}\mathbb{Z}_2; \mathbb{F}_2)$$

with a property that that composition

$$(p^*)^{-1} \circ \pi^* \circ \mathrm{tr} \circ p^*\colon H^*(X \times X; \mathbb{F}_2) \longrightarrow H^*(X \times X; \mathbb{F}_2)$$

is the map

$$v_i \otimes v_j \longmapsto v_i \otimes v_j + v_j \otimes v_i = (1 + t) \cdot (v_i \otimes v_j),$$

where $v_i, v_j \in H^*(X; \mathbb{F}_2)$, t is a generator of $\mathbb{Z}_2$, and $1 + t \in \mathbb{Z}[\mathbb{Z}_2]$. Thus, the image $\mathrm{im}((p^*)^{-1} \circ \pi^* \circ \mathrm{tr} \circ p^*)$ is contained in $H^*(X \times X; \mathbb{F}_2)^{\mathbb{Z}_2}$ and each element is associated to an invariant element $v_i \otimes v_j + v_j \otimes v_i$, where $v_i \neq v_j$, of a free $\mathbb{F}_2[\mathbb{Z}_2]$-module in the decomposition of the cohomology $H^*(X \times X; \mathbb{F}_2)$. Since the composition $(p^*)^{-1} \circ \pi^* \circ \mathrm{tr} \circ p^*$ factors through the cohomology $H^*((X \times X) \times_{\mathbb{Z}_2} \mathrm{E}\mathbb{Z}_2; \mathbb{F}_2)$ all these elements survive to the E_∞-term.

(B) Let v be a cohomology class of dimension n from the basis $\mathcal{B}$. In the second step we prove that all the elements (of the form $v \otimes v$) in $E_2^{0,s}$ associated to the invariants of trivial $\mathbb{F}_2[\mathbb{Z}_2]$-modules (generated by $v \otimes v$) in the decomposition of $H^*(X \times X; \mathbb{F}_2)$ survive to the E_∞-term. Using the bijective correspondence

$$H^n(X; \mathbb{F}_2) \longleftrightarrow [X, K(\mathbb{Z}_2, n)] \tag{3.13}$$

we can present the cohomology class v as the image of the fundamental class $\iota_n \in H^*(K(\mathbb{Z}_2, n); \mathbb{F}_2)$ along the map $\nu\colon X \longrightarrow K(\mathbb{Z}_2, n)$ which is associated to v via the correspondence (3.13), that is $v = \nu^*(\iota_n)$. For more details about the correspondence (3.13) consult for example [81, Thm. 1, page 3]. The map ν induces the following $\mathbb{Z}_2$-equivariant map

$$\nu \times \nu\colon X \times X \longrightarrow K(\mathbb{Z}_2, n) \times K(\mathbb{Z}_2, n),$$

and consequently a morphism of Borel construction fibrations:

$$\begin{array}{ccc} (X \times X) \times_{\mathbb{Z}_2} E\mathbb{Z}_2 & \xrightarrow{(\nu\times\nu)\times_{\mathbb{Z}_2}\mathrm{id}} & (K(\mathbb{Z}_2, n) \times K(\mathbb{Z}_2, n)) \times_{\mathbb{Z}_2} E\mathbb{Z}_2 \\ \downarrow & & \downarrow \\ B\mathbb{Z}_2 & \longrightarrow & B\mathbb{Z}_2. \end{array}$$

This morphism of fibrations induces a morphism between associated Serre spectral sequences. In particular, the map between $E_2^{0,2n}$ entries sends the class $\iota_n \otimes \iota_n$ to the class $v \otimes v$, see Fig. 3.4. Consequently, if the class $\iota_n \otimes \iota_n$ survives to the E_∞-term (all differentials evaluated at $\iota_n \otimes \iota_n$ are zero), then the class $v \otimes v$ also survives to the E_∞-term.

Hence, we prove that the class $\iota_n \otimes \iota_n$ survives to the E_∞-term in the Serre spectral sequences associated to the Borel construction fibration

$$K(\mathbb{Z}_2, n) \times K(\mathbb{Z}_2, n) \longrightarrow (K(\mathbb{Z}_2, n) \times K(\mathbb{Z}_2, n)) \times_{\mathbb{Z}_2} E\mathbb{Z}_2 \xrightarrow{\ q\ } B\mathbb{Z}_2. \tag{3.14}$$

The E_2-term of this spectral sequence is of the form

$$E_2^{r,s} = H^r(\mathbb{Z}_2; H^s(K(\mathbb{Z}_2, n) \times K(\mathbb{Z}_2, n); \mathbb{F}_2)). \tag{3.15}$$

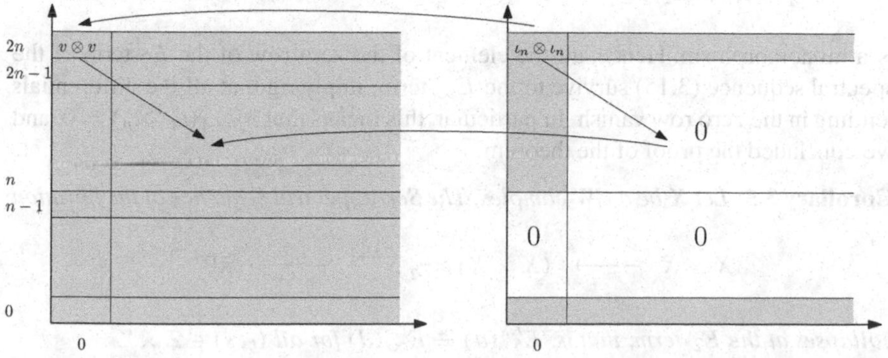

Fig. 3.4 The morphism between Serre spectral sequences induced by the map $\mathrm{id} \times_{\mathbb{Z}_2} (\nu \times \nu)$

Since the Eilenberg–Mac Lane space $K(\mathbb{Z}_2, n)$ is $(n-1)$-connected the Künneth formula [23, Thm. VI.3.2] implies that

$$H^s(K(\mathbb{Z}_2, n)\times K(\mathbb{Z}_2, n); \mathbb{F}_2) = \begin{cases} \mathbb{F}_2, & \text{for } s = 0, \\ 0, & \text{for } 1 \le s \le n-1, \\ \text{free } \mathbb{F}_2[\mathbb{Z}_2]\text{-module}, & \text{for } n \le s \le 2n-1, \\ \text{not relevant for our proof}, & \text{otherwise.} \end{cases}$$

Here we used that for $n \le s \le 2n-1$ the following holds:

$$\begin{aligned} &H^s(K(\mathbb{Z}_2, n) \times K(\mathbb{Z}_2, n); \mathbb{F}_2) \\ &\quad = \big(H^s(K(\mathbb{Z}_2, n); \mathbb{F}_2) \otimes H^0(K(\mathbb{Z}_2, n); \mathbb{F}_2)\big) \\ &\quad\quad \oplus \big(H^0(K(\mathbb{Z}_2, n); \mathbb{F}_2) \otimes H^s(K(\mathbb{Z}_2, n); \mathbb{F}_2)\big). \end{aligned}$$

Thus, a part of the E_2-term vanishes, meaning

$$E_2^{r,s} = H^r(\mathbb{Z}_2; H^s(K(\mathbb{Z}_2, n) \times K(\mathbb{Z}_2, n); \mathbb{F}_2)) = 0,$$

for all $r \ge 1$ and $1 \le s \le 2n-1$. Therefore the element $\iota_n \otimes \iota_n \in E_2^{0,2n}$ survives to the E_∞-term if and only if $\partial_{2n+1}(\iota_n \otimes \iota_n) = 0$. It suffices to prove that no non-zero differential lands in the zero row of the spectral sequence.

The fibration (3.14) has a section

$$\sigma : \mathrm{B}\mathbb{Z}_2 \longrightarrow (K(\mathbb{Z}_2, n) \times K(\mathbb{Z}_2, n)) \times_{\mathbb{Z}_2} \mathrm{E}\mathbb{Z}_2$$

induced by the $\mathbb{Z}_2$-map $\mathrm{E}\mathbb{Z}_2 \longrightarrow (K(\mathbb{Z}_2, n) \times K(\mathbb{Z}_2, n)) \times \mathrm{E}\mathbb{Z}_2$ given by $e \longmapsto (x_0, x_0, e)$ where $e \in \mathrm{E}\mathbb{Z}_2$ and $x_0 \in K(\mathbb{Z}_2, n)$ is an arbitrary point that we fixed. Consequently, $q \circ \sigma = \mathrm{id}_{\mathrm{B}\mathbb{Z}_2}$. Passing to cohomology we get $\sigma^* \circ q^* = \mathrm{id}_{H^*(\mathrm{B}\mathbb{Z}_2;\mathbb{F}_2)}$ implying that

$$q^* \colon H^*(\mathrm{B}\mathbb{Z}_2; \mathbb{F}_2) \longrightarrow H^*((K(\mathbb{Z}_2, n) \times K(\mathbb{Z}_2, n)) \times_{\mathbb{Z}_2} \mathrm{E}\mathbb{Z}_2; \mathbb{F}_2)$$

is a monomorphism. Hence, all the element of the zero row of the E_2-term of the spectral sequence (3.15) survive to the E_∞-term, implying that all the differentials lending in the zero row vanish. In particular, this means that $\partial_{2n+1}(\iota_n \otimes \iota_n) = 0$, and we concluded the proof of the theorem. □

Corollary 3.5 *Let X be a CW-complex. The Serre spectral sequence of the fibration*

$$X \times X \longrightarrow (X \times X) \times_{\mathbb{Z}_2} S^{d-1} \longrightarrow \mathbb{R}\mathrm{P}^{d-1}$$

collapses at the E_2*-term, that is,* $E_2^{r,s}(d) \cong E_\infty^{r,s}(d)$ *for all* $(r, s) \in \mathbb{Z} \times \mathbb{Z}$.

Proof We prove that all differentials of the spectral sequence $E_*^{*,*}(d)$ vanish. In (3.8), using S^∞ as a model for $\mathrm{E}\mathbb{Z}_2$, we have defined the map

$$i_d\colon (X \times X) \times_{\mathbb{Z}_2} S^{d-1} \longrightarrow (X \times X) \times_{\mathbb{Z}_2} \mathrm{E}S^\infty \cong (X \times X) \times_{\mathbb{Z}_2} \mathrm{E}\mathbb{Z}_2.$$

The map i_d induces a morphism of the fiber bundles:

$$\begin{array}{ccc} (X \times X) \times_{\mathbb{Z}_2} S^{d-1} & \xrightarrow{\ i_d\ } & (X \times X) \times_{\mathbb{Z}_2} S^\infty \cong (X \times X) \times_{\mathbb{Z}_2} \mathrm{E}\mathbb{Z}_2 \\ \downarrow & & \downarrow \\ \mathbb{R}\mathrm{P}^{d-1} & \longrightarrow & \mathbb{R}\mathrm{P}^\infty \cong \mathrm{B}\mathbb{Z}_2, \end{array}$$

that in turn gives a morphism between the corresponding Serre spectral sequences:

$$E_2^{r,s}(d) = H^r(\mathbb{R}\mathrm{P}^{d-1}; \mathcal{H}^s(X \times X; \mathbb{F}_2)) \xleftarrow{E_2^{r,s}(i_d)} E_2^{r,s}(\infty) = H^r(\mathbb{R}\mathrm{P}^\infty; \mathcal{H}^s(X \times X; \mathbb{F}_2)).$$

Since by Theorem 3.4 the spectral sequence $E_2^{r,s}(\infty)$ collapses at the E_2-term all the differential of this spectral sequence vanish. Consequently, all the elements in the image $\mathrm{im}\left(E_2^{r,s}(i_d)\right) \subseteq E_2^{r,s}(d)$ survive to the infinity term. The only elements of $E_2^{r,s}(d)$ not contained in $\mathrm{im}\left(E_2^{r,s}(i_d)\right)$ belong to the $(d-1)$-column, correspond to $\mathbb{F}_2[\mathbb{Z}_2]$-free summands in the cohomology $H^*(X \times X; \mathbb{F}_2)$, and are of the form (something) $\otimes_{\mathbb{Z}_2} z_{d-1}$, consult Sect. 10.4. Because all differentials emanating from the $(d-1)$-column of the spectral sequence $E_*^{*,*}(d)$ are zero and all differentials arriving at the $(d-1)$-column have to be zero we conclude that all differentials in $E_*^{*,*}(d)$ indeed vanish. □

In summary, when an additive basis of the $\mathbb{F}_2$ vector space $H^*(X; \mathbb{F}_2)$ is given, then we can describe a basis of the cohomology of the spaces $(X \times X) \times_{\mathbb{Z}_2} S^{d-1}$ and $(X \times X) \times_{\mathbb{Z}_2} \mathrm{E}\mathbb{Z}_2$ as follows. Keep in mind notation introduced in Sect. 10.4.

Theorem 3.6 *Let $d \geq 2$ be integer or $d = \infty$, let X be a CW-complex, and let $\mathcal{B}_X$ be an additive basis of $H^*(X; \mathbb{F}_2)$. Denote by f^j for $0 \leq j \leq d-1$ the additive generator of the group $H^j(\mathbb{R}\mathrm{P}^{d-1}; \mathbb{F}_2) \cong \mathbb{F}_2$, where $f \in H^1(\mathbb{R}\mathrm{P}^{d-1}; \mathbb{F}_2)$ is the multiplicative generator of the cohomology ring $H^*(\mathbb{R}\mathrm{P}^{d-1}; \mathbb{F}_2)$. Here we assume that $1 = f^0 \in H^0(\mathbb{R}\mathrm{P}^{d-1}; \mathbb{F}_2)$. An additive basis $\mathcal{B}$ of the cohomology*

$$H^*((X \times X) \times_{\mathbb{Z}_2} S^{d-1}; \mathbb{F}_2)$$

can be given the following way:

(1) *If $v \in \mathcal{B}_X$, then $(v \otimes v) \otimes_{\mathbb{Z}_2} f^j \in \mathcal{B}$ for $0 \leq j \leq d-1$ with* $\deg((v \otimes v) \otimes_{\mathbb{Z}_2} f^j) = 2\deg(v) + \deg(f^j) = 2\deg(v) + j$;

(2) *If* $u, v \in \mathcal{B}_X$ *and* $u \neq v$, *then* $(u \otimes v) \otimes_{\mathbb{Z}_2} 1 \in \mathcal{B}'$ *with* $\deg((u \otimes v) \otimes_{\mathbb{Z}_2} 1) = \deg(u) + \deg(v)$; *and*
(3) *If* $u, v \in \mathcal{B}_X$ *and* $u \neq v$ *with* $d < \infty$, *then we set* $(u \otimes v) \otimes_{\mathbb{Z}_2} z_{d-1} \in \mathcal{B}$ *where* $\deg((u \otimes v) \otimes_{\mathbb{Z}_2} z_{d-1}) = \deg(u) + \deg(v) + d - 1$.

Proof The spectral sequences (3.10) and (3.11), depending whether $d < \infty$ or $d = \infty$, converge to the cohomology $H^*((X \times X) \times_{\mathbb{Z}_2} S^{d-1}; \mathbb{F}_2)$. Since by Theorem 3.4 and Corollary 3.5 both spectral sequences collapse at the E_2-term it suffices to find a basis of E_2-terms. Thus, the proof is concluded by a direct application of Lemmas 3.2 and 3.3. □

The calculation of the cohomology of $(X \times X) \times_{\mathbb{Z}_2} S^{d-1}$ we presented was done with coefficients in the field $\mathbb{F}_2$. The Universal Coefficient theorem transcribes the arguments for cohomology into homology arguments, implying the following claim.

Theorem 3.7 *Let* $d \geq 2$ *be integer or* $d = \infty$, *let* X *be a CW-complex, and let* $\mathcal{B}'_X$ *be an additive basis of the homology* $H_*(X; \mathbb{F}_2)$. *Denote by* f_j *for* $0 \leq j \leq d-1$ *the generator of the group* $H_j(\mathbb{RP}^{d-1}; \mathbb{F}_2) \cong \mathbb{F}_2$. *An additive basis* $\mathcal{B}'$ *of the homology*

$$H_*((X \times X) \times_{\mathbb{Z}_2} S^{d-1}; \mathbb{F}_2)$$

can be given the following way:

(1) *If* $v \in \mathcal{B}'_X$, *then* $(v \otimes v) \otimes_{\mathbb{Z}_2} f_j \in \mathcal{B}'$ *for* $1 \leq j \leq d-1$ *with* $\deg((v \otimes v) \otimes_{\mathbb{Z}_2} f_j) = 2\deg(v) + \deg(f_j) = 2\deg(v) + j$;
(2) *If* $u, v \in \mathcal{B}'_X$ *and* $u \neq v$, *then* $(u \otimes v) \otimes_{\mathbb{Z}_2} 1 \in \mathcal{B}'$ *with* $\deg((u \otimes v) \otimes_{\mathbb{Z}_2} 1) = \deg(u) + \deg(v)$; *and*
(3) *If* $u, v \in \mathcal{B}'_X$ *and* $u \neq v$ *with* $d < \infty$, *then we set* $(u \otimes v) \otimes_{\mathbb{Z}_2} h_{d-1} \in \mathcal{B}'$ *where* $\deg((u \otimes v) \otimes_{\mathbb{Z}_2} h_{d-1}) = \deg(u) + \deg(v) + d - 1$.

A useful consequence of Theorem 3.6 is the following fact.

Corollary 3.8 *Let* $d \geq 2$ *be integer or* $d = \infty$, *let* X *and* Y *be CW-complexes, and let* $f : X \longrightarrow Y$ *be a continuous map. If the induced map in cohomology*

$$f^* : H^*(Y; \mathbb{F}_2) \longrightarrow H^*(X; \mathbb{F}_2)$$

is injective, then the induced map

$$((f \times f) \times_{\mathbb{Z}_2} \mathrm{id})^* : H^*((Y \times Y) \times_{\mathbb{Z}_2} S^{d-1}; \mathbb{F}_2) \longrightarrow H^*((X \times X) \times_{\mathbb{Z}_2} S^{d-1}; \mathbb{F}_2),$$

is also injective.

Proof The induced map $(f \times f) \times_{\mathbb{Z}_2} \mathrm{id} : (X \times X) \times_{\mathbb{Z}_2} S^{d-1} \longrightarrow (Y \times Y) \times_{\mathbb{Z}_2} S^{d-1}$ is covering the identity map in the following bundle morphism:

$$\begin{array}{ccc} (X \times X) \times_{\mathbb{Z}_2} S^{d-1} & \xrightarrow{(f \times f) \times_{\mathbb{Z}_2} \mathrm{id}} & (Y \times Y) \times_{\mathbb{Z}_2} S^{d-1} \\ \downarrow & & \downarrow \\ S^{d-1}/\mathbb{Z}_2 & \xrightarrow{\mathrm{id}} & S^{d-1}/\mathbb{Z}_2. \end{array}$$

The bundle morphism induces the morphism between the corresponding Serre spectral sequences that corresponds to the homomorphism $((f \times f) \times_{\mathbb{Z}_2} \mathrm{id})^*$. Since both spectral sequences collapse at E_2-term and there is no extension problem Theorem 3.6 in combination with the assumption that f^* is injective yields the injectivity of $((f \times f) \times_{\mathbb{Z}_2} \mathrm{id})^*$. This concludes the proof. □

The following dual version of the previous corollary also holds.

Corollary 3.9 *Let $d \geq 2$ be integer or $d = \infty$, let X and Y be CW-complexes, and let $f : X \longrightarrow Y$ be a continuous map. If the induced map in homology*

$$f_* : H_*(X; \mathbb{F}_2) \longrightarrow H_*(Y; \mathbb{F}_2)$$

is surjective, then the induced map

$$((f \times f) \times_{\mathbb{Z}_2} \mathrm{id})_* : H^*((X \times X) \times_{\mathbb{Z}_2} S^{d-1}; \mathbb{F}_2) \longrightarrow H^*((Y \times Y) \times_{\mathbb{Z}_2} S^{d-1}; \mathbb{F}_2),$$

is also surjective.

3.4 The Induction Step

In this section we take $d \geq 2$ to be an integer or $d = \infty$. Let us assume that for $m = k$ the cohomology

$$H^*(\mathrm{Pe}(\mathbb{R}^d, 2^m)/\mathcal{S}_{2^m}; \mathbb{F}_2) = H^*(\mathrm{Pe}(\mathbb{R}^d, 2^k)/\mathcal{S}_{2^k}; \mathbb{F}_2)$$

is determined by specifying a basis $\mathcal{B}(\mathbb{R}^d, 2^k)$ and its associated partition

$$\mathcal{B}(\mathbb{R}^d, 2^k) = \mathcal{B}_a(\mathbb{R}^d, 2^k) \cup \mathcal{B}_i(\mathbb{R}^d, 2^k)$$

in such a way that

$$\mathrm{A}^*(\mathbb{R}^d, 2^k) := \langle \mathcal{B}_a(\mathbb{R}^d, 2^k) \rangle \cong \mathbb{F}_2[V_{k,1}, \ldots, V_{k,k}]/\langle V_{k,1}^d, \ldots, V_{k,k}^d \rangle$$

is a subalgebra and $\mathrm{I}^*(\mathbb{R}^d, 2^k) := \langle \mathcal{B}_i(\mathbb{R}^d, 2^k) \rangle$ is an ideal of the cohomology ring $H^*(\mathrm{Pe}(\mathbb{R}^d, 2^k)/\mathcal{S}_{2^k}; \mathbb{F}_2)$, and in addition

$$\begin{aligned} H^*(\mathrm{Pe}(\mathbb{R}^d, 2^k)/\mathcal{S}_{2^k}; \mathbb{F}_2) &\cong \mathrm{A}^*(\mathbb{R}^d, 2^k) \oplus \mathrm{I}^*(\mathbb{R}^d, 2^k) \\ &\cong \mathbb{F}_2[V_{k,1}, \ldots, V_{k,k}]/\langle V_{k,1}^d, \ldots, V_{k,k}^d \rangle \oplus \mathrm{I}^*(\mathbb{R}^d, 2^k), \end{aligned} \tag{3.16}$$

where $\deg(V_{k,r}) = 2^{r-1}$ for $1 \leq r \leq k$.

Now, for $m = k + 1$ we study the cohomology

$$H^*(\mathrm{Pe}(\mathbb{R}^d, 2^{k+1})/\mathcal{S}_{2^{k+1}}; \mathbb{F}_2).$$

According to Definition 2.1 we have that

$$\mathrm{Pe}(\mathbb{R}^d, 2^{k+1}) = (\mathrm{Pe}(\mathbb{R}^d, 2^k) \times \mathrm{Pe}(\mathbb{R}^d, 2^k)) \times S^{d-1} \quad \text{and} \quad \mathcal{S}_{2^{k+1}} = (\mathcal{S}_{2^k} \times \mathcal{S}_{2^k}) \rtimes \mathbb{Z}_2.$$

Using the nature of the $\mathcal{S}_{2^{k+1}}$-action on the spaces $\mathrm{Pe}(\mathbb{R}^d, 2^{k+1})$ we have that

$$\begin{aligned}
\mathrm{Pe}(\mathbb{R}^d, 2^{k+1})/\mathcal{S}_{2^{k+1}} &\cong \big((\mathrm{Pe}(\mathbb{R}^d, 2^k) \times \mathrm{Pe}(\mathbb{R}^d, 2^k)) \times S^{d-1}\big)/\mathcal{S}_{2^{k+1}} \\
&\cong \big((\mathrm{Pe}(\mathbb{R}^d, 2^k)/\mathcal{S}_{2^k} \times \mathrm{Pe}(\mathbb{R}^d, 2^k)/\mathcal{S}_{2^k}) \times S^{d-1}\big)/\mathbb{Z}_2 \\
&=: \big(\mathrm{Pe}(\mathbb{R}^d, 2^k)/\mathcal{S}_{2^k} \times \mathrm{Pe}(\mathbb{R}^d, 2^k)/\mathcal{S}_{2^k}\big) \times_{\mathbb{Z}_2} S^{d-1}.
\end{aligned}$$

Since the additive basis $\mathcal{B}(\mathbb{R}^d, 2^k)$ for the cohomology $H^*(\mathrm{Pe}(\mathbb{R}^d, 2^k)/\mathcal{S}_{2^k}; \mathbb{F}_2)$ is already fixed we can now use Theorem 3.6 and get the basis $\mathcal{B}(\mathbb{R}^d, 2^{k+1})$ for the cohomology

$$H^*(\mathrm{Pe}(\mathbb{R}^d, 2^{k+1})/\mathcal{S}_{2^{k+1}}; \mathbb{F}_2)$$

as follows:

(i) If $v \in \mathcal{B}(\mathbb{R}^d, 2^k)$, then for all $0 \le j \le d-1$

$$(v \otimes v) \otimes_{\mathbb{Z}_2} f^j \in \mathcal{B}(\mathbb{R}^d, 2^{k+1}),$$

with $\deg((v \otimes v) \otimes_{\mathbb{Z}_2} f^j) = 2\deg(v) + \deg(f^j) = 2\deg(v) + j$;

(ii) If $u, v \in \mathcal{B}(\mathbb{R}^d, 2^k)$ and $u \ne v$, then

$$(u \otimes v) \otimes_{\mathbb{Z}_2} 1 \in \mathcal{B}(\mathbb{R}^d, 2^{k+1}),$$

with $\deg((u \otimes v) \otimes_{\mathbb{Z}_2} 1) = \deg(u) + \deg(v)$; and

(iii) If $u, v \in \mathcal{B}(\mathbb{R}^d, 2^k)$ and $u \ne v$ and $d < \infty$, then

$$(u \otimes v) \otimes_{\mathbb{Z}_2} z_{d-1} \in \mathcal{B}(\mathbb{R}^d, 2^{k+1}),$$

with $\deg((u \otimes v) \otimes_{\mathbb{Z}_2} z_{d-1}) = \deg(u) + \deg(v) + d - 1$.

The partition of the basis $\mathcal{B}(\mathbb{R}^d, 2^{k+1})$ just introduced is given by

$$(v \otimes v) \otimes_{\mathbb{Z}_2} f^j \in \mathcal{B}_a(\mathbb{R}^d, 2^{k+1}),$$

for $v \in \mathcal{B}_a(\mathbb{R}^d, 2^k)$ and $0 \le j \le d-1$, that is,

$$\mathcal{B}_a(\mathbb{R}^d, 2^{k+1}) := \{(v \otimes v) \otimes_{\mathbb{Z}_2} f^j : v \in \mathcal{B}_a(\mathbb{R}^d, 2^k),\ 0 \le j \le d-1\}.$$

Then we set $\mathcal{B}_i(\mathbb{R}^d, 2^{k+1}) = \mathcal{B}(\mathbb{R}^d, 2^{k+1}) \backslash \mathcal{B}_a(\mathbb{R}^d, 2^{k+1})$. As before, we denote by

$$\mathrm{A}^*(\mathbb{R}^d, 2^{k+1}) := \langle \mathcal{B}_a(\mathbb{R}^d, 2^{k+1}) \rangle \qquad \text{and} \qquad \mathrm{I}^*(\mathbb{R}^d, 2^{k+1}) := \langle \mathcal{B}_i(\mathbb{R}^d, 2^{k+1}) \rangle.$$

Then we have the following additive decomposition of the cohomology

$$H^*(\mathrm{Pe}(\mathbb{R}^d, 2^{k+1})/\mathcal{S}_{2^{k+1}}; \mathbb{F}_2) \cong \mathrm{A}^*(\mathbb{R}^d, 2^{k+1}) \oplus \mathrm{I}^*(\mathbb{R}^d, 2^{k+1}).$$

Lemma 3.10 *Let* $d \ge 2$ *be an integer or* $d = \infty$. *Then* $\mathrm{A}^*(\mathbb{R}^d, 2^{k+1})$ *is a subalgebra and* $\mathrm{I}^*(\mathbb{R}^d, 2^{k+1})$ *is an ideal of the cohomology algebra* $H^*(\mathrm{Pe}(\mathbb{R}^d, 2^{k+1})/\mathcal{S}_{2^{k+1}}; \mathbb{F}_2)$. *Moreover,*

$$\mathrm{A}^*(\mathbb{R}^d, 2^{k+1}) \cong \mathbb{F}_2[V_{k+1,1}, \dots, V_{k+1,k+1}]/\langle V_{k+1,1}^d, \dots, V_{k+1,k+1}^d \rangle, \tag{3.17}$$

and $\deg(V_{k+1,r}) = 2^{r-1}$ *for* $1 \le r \le k+1$.

Proof Since the description of the cohomology $H^*(\mathrm{Pe}(\mathbb{R}^d, 2^{k+1})/\mathcal{S}_{2^{k+1}}; \mathbb{F}_2)$ is derived from a spectral sequence with appropriate multiplication structure it follows directly that $\mathrm{A}^*(\mathbb{R}^d, 2^{k+1})$ is a subalgebra, and $\mathrm{I}^*(\mathbb{R}^d, 2^{k+1})$ is an ideal, consult Theorem 3.6. It remains to establish the isomorphism (3.17).

Let us set

$$V_{k+1,1} := (1 \otimes 1) \otimes_{\mathbb{Z}_2} f, \tag{3.18}$$

and for all $2 \le r \le k+1$ let

$$V_{k+1,r} := (V_{k,r-1} \otimes V_{k,r-1}) \otimes_{\mathbb{Z}_2} 1. \tag{3.19}$$

Then

$$\deg(V_{k+1,1}) = 2\deg(1) + \deg(f) = 2 \cdot 0 + 1 = 1,$$

and

$$\deg(V_{k+1,r}) = 2\deg(V_{k,r-1}) + \deg(1) = 2 \cdot 2^{r-2} + 0 = 2^{r-1}$$

for $2 \le r \le k+1$. Therefore, from the construction of the set $\mathcal{B}_a(\mathbb{R}^d, 2^k)$ and the assumption about the structure of the subalgebra

$$\mathrm{A}^*(\mathbb{R}^d, 2^k) \cong \mathbb{F}_2[V_{k,1}, \dots, V_{k,k}]/\langle V_{k,1}^d, \dots, V_{k,k}^d \rangle,$$

directly follows that the isomorphism (3.17) holds. □

The calculations of this section establish the following theorem.

Theorem 3.11 *Let $d \geq 2$ be an integer or $d = \infty$, and let $m \geq 0$ be an integer. Then*

$$\begin{aligned} & H^*(\mathrm{Pe}(\mathbb{R}^d, 2^m)/\mathcal{S}_{2^m}; \mathbb{F}_2) \\ & \quad \cong \mathbb{F}_2[V_{m,1}, \ldots, V_{m,m}]/\langle V_{m,1}^d, \ldots, V_{m,m}^d \rangle \oplus \mathrm{I}^*(\mathbb{R}^d, 2^m), \end{aligned} \tag{3.20}$$

where $\mathrm{I}^*(\mathbb{R}^d, 2^m)$ is an ideal, and $\deg(V_{m,r}) = 2^{r-1}$ for $1 \leq r \leq m$. In particular, for $d = \infty$ we have

$$H^*(\mathrm{Pe}(\mathbb{R}^\infty, 2^m)/\mathcal{S}_{2^m}; \mathbb{F}_2) \cong \mathbb{F}_2[V_{m,1}, \ldots, V_{m,m}] \oplus \mathrm{I}^*(\mathbb{R}^\infty, 2^m). \tag{3.21}$$

3.5 The Restriction Homomorphisms – Three Aspects

Let $m \geq 0$ be an integer. Consider the sequence of inclusions

$$\mathcal{E}_m \longrightarrow \mathcal{S}_{2^m} \xrightarrow{\iota_{2^m}} \mathfrak{S}_{2^m} \longrightarrow \mathrm{O}(2^m)$$

where the last inclusion is the embedding give via the permutation representation. The corresponding sequence of maps between classifying spaces

$$\mathrm{B}\mathcal{E}_m \longrightarrow \mathrm{B}\mathcal{S}_{2^m} \longrightarrow \mathrm{B}\mathfrak{S}_{2^m} \longrightarrow \mathrm{BO}(2^m)$$

induces the following sequence of restriction homomorphisms:

$$H^*(\mathrm{O}(2^m); \mathbb{F}_2) \xrightarrow{\mathrm{res}^{\mathrm{O}(2^m)}_{\mathfrak{S}_{2^m}}} H^*(\mathfrak{S}_{2^m}; \mathbb{F}_2) \xrightarrow{\mathrm{res}^{\mathfrak{S}_{2^m}}_{\mathcal{S}_{2^m}}} H^*(\mathcal{S}_{2^m}; \mathbb{F}_2) \xrightarrow{\mathrm{res}^{\mathcal{S}_{2^m}}_{\mathcal{E}_m}} H^*(\mathcal{E}_m; \mathbb{F}_2).$$

In this section we study various aspects of these restriction homomorphisms.

3.5.1 A Restriction Homomorphism and the Mùi Invariants

For $d = \infty$ the isomorphism (3.20) gives the following decomposition of the cohomology of the group $\mathcal{S}_{2^m}$:

$$\begin{aligned} H^*(\mathcal{S}_{2^m}; \mathbb{F}_2) & \cong H^*(\mathrm{Pe}(\mathbb{R}^\infty, 2^m)/\mathcal{S}_{2^m}; \mathbb{F}_2) \\ & \cong \mathbb{F}_2[V_{m,1}, \ldots, V_{m,m}] \oplus \mathrm{I}^*(\mathbb{R}^\infty, 2^m), \end{aligned} \tag{3.22}$$

where $\deg(V_{m,r}) = 2^{r-1}$ for $1 \leq r \leq m$.

Recall that in Definition 2.7 we have specified the elementary abelian group $\mathcal{E}_m \cong \mathbb{Z}_2^{\oplus m}$ as a subgroup of $\mathcal{S}_{2^m}$. First we study the restriction map

$$\mathrm{res}_{\mathcal{E}_m}^{\mathcal{S}_{2^m}} : H^*(\mathcal{S}_{2^m}; \mathbb{F}_2) \longrightarrow H^*(\mathcal{E}_m; \mathbb{F}_2).$$

From the definition of the basis $\mathcal{B}(\mathbb{R}^d, 2^m)$, its partition into subsets $\mathcal{B}_a(\mathbb{R}^d, 2^m)$ and $\mathcal{B}_i(\mathbb{R}^d, 2^m)$, and the definition of the element $V_{m,1}$ follows that

$$V_{m,1} \cdot I^*(\mathbb{R}^\infty, 2^m) = 0.$$

In addition, if we recall how we introduced the subgroup $\mathcal{E}_m$ of $\mathcal{S}_{2^m}$ (see Definition 2.7), and observe that multiplication by $\mathrm{res}_{\mathcal{E}_m}^{\mathcal{S}_{2^m}}(V_{m,1})$ in $H^*(\mathcal{E}_m; \mathbb{F}_2)$ is injective, we can conclude that

$$\mathrm{res}_{\mathcal{E}_m}^{\mathcal{S}_{2^m}}(\mathrm{I}^*(\mathbb{R}^\infty, 2^m)) = 0 \qquad \Longleftrightarrow \qquad \mathrm{I}^*(\mathbb{R}^\infty, 2^m) \subseteq \ker(\mathrm{res}_{\mathcal{E}_m}^{\mathcal{S}_{2^m}}). \tag{3.23}$$

Now, we shift our interest to the generators $V_{m,1}, \ldots, V_{m,m}$ of the polynomial subalgebra in the decomposition (3.22) and will identify its images under the restriction $\mathrm{res}_{\mathcal{E}_m}^{\mathcal{S}_{2^m}}(V_{m,1}), \ldots, \mathrm{res}_{\mathcal{E}_m}^{\mathcal{S}_{2^m}}(V_{m,m})$. Even the definitions of the group $\mathcal{S}_{2^m}$ and its subgroup $\mathcal{E}_m$ were inductive our approach to the description of the image of $\mathrm{im}(\mathrm{res}_{\mathcal{E}_m}^{\mathcal{S}_{2^m}})$ is not be inductive. For that we follow [64, p. 266] and first note that according to Lemma 10.5 the restriction image is contained in the ring of invariants of the corresponding Weyl group

$$\begin{aligned}\mathrm{im}\left(\mathrm{res}_{\mathcal{E}_m}^{\mathcal{S}_{2^m}} : H^*(\mathcal{S}_{2^m}; \mathbb{F}_2) \longrightarrow H^*(\mathcal{E}_m; \mathbb{F}_2)\right) &\subseteq H^*(\mathcal{E}_m; \mathbb{F}_2)^{W_{\mathcal{S}_{2^m}}(\mathcal{E}_m)} \\ &= \mathbb{F}_2[y_1, \ldots, y_m]^{W_{\mathcal{S}_{2^m}}(\mathcal{E}_m)}.\end{aligned}$$

From Lemma 10.6 we know that $W_{\mathcal{S}_{2^m}}(\mathcal{E}_m) = \mathrm{L}_m(\mathbb{F}_2)$ is the Sylow 2-subgroup of $\mathrm{GL}_m(\mathbb{F}_2)$ of all lower triangular matrices with 1's on the main diagonal. Therefore all polynomials

$$v_{m,1} := \mathrm{res}_{\mathcal{E}_m}^{\mathcal{S}_{2^m}}(V_{m,1}), \ \ldots, \ v_{m,m} := \mathrm{res}_{\mathcal{E}_m}^{\mathcal{S}_{2^m}}(V_{m,m})$$

are $\mathrm{L}_m(\mathbb{F}_2)$ invariant polynomials—exactly the corresponding Mùi invariants. Each polynomial $v_{m,r}$ has y_{m-r+1} as a factor by (3.18) and (3.19), $\deg(v_{m,r}) = 2^{r-1}$, and $v_{m,r}$ is $\mathrm{L}_m(\mathbb{F}_2)$ invariant, consequently

$$v_{m,r} = \prod_{(\lambda_m, \ldots, \lambda_{m-r+2}) \in \mathbb{F}_2^{r-1}} \left(\lambda_m\, y_m + \cdots + \lambda_{m-r+2}\, y_{m-r+2} + y_{m-r+1}\right), \tag{3.24}$$

as stated in [62, (2.14)]. In particular, the polynomials $v_{m,1}, \ldots, v_{m,m}$ are algebraically independent. In the notation of Theorem 8.1 we have that $v_{m,r} = h_r$ are Mùi invariants, with $m = n$ and $y_1 = x_1, \ldots, y_m = x_m$.

3.5.2 A Restriction Homomorphism and the Dickson Invariants

Let us now consider the restriction homomorphism

$$\mathrm{res}^{\mathfrak{S}_{2^m}}_{\mathcal{E}_m} : H^*(\mathfrak{S}_{2^m}; \mathbb{F}_2) \longrightarrow H^*(\mathcal{E}_m; \mathbb{F}_2).$$

Like in the previous case using Lemma 10.5 we get that

$$\begin{aligned}\mathrm{im}\left(\mathrm{res}^{\mathfrak{S}_{2^m}}_{\mathcal{E}_m} : H^*(\mathfrak{S}_{2^m}; \mathbb{F}_2) \longrightarrow H^*(\mathcal{E}_m; \mathbb{F}_2)\right) &\subseteq H^*(\mathcal{E}_m; \mathbb{F}_2)^{W_{\mathfrak{S}_{2^m}}(\mathcal{E}_m)} \\ &= \mathbb{F}_2[y_1, \dots, y_m]^{W_{\mathfrak{S}_{2^m}}(\mathcal{E}_m)}.\end{aligned}$$

From Lemma 10.7 we get that $W_{\mathfrak{S}_{2^m}}(\mathcal{E}_m) \cong \mathrm{GL}_m(\mathbb{F}_2)$, and consequently

$$\begin{aligned}\mathrm{im}\left(\mathrm{res}^{\mathfrak{S}_{2^m}}_{\mathcal{E}_m} : H^*(\mathfrak{S}_{2^m}; \mathbb{F}_2) \longrightarrow H^*(\mathcal{E}_m; \mathbb{F}_2)\right) &\subseteq H^*(\mathcal{E}_m; \mathbb{F}_2)^{W_{\mathfrak{S}_{2^m}}(\mathcal{E}_m)} \\ &= \mathbb{F}_2[y_1, \dots, y_m]^{\mathrm{GL}_m(\mathbb{F}_2)}.\end{aligned}$$

Now the description of the ring of invariants given in Theorem 8.2 yields that

$$\begin{aligned}&\mathrm{im}\left(\mathrm{res}^{\mathfrak{S}_{2^m}}_{\mathcal{E}_m} : H^*(\mathfrak{S}_{2^m}; \mathbb{F}_2) \longrightarrow H^*(\mathcal{E}_m; \mathbb{F}_2)\right) \subseteq H^*(\mathcal{E}_m; \mathbb{F}_2)^{W_{\mathfrak{S}_{2^m}}(\mathcal{E}_m)} \\ &= \mathbb{F}_2[y_1, \dots, y_m]^{\mathrm{GL}_m(\mathbb{F}_2)} = \mathbb{F}_2[d_{m,0}, \dots, d_{m,m-1}], \qquad (3.25)\end{aligned}$$

where $d_{m,0}, \dots, d_{m,m-1}$ are the Dickson invariants.

Next we recall that

$$H^*(\mathrm{BO}(2^m); \mathbb{F}_2) = \mathbb{F}[w_1, \dots, w_{2^m}],$$

where w_i, for $1 \le i \le 2^m$, denotes the ith Stiefel–Whitney class of the tautological vector bundle γ_{2^m} over $\mathrm{BO}(2^m)$, see [80, Thm. 7.1]. Let us introduce the following notation

$$w_{2^m-2^r} \xmapsto{\mathrm{res}^{\mathrm{O}(2^m)}_{\mathfrak{S}_{2^m}}} w_{m,r} \xmapsto{\mathrm{res}^{\mathfrak{S}_{2^m}}_{S_{2^m}}} D_{m,r} \xmapsto{\mathrm{res}^{S_{2^m}}_{\mathcal{E}_m}} d_{m,r},$$

where $0 \le r \le m-1$. From Theorem 8.4 we have that indeed the classes $d_{m,0}, \dots, d_{m,m-1}$ are Dickson invariants, and furthermore

$$\mathrm{res}^{\mathrm{O}(2^m)}_{\mathcal{E}_m}(w_i) = \begin{cases} d_{m,r} = \mathrm{res}^{\mathfrak{S}_{2^m}}_{\mathcal{E}_m}(w_{m,r}), & i = 2^m - 2^r,\ 0 \le r \le m-1, \\ 1, & i = 0, \\ 0, & \text{otherwise.} \end{cases}$$

Hence, from (3.25) we conclude that

$$\operatorname{im}\left(\operatorname{res}^{\mathfrak{S}_{2^m}}_{\mathcal{E}_m} : H^*(\mathfrak{S}_{2^m};\mathbb{F}_2) \longrightarrow H^*(\mathcal{E}_m;\mathbb{F}_2)\right) = H^*(\mathcal{E}_m;\mathbb{F}_2)^{W_{\mathfrak{S}_{2^m}}(\mathcal{E}_m)}$$
$$= \mathbb{F}_2[y_1,\dots,y_m]^{\mathrm{GL}_m(\mathbb{F}_2)} = \mathbb{F}_2[d_{m,0},\dots,d_{m,m-1}].$$

To summarize, we have proved the following lemma.

Lemma 3.12 *Let $m \geq 0$ be an integer. Then*

$$\operatorname{im}\left(\operatorname{res}^{\mathfrak{S}_{2^m}}_{\mathcal{E}_m} : H^*(\mathfrak{S}_{2^m};\mathbb{F}_2) \longrightarrow H^*(\mathcal{E}_m;\mathbb{F}_2)\right) = \mathbb{F}_2[y_1,\dots,y_m]^{\mathrm{GL}_m(\mathbb{F}_2)}$$
$$= \mathbb{F}_2[d_{m,0},\dots,d_{m,m-1}], \quad (3.26)$$

where $d_{m,0},\dots,d_{m,m-1}$ are the Dickson invariants. Consequently,

$$H^*(\mathfrak{S}_{2^m};\mathbb{F}_2) \cong \operatorname{im}\left(\operatorname{res}^{\mathfrak{S}_{2^m}}_{\mathcal{E}_m} : H^*(\mathfrak{S}_{2^m};\mathbb{F}_2) \longrightarrow H^*(\mathcal{E}_m;\mathbb{F}_2)\right) \oplus \ker(\operatorname{res}^{\mathfrak{S}_{2^m}}_{\mathcal{E}_m})$$
$$\cong \mathbb{F}_2[d_{m,0},\dots,d_{m,m-1}] \oplus \ker(\operatorname{res}^{\mathfrak{S}_{2^m}}_{\mathcal{E}_m})$$
$$\cong \mathbb{F}_2[w_{m,0},\dots,w_{m,m-1}] \oplus \ker(\operatorname{res}^{\mathfrak{S}_{2^m}}_{\mathcal{E}_m}). \quad (3.27)$$

Let us reflect on the facts we obtained so far. For $0 \leq r \leq m-1$ we specified the following elements under the restriction maps:

$$\begin{array}{ccccccc}
H^*(\mathrm{O}(2^m);\mathbb{F}_2) & \xrightarrow{\operatorname{res}^{\mathrm{O}(2^m)}_{\mathfrak{S}_{2^m}}} & H^*(\mathfrak{S}_{2^m};\mathbb{F}_2) & \xrightarrow{\operatorname{res}^{\mathfrak{S}_{2^m}}_{S_{2^m}}} & H^*(S_{2^m};\mathbb{F}_2) & \xrightarrow{\operatorname{res}^{S_{2^m}}_{\mathcal{E}_m}} & H^*(\mathcal{E}_m;\mathbb{F}_2) \\
w_{2^m-2^r} & \xmapsto{\operatorname{res}^{\mathrm{O}(2^m)}_{\mathfrak{S}_{2^m}}} & w_{m,r} & \xmapsto{\operatorname{res}^{\mathfrak{S}_{2^m}}_{S_{2^m}}} & D_{m,r} & \xmapsto{\operatorname{res}^{S_{2^m}}_{\mathcal{E}_m}} & d_{m,r} \\
 & & & & V_{m,r+1} & \xmapsto{\operatorname{res}^{S_{2^m}}_{\mathcal{E}_m}} & v_{m,r+1}.
\end{array} \quad (3.28)$$

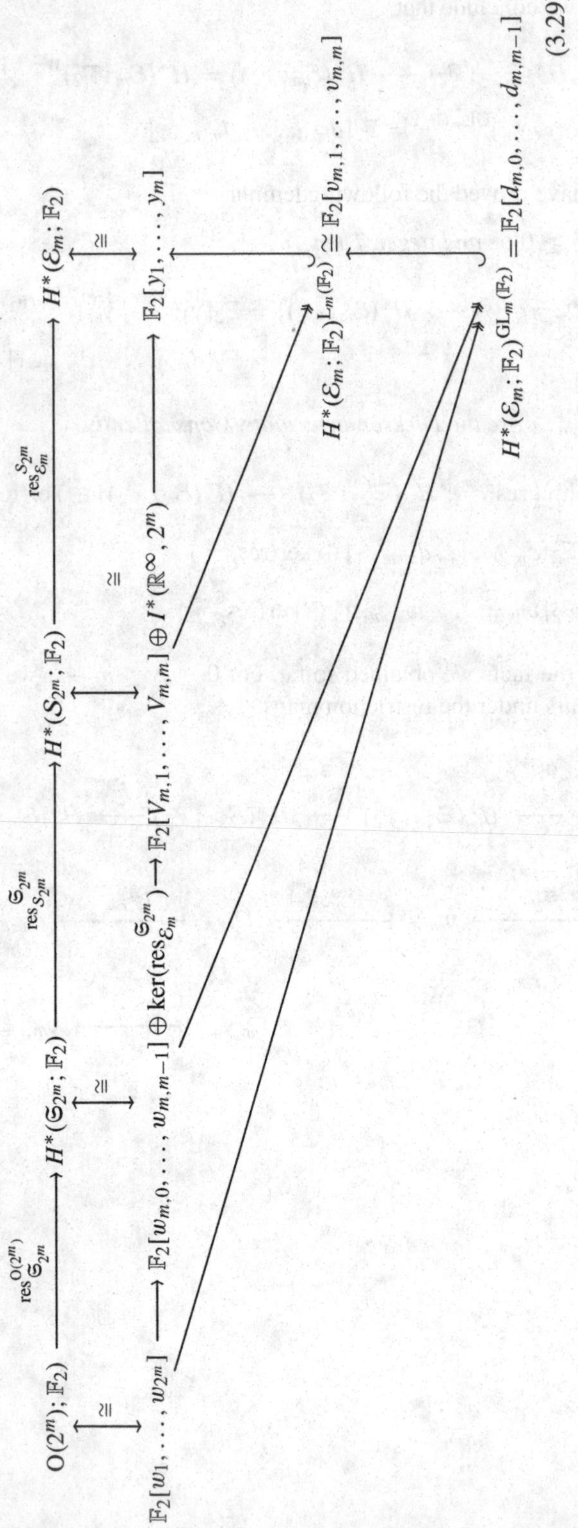
$\mathrm{O}(2^m); \mathbb{F}_2)$
$\mathrm{res}^{\mathrm{O}(2^m)}_{\mathfrak{S}_{2^m}}$
$H^*(\mathfrak{S}_{2^m}; \mathbb{F}_2)$
$\mathrm{res}^{\mathfrak{S}_{2^m}}_{S_{2^m}}$
$H^*(S_{2^m}; \mathbb{F}_2)$
$\mathrm{res}^{S_{2^m}}_{\mathcal{E}_m}$
$H^*(\mathcal{E}_m; \mathbb{F}_2)$
$\cong$
$\mathbb{F}_2[w_1, \ldots, w_{2^m}]$
$\mathbb{F}_2[w_{m,0}, \ldots, w_{m,m-1}] \oplus \ker(\mathrm{res}^{\mathfrak{S}_{2^m}}_{\mathcal{E}_m})$
$\mathbb{F}_2[V_{m,1}, \ldots, V_{m,m}] \oplus I^*(\mathbb{R}^\infty, 2^m)$
$\mathbb{F}_2[y_1, \ldots, y_m]$
$H^*(\mathcal{E}_m; \mathbb{F}_2)^{\mathrm{L}_m(\mathbb{F}_2)} \cong \mathbb{F}_2[v_{m,1}, \ldots, v_{m,m}]$
$H^*(\mathcal{E}_m; \mathbb{F}_2)^{\mathrm{GL}_m(\mathbb{F}_2)} = \mathbb{F}_2[d_{m,0}, \ldots, d_{m,m-1}]$

(3.29)

Furthermore, we have proved the factorizations of the restriction homomorphism which are described by the diagram (3.29) on the next page. In other words,

– the restriction homomorphism $\mathrm{res}^{\mathrm{O}(2^m)}_{\mathcal{E}_m}$ factors as follows:

$$\begin{array}{ccccc} H^*(\mathrm{O}(2^m);\mathbb{F}_2) & \twoheadrightarrow & H^*(\mathcal{E}_m;\mathbb{F}_2)^{\mathrm{GL}_m(\mathbb{F}_2)} & \hookrightarrow & H^*(\mathcal{E}_m;\mathbb{F}_2) \\ \updownarrow\cong & & \updownarrow\cong & & \updownarrow\cong \\ \mathbb{F}_2[w_1,\dots,w_{2^m}] & \twoheadrightarrow & \mathbb{F}_2[d_{m,0},\dots,d_{m,m-1}] & \hookrightarrow & \mathbb{F}[y_1,\dots,y_m], \end{array}$$

– the restriction homomorphism $\mathrm{res}^{\mathfrak{S}_{2^m}}_{\mathcal{E}_m}$ factors as follows

$$\begin{array}{ccccc} H^*(\mathfrak{S}_{2^m};\mathbb{F}_2) & \twoheadrightarrow & H^*(\mathcal{E}_m;\mathbb{F}_2)^{\mathrm{GL}_m(\mathbb{F}_2)} & \hookrightarrow & H^*(\mathcal{E}_m;\mathbb{F}_2) \\ \updownarrow\cong & & \updownarrow\cong & & \updownarrow\cong \\ \mathbb{F}_2[w_{m,0},\dots,w_{m,m-1}]\oplus\ker(\mathrm{res}^{\mathfrak{S}_{2^m}}_{\mathcal{E}_m}) & \twoheadrightarrow & \mathbb{F}_2[d_{m,0},\dots,d_{m,m-1}] & \hookrightarrow & \mathbb{F}[y_1,\dots,y_m], \end{array}$$

– the restriction homomorphism $\mathrm{res}^{\mathcal{S}_{2^m}}_{\mathcal{E}_m}$ factors as follows:

$$\begin{array}{ccccc} H^*(\mathcal{S}_{2^m};\mathbb{F}_2) & \twoheadrightarrow & H^*(\mathcal{E}_m;\mathbb{F}_2)^{\mathrm{L}_m(\mathbb{F}_2)} & \hookrightarrow & H^*(\mathcal{E}_m;\mathbb{F}_2) \\ \updownarrow\cong & & \updownarrow\cong & & \updownarrow\cong \\ \mathbb{F}_2[V_{m,1},\dots,V_{m,m}]\oplus I^*(\mathbb{R}^\infty,2^m) & \twoheadrightarrow & \mathbb{F}_2[v_{m,1},\dots,v_{m,m}] & \hookrightarrow & \mathbb{F}[y_1,\dots,y_m]. \end{array}$$

Now from Proposition 8.3 we get a connection between the Dickson invariant polynomials $d_{m,0},\dots,d_{m,m-1}$ and the Mùi invariant polynomials, $v_{m,1},\dots,v_{m,m}$. For $0\le r\le m-1$ and $d_{m-1,-1}=0$ holds:

$$d_{m,r}=(\chi_m d_{m-1,r})\,v_{m,m}+(\chi_m d_{m-1,r-1})^2. \tag{3.30}$$

Here $\chi_m\in\mathrm{GL}_m(\mathbb{F}_2)$ is the variable change given by the matrix

$$\begin{pmatrix} 0\,0\cdots 0\,1 \\ 0\,0\cdots 1\,0 \\ \cdots \\ 0\,1\cdots 0\,0 \\ 1\,0\cdots 0\,0 \end{pmatrix}.$$

3.5.3 Two Lemmas

In the final part of this section we prove the following two lemmas. For the next lemma see also [64, Lem. 3.14].

Lemma 3.13 *Let $m \geq 0$ be an integer. Then*

$$\operatorname{res}_{\mathcal{S}_{2^m}}^{\mathfrak{S}_{2^m}}\big(\langle w_{m,0}\rangle\big) \subseteq \mathbb{F}_2[V_{m,1}, \ldots, V_{m,m}], \tag{3.31}$$

where $\langle w_{m,0}\rangle$ denotes the principal ideal generated by the class $w_{m,0}$ in $H^*(\mathfrak{S}_{2^m}; \mathbb{F}_2)$.

Proof Recall that in (3.22) we concluded that

$$H^*(\mathcal{S}_{2^m}; \mathbb{F}_2) \cong \mathbb{F}_2[V_{m,1}, \ldots, V_{m,m}] \oplus \mathrm{I}^*(\mathbb{R}^\infty, 2^m).$$

In order to prove (3.31) it suffices to show that

$$D_{m,0} \cdot \mathrm{I}^*(\mathbb{R}^\infty, 2^m) = 0, \qquad \text{and} \qquad D_{m,0} = V_{m,1} \cdots V_{m,m}. \tag{3.32}$$

Indeed, if the equalities (3.32) hold, and because $\operatorname{res}_{\mathcal{S}_{2^m}}^{\mathfrak{S}_{2^m}}(w_{m,0}) = D_{m,0}$, we have that

$$\begin{aligned}
\operatorname{res}_{\mathcal{S}_{2^m}}^{\mathfrak{S}_{2^m}}\big(\langle w_{m,0}\rangle\big) &= \operatorname{res}_{\mathcal{S}_{2^m}}^{\mathfrak{S}_{2^m}}\big(w_{m,0} \cdot H^*(\mathfrak{S}_{2^m}; \mathbb{F}_2)\big) \\
&\subseteq D_{m,0} \cdot H^*(\mathcal{S}_{2^m}; \mathbb{F}_2) \\
&= D_{m,0} \cdot \big(\mathbb{F}_2[V_{m,1}, \ldots, V_{m,m}] \oplus \mathrm{I}^*(\mathbb{R}^\infty, 2^m)\big) \\
&= D_{m,0} \cdot \mathbb{F}_2[V_{m,1}, \ldots, V_{m,m}] \subseteq \mathbb{F}_2[V_{m,1}, \ldots, V_{m,m}].
\end{aligned}$$

Thus, in order to finish the proof of the lemma it remains to verify equalities (3.32).

First we verify that $D_{m,0} \cdot \mathrm{I}^*(\mathbb{R}^\infty, 2^m) = 0$. For that we use a classical result of Quillen [90] about detection of group cohomology, see Sect. 10.1. In particular, according to Theorem 10.2 we have that the cohomology $H^*(\mathcal{S}_{2^m}; \mathbb{F}_2)$ of the Sylow 2-subgroup $\mathcal{S}_{2^m}$ modulo $\mathbb{F}_2$ is detected by the subgroups $\mathcal{E}_m$ and $\mathcal{S}_{2^{m-1}} \times \mathcal{S}_{2^{m-1}}$. From (3.23) we have that $\operatorname{res}_{\mathcal{E}_m}^{\mathcal{S}_{2^m}}(\mathrm{I}^*(\mathbb{R}^\infty, 2^m)) = 0$. Consequently, if we prove that $\operatorname{res}_{\mathcal{S}_{2^{m-1}} \times \mathcal{S}_{2^{m-1}}}^{\mathcal{S}_{2^m}}(D_{m,0}) = 0$ it would follow that $D_{m,0} \cdot \mathrm{I}^*(\mathbb{R}^\infty, 2^m) = 0$. To see that this restriction of $D_{m,0}$ vanishes we first recall that $D_{m,0}$ is a $(2^m - 1)$-Stiefel–Whitney class of the vector bundle η_{2^m}:

$$\mathbb{R}^{2^m} \longrightarrow \mathrm{E}\mathcal{S}_{2^m} \times_{\mathcal{S}_{2^m}} \mathbb{R}^{2^m} \longrightarrow \mathrm{B}\mathcal{S}_{2^m},$$

see Sect. 8.2. The vector bundle η_{2^m} can be decomposed into a Whitney sum of two vector bundles where one of them is a trivial line bundle. Consequently, the 2^m-Stiefel–Whitney class of the bundle vanishes. The trivial line subbundle is determined by the trivial $\mathcal{S}_{2^m}$ subrepresentation $\{(x_1, \dots, x_{2^m}) \in \mathbb{R}^{2^m} : x_1 = \dots = x_{2^m}\}$ of $\mathbb{R}^{2^m}$. Using the naturality property of Stiefel–Whitney classes [80, Ax. 2, p. 35] we have that $\operatorname{res}^{\mathcal{S}_{2^m}}_{\mathcal{S}_{2^{m-1}} \times \mathcal{S}_{2^{m-1}}}(D_{m,0})$ is the $(2^m - 1)$-Stiefel–Whitney class of the pull-back vector bundle:

$$\begin{array}{ccc} \mathrm{E}(\mathcal{S}_{2^{m-1}} \times \mathcal{S}_{2^{m-1}}) \times_{(\mathcal{S}_{2^{m-1}} \times \mathcal{S}_{2^{m-1}})} \mathbb{R}^{2^m} & \longrightarrow & \mathrm{E}\mathcal{S}_{2^m} \times_{\mathcal{S}_{2^m}} \mathbb{R}^{2^m} \\ \omega_{2^m} \downarrow & & \downarrow \eta_{2^m} \\ \mathrm{B}(\mathcal{S}_{2^{m-1}} \times \mathcal{S}_{2^{m-1}}) & \longrightarrow & \mathrm{B}\mathcal{S}_{2^m}. \end{array}$$

The pull-back vector bundle ω_{2^m} can be decomposed into a Whitney sum of two vector bundles where one of them is a two dimensional trivial vector bundle. This trivial vector subbundle is determined by the trivial $\mathcal{S}_{2^{m-1}} \times \mathcal{S}_{2^{m-1}}$ subrepresentation $\{(x_1, \dots, x_{2^m}) \in \mathbb{R}^{2^m} : x_1 = \dots = x_{2^{m-1}}, x_{2^{m-1}+1} = \dots = x_{2^m}\}$ of $\mathbb{R}^{2^m}$. Hence, $(2^m - 1)$-Stiefel–Whitney class of this bundle $w_{2^m-1}(\omega_{2^m}) = \operatorname{res}^{\mathcal{S}_{2^m}}_{\mathcal{S}_{2^{m-1}} \times \mathcal{S}_{2^{m-1}}}(D_{m,0})$ has to vanish. This completes the proof of the first equality $D_{m,0} \cdot \mathrm{I}^*(\mathbb{R}^\infty, 2^m) = 0$ in (3.32).

Next we prove that $D_{m,0} = V_{m,1} \cdots V_{m,m}$. Once again we use the fact that $H^*(\mathcal{S}_{2^m}; \mathbb{F}_2)$ is detected by the subgroups $\mathcal{E}_m$ and $\mathcal{S}_{2^{m-1}} \times \mathcal{S}_{2^{m-1}}$. Since we showed that $\operatorname{res}^{\mathcal{S}_{2^m}}_{\mathcal{S}_{2^{m-1}} \times \mathcal{S}_{2^{m-1}}}(D_{m,0}) = 0$ it suffices to show that

$$\operatorname{res}^{\mathcal{S}_{2^m}}_{\mathcal{E}_m}(D_{m,0}) = \operatorname{res}^{\mathcal{S}_{2^m}}_{\mathcal{E}_m}(V_{m,1} \cdots V_{m,m}) \quad \Longleftrightarrow \quad d_{m,0} = v_{m,1} \cdots v_{m,m}.$$

Indeed, the equality $d_{m,0} = v_{m,1} \cdots v_{m,m}$ can be established by direct computation using the induction on m in combination with relations (3.24) and (3.30), and observation that

$$\chi_m(v_{m-1,1} \cdots v_{m-1,m-1}) = v_{m,1} \cdots v_{m,m-1}.$$

Hence, we showed that $D_{m,0} = V_{m,1} \cdots V_{m,m}$ and consequently verification of the second equality in (3.32). This completes the proof of the lemma. □

In the final lemma of this section we describe the kernel of the restriction homomorphism $\operatorname{res}^{\mathfrak{S}_{2^m}}_{\mathfrak{S}_{2^{m-1}} \times \mathfrak{S}_{2^{m-1}}}$.

Lemma 3.14 *Let $m \geq 0$ be an integer. Then*

$$\ker\left(\operatorname{res}^{\mathfrak{S}_{2^m}}_{\mathfrak{S}_{2^{m-1}} \times \mathfrak{S}_{2^{m-1}}}\right) = \langle w_{m,0} \rangle \subseteq H^*(\mathfrak{S}_{2^m}; \mathbb{F}_2). \tag{3.33}$$

Proof For the proof of the lemma we use again a classical result of Quillen on the detection of group cohomology, see Sect. 10.1. From Theorem 10.2 we have that the cohomology $H^*(\mathfrak{S}_{2^m}; \mathbb{F}_2)$ of the symmetric group $\mathfrak{S}_{2^m}$ modulo $\mathbb{F}_2$ is detected by the subgroups $\mathcal{E}_m$ and $\mathfrak{S}_{2^{m-1}} \times \mathfrak{S}_{2^{m-1}}$. This means that the homomorphism

$$H^*(\mathfrak{S}_{2^m}; \mathbb{F}_2) \xrightarrow{\operatorname{res}^{\mathfrak{S}_{2^m}}_{\mathcal{E}_m} \times \operatorname{res}^{\mathfrak{S}_{2^m}}_{\mathfrak{S}_{2^{m-1}} \times \mathfrak{S}_{2^{m-1}}}} H^*(\mathcal{E}_m; \mathbb{F}_2) \times H^*(\mathfrak{S}_{2^{m-1}} \times \mathfrak{S}_{2^{m-1}}; \mathbb{F}_2)$$

is a monomorphism. Thus,

$$0 \neq x \in \ker\left(\operatorname{res}^{\mathfrak{S}_{2^m}}_{\mathfrak{S}_{2^{m-1}} \times \mathfrak{S}_{2^{m-1}}}\right) \quad \Longrightarrow \quad \operatorname{res}^{\mathfrak{S}_{2^m}}_{\mathcal{E}_m}(x) \neq 0.$$

Further on, using the decomposition (3.27) we get the implication

$$0 \neq x \in \ker\left(\operatorname{res}^{\mathfrak{S}_{2^m}}_{\mathfrak{S}_{2^{m-1}} \times \mathfrak{S}_{2^{m-1}}}\right) \Longrightarrow x \in \langle w_{m,0}, \ldots, w_{m,m-1} \rangle.$$

Like in the proof of the previous lemma we consider the vector bundle ξ_{2^m} and its pull-back θ_{2^m} introduced by the following pull-back diagram:

$$\begin{array}{ccc} \mathrm{E}(\mathfrak{S}_{2^{m-1}} \times \mathfrak{S}_{2^{m-1}}) \times_{(\mathfrak{S}_{2^{m-1}} \times \mathfrak{S}_{2^{m-1}})} \mathbb{R}^{2^m} & \longrightarrow & \mathrm{E}\mathfrak{S}_{2^m} \times_{\mathfrak{S}_{2^m}} \mathbb{R}^{2^m} \\ \downarrow{\scriptstyle \theta_{2^m}} & & \downarrow{\scriptstyle \xi_{2^m}} \\ \mathrm{B}(\mathfrak{S}_{2^{m-1}} \times \mathfrak{S}_{2^{m-1}}) & \longrightarrow & \mathrm{B}\mathfrak{S}_{2^m}. \end{array}$$

As we know the classes $w_{m,0}, \ldots, w_{m,m-1}$ are the Stiefel–Whitney classes of the vector bundle ξ_{2^m} in dimensions $2^m - 2^0, \ldots, 2^m - 2^{m-1}$, respectively. The pull-back vector bundle θ_{2^m} can be decomposed into a Whitney sum of two vector bundles where one of them is a two dimensional trivial vector bundle. The trivial vector subbundle is determined by the trivial $\mathfrak{S}_{2^{m-1}} \times \mathfrak{S}_{2^{m-1}}$ subrepresentation $\{(x_1, \ldots, x_{2^m}) \in \mathbb{R}^{2^m} : x_1 = \cdots = x_{2^{m-1}}, x_{2^{m-1}+1} = \cdots = x_{2^m}\}$ of $\mathbb{R}^{2^m}$. Hence, $(2^m - 1)$-Stiefel–Whitney class of this bundle $\operatorname{res}^{\mathfrak{S}_{2^m}}_{\mathfrak{S}_{2^{m-1}} \times \mathfrak{S}_{2^{m-1}}}(w_{m,0})$ has to vanish, or equivalently

$$w_{m,0} \in \ker\left(\operatorname{res}^{\mathfrak{S}_{2^m}}_{\mathfrak{S}_{2^{m-1}} \times \mathfrak{S}_{2^{m-1}}}\right).$$

On the other hand the pull-back vector bundle θ_{2^m} is isomorphic to the vector bundle $\xi_{2^{m-1}} \times \xi_{2^{m-1}}$. Therefore,

$$\operatorname{res}^{\mathfrak{S}_{2^m}}_{\mathfrak{S}_{2^{m-1}} \times \mathfrak{S}_{2^{m-1}}}\left(w(\xi_{2^m})\right) = w(\theta_{2^m}) = w(\xi_{2^{m-1}} \times \xi_{2^{m-1}}) = w(\xi_{2^{m-1}}) \times w(\xi_{2^{m-1}}),$$

and consequently for $1 \le r \le m-1$ we have

$$\begin{aligned}
\operatorname{res}^{\mathfrak{S}_{2^m}}_{\mathfrak{S}_{2^{m-1}}\times\mathfrak{S}_{2^{m-1}}}(w_{m,r}) &= \operatorname{res}^{\mathfrak{S}_{2^m}}_{\mathfrak{S}_{2^{m-1}}\times\mathfrak{S}_{2^{m-1}}}(w_{2^m-2^r}(\xi_{2^m})) \\
&= w_{2^m-2^r}(\xi_{2^{m-1}} \times \xi_{2^{m-1}}) \\
&= \sum_{i=0}^{2^m-2^r} w_i(\xi_{2^{m-1}}) \times w_{2^m-2^r-i}(\xi_{2^{m-1}}) \\
&= \sum_{i=0}^{2^m-2^r} w_i(\xi_{2^{m-1}}) \otimes w_{2^m-2^r-i}(\xi_{2^{m-1}}) \\
&\in \bigoplus_{i=0}^{2^m-2^r} H^i(\mathfrak{S}_{2^{m-1}};\mathbb{F}_2) \otimes H^{2^m-2^r-i}(\mathfrak{S}_{2^{m-1}};\mathbb{F}_2).
\end{aligned}$$

Here we silently use the Eilenberg–Zilber isomorphism [23, Th. VI.3.2]. In particular, we can isolate a concrete (direct) summand in the decomposition as follows:

$$\begin{aligned}
\operatorname{res}^{\mathfrak{S}_{2^m}}_{\mathfrak{S}_{2^{m-1}}\times\mathfrak{S}_{2^{m-1}}}(w_{m,r}) &= \sum_{i=0}^{2^m-2^r} w_i(\xi_{2^{m-1}}) \otimes w_{2^m-2^r-i}(\xi_{2^{m-1}}) \\
&= w_{2^{m-1}-2^{r-1}}(\xi_{2^{m-1}}) \otimes w_{2^{m-1}-2^{r-1}}(\xi_{2^{m-1}}) \\
&\quad + \sum_{i\neq 2^{m-1}-2^{r-1}} w_i(\xi_{2^{m-1}}) \otimes w_{2^m-2^r-i}(\xi_{2^{m-1}}).
\end{aligned}$$

Now we use the fact that the Stiefel–Whitney classes

$$w_{2^{m-1}-2^{m-2}}(\xi_{2^{m-1}}), \ldots, w_{2^{m-1}-2^0}(\xi_{2^{m-1}})$$

are algebraically independent. Indeed, they restrict to the corresponding Dickson invariants $d_{m-1,0}, \ldots, d_{m-1,m-2}$ for which we know to be algebraically independent. Consequently, the restricted homomorphism

$$\operatorname{res}^{\mathfrak{S}_{2^m}}_{\mathfrak{S}_{2^{m-1}}\times\mathfrak{S}_{2^{m-1}}}|_{\langle w_{m,1},\ldots,w_{m,m-1}\rangle}$$

has to be a monomorphism. Hence, $\ker(\operatorname{res}^{\mathfrak{S}_{2^m}}_{\mathfrak{S}_{2^{m-1}}\times\mathfrak{S}_{2^{m-1}}}) = \langle w_{m,0}\rangle$, and the proof of the lemma is complete. $\square$

Chapter 4
Hu'ng's Injectivity Theorem

Let $d \geq 2$ be an integer or $d = \infty$, and let $m \geq 0$ be an integer. Consider the composition map $\rho_{d,2^m} := \mathrm{id}/\mathfrak{S}_{2^m} \circ \mathrm{ecy}_{d,2^m}/S_{2^m}$ between the quotient spaces

$$\mathrm{Pe}(\mathbb{R}^d, 2^m)/S_{2^m} \xrightarrow{\mathrm{ecy}_{d,2^m}/S_{2^m}} \mathrm{F}(\mathbb{R}^d, 2^m)/S_{2^m} \xrightarrow{\mathrm{id}/\mathfrak{S}_{2^m}} \mathrm{F}(\mathbb{R}^d, 2^m)/\mathfrak{S}_{2^m}, \tag{4.1}$$

where the first map is induced by the S_{2^m}-equivariant map $\mathrm{ecy}_{d,2^m}\colon \mathrm{Pe}(\mathbb{R}^d, 2^m) \longrightarrow \mathrm{F}(\mathbb{R}^d, 2^m)$, and the second map is induced by the identity.

The central objective of this chapter is to present a new and complete proof of the following claim, but first it is necessary to explain in detail several critical gaps in the published proof of this result, [62, Thm. 3.1].

Theorem 4.1 *Let $d \geq 2$ be an integer or $d = \infty$, and let $m \geq 0$ be an integer. Then the homomorphism*

$$\rho^*_{d,2^m}\colon H^*(\mathrm{F}(\mathbb{R}^d, 2^m)/\mathfrak{S}_{2^m}; \mathbb{F}_2) \longrightarrow H^*(\mathrm{Pe}(\mathbb{R}^d, 2^m)/S_{2^m}; \mathbb{F}_2) \tag{4.2}$$

is a monomorphism.

Remark 4.2 The homomorphism $\rho^*_{d,2^m}$ decomposes into the composition

$$(\mathrm{ecy}_{d,2^m}/S_{2^m})^* \circ (\mathrm{id}/\mathfrak{S}_{2^m})^*.$$

Since the map $\mathrm{id}/\mathfrak{S}_{2^m}\colon \mathrm{F}(\mathbb{R}^d, 2^m)/S_{2^m} \longrightarrow \mathrm{F}(\mathbb{R}^d, 2^m)/\mathfrak{S}_{2^m}$ is a covering map then the composition homomorphism

$$\begin{array}{ccc} H^*(\mathrm{F}(\mathbb{R}^d, 2^m)/\mathfrak{S}_{2^m}; \mathbb{F}_2) & \xrightarrow{(\mathrm{id}/\mathfrak{S}_{2^m})^*} & H^*(\mathrm{F}(\mathbb{R}^d, 2^m)/S_{2^m}; \mathbb{F}_2) \\ & \searrow^{[\mathfrak{S}_{2^m}:S_{2^m}]\cdot} & \downarrow \mathrm{tr} \\ & & H^*(\mathrm{F}(\mathbb{R}^d, 2^m)/\mathfrak{S}_{2^m}; \mathbb{F}_2) \end{array}$$

P. V. M. Blagojević et al., *Equivariant Cohomology of Configuration Spaces Mod 2*, Lecture Notes in Mathematics 2282, https://doi.org/10.1007/978-3-030-84138-6_4

is the multiplication with the index $[\mathfrak{S}_{2^m} : \mathcal{S}_{2^m}]$. Here tr denotes the classical transfer homomorphism, consult for example [60, Sec. 3.G]. Since $\mathcal{S}_{2^m}$ is a Sylow 2-subgroup the index $[\mathfrak{S}_{2^m} : \mathcal{S}_{2^m}]$ has to be odd. Hence the composition $\mathrm{tr} \circ (\mathrm{id}/\mathfrak{S}_{2^m})^*$ is an isomorphism implying that $(\mathrm{id}/\mathfrak{S}_{2^m})^*$ is a monomorphism. This means that in order to prove Theorem 4.1 it suffices to show that the homomorphism

$$(\mathrm{ecy}_{d,2^m}/\mathcal{S}_{2^m})^*\colon H^*(\mathrm{F}(\mathbb{R}^d, 2^m)/\mathcal{S}_{2^m}; \mathbb{F}_2) \longrightarrow H^*(\mathrm{Pe}(\mathbb{R}^d, 2^m)/\mathcal{S}_{2^m}; \mathbb{F}_2) \tag{4.3}$$

is a monomorphism.

Remark 4.3 In the case when $d = \infty$ the homomorphism $\rho^*_{\infty,2^m}$ becomes the restriction homomorphism $\mathrm{res}^{\mathfrak{S}_{2^m}}_{\mathcal{S}_{2^m}}$. Since we are working in the field $\mathbb{F}_2$ and $\mathcal{S}_{2^m}$ is a Sylow 2-subgroup the restriction map $\mathrm{res}^{\mathfrak{S}_{2^m}}_{\mathcal{S}_{2^m}}$ is injective, see for example [25, Prop. III.9.5(ii) and Thm. III.10.3]. Thus, Theorem 4.1 holds for $d = \infty$.

4.1 Critical Points in Hưng's Proof of His Injectivity Theorem

In order to simplify the comparison with the work of Hưng we begin with a dictionary that translates between our notation and the notation used in [64].

Paper [64]	This book	
$\mathfrak{S}_{2^m}$	$\mathfrak{S}_{2^m}$	The symmetric group on the set $\mathbb{Z}_2^{\oplus m}$
$\mathfrak{S}_{2^m,2}$	$\mathcal{S}_{2^m}$	The Sylow 2-subgroup of $\mathfrak{S}_{2^m}$ that contains $\mathcal{E}_m$
E^m	$\mathcal{E}_m$	$\mathbb{Z}_2^{\oplus m}$ regularly embedded elementary abelian group in $\mathfrak{S}_{2^m}$
$F(X, n)$	$\mathrm{F}(X, n)$	The ordered configuration space of n distinct points in X
$\widetilde{M}(d, m)$	$\mathrm{Pe}(\mathbb{R}^d, 2^m)$	The ordered Ptolemaic epicycles space $(S^{d-1})^{2^m-1}$
$M(d, m)$	$\mathrm{Pe}(\mathbb{R}^d, 2^m)/\mathcal{S}_{2^m}$	The unordered Ptolemaic epicycles space
$\tilde{i}(d, m)$	$\mathrm{ecy}_{d,2^m}$	The map from Definition 2.1 with $d \geq 1$ integer or $d = \infty$
$i(d, m)$	$\rho_{d,2^m}$	The map introduced in (4.1) with $d \geq 1$ integer or $d = \infty$
$i(M, d)$	$\kappa_{d,m}/\mathcal{S}_{2^m}$	The map $\mathrm{Pe}(\mathbb{R}^d, m)/\mathcal{S}_{2^m} \longrightarrow \mathrm{Pe}(\mathbb{R}^\infty, m)/\mathcal{S}_{2^m}$
$i(F, d)$		The map $F(\mathbb{R}^d, 2^m) \longrightarrow F(\mathbb{R}^\infty, 2^m)$
$W_{m,r}$	$w_{m,r}$	Restriction of the Stiefel–Whitney class $w_{2^m-2^r}$ to $H^*(\mathfrak{S}_{2^m})$
$\overline{Q}_{m,r}$	$D_{m,r}$	Restriction of the Stiefel–Whitney class $w_{2^m-2^r}$ to $H^*(\mathcal{S}_{2^m})$
$Q_{m,r}$	$d_{m,r}$	Restriction of the Stiefel–Whitney class $w_{2^m-2^r}$ to $H^*(\mathcal{E}_m)$, or the Dickson invariant
$\overline{V}_{m,r}$	$V_{m,r}$	Elements of $H^*(\mathcal{S}_{2^m})$ defined in Theorem 3.11
$V_{m,r}$	$v_{m,r}$	Elements of $H^*(\mathcal{E}_m)$ given by restriction $\mathrm{res}^{\mathcal{S}_{2^m}}_{\mathcal{E}_m}(V_{m,r})$

The statement of Theorem 4.1 in [64, Thm. 3.1] is written as follows; the coefficient field $\mathbb{F}_2$ is always to be assumed.

3.1. THEOREM *$i^*(q,n)\colon H^*(F(\mathbb{R}^q,2^n)/\mathfrak{S}_{2^n}) \longrightarrow H^*(M(q,n))$ is a monomorphism for $q \geq 1, n \geq 0$.*

The proof of [64, Thm. 3.1] presented in Hưng's paper is by induction on n. It starts on the page 269 and ends on page 271. This proof relies on [64, Prop. 3.5]. The claim of [64, Prop. 3.5] is proven on page 275 and relies on [64, Lem. 3.14, Lem. 3.19].

Now we outline the proof given by Hưng and exhibit two critical points that we have identified. The proof is by induction on n. For $n = 0$ the statement is easy to verify since $F(\mathbb{R}^q, 2^n) = \mathbb{R}^q$ and $\widetilde{M}(q,n) = \text{pt}$. Let us assume that $i^*(q, n-1)$ is injective. Before we make the next step in the proof we define the maps $\mu_{m,n}$ introduced in [64, (3.2)], and the map φ_{n-1} defined in [64, (2.3)].

For integers $m \geq 1$ and $n \geq 1$ consider the map

$$\mu_{m,n}\colon F(\mathbb{R}^q,m)/\mathfrak{S}_m \times F(\mathbb{R}^q,n)/\mathfrak{S}_n \longrightarrow F(\mathbb{R}^q,m+n)/\mathfrak{S}_{m+n},$$

given by

$$[(x_1,\dots,x_m)] \times [(y_1,\dots,y_n)] \longmapsto [(x_1,\dots,x_m,y_1+z,\dots,y_n+z)].$$

Here for

$$R_1 = \max_{1\leq k\leq m}\left\| x_k - \frac{1}{m}\sum_{i=1}^m x_i\right\| \quad\text{and}\quad R_2 = \max_{1\leq k\leq n}\left\| y_k - \frac{1}{n}\sum_{j=1}^n y_j\right\|$$

we define

$$z = \frac{1}{m}\sum_{i=1}^m x_i - \frac{1}{n}\sum_{j=1}^n y_j - (R_1+R_2+1,0,\dots,0).$$

Next for any $n \geq 1$ we introduce the following map

$$\varphi_{n-1}\colon M(q,n-1)\times M(q,n-1) \longrightarrow M(q,n), \qquad (x,y) \longmapsto [(x,y,*)].$$

Let us now consider the following diagram that commutes up to a homotopy

$$\begin{array}{ccc} F(\mathbb{R}^q,2^n)/\mathfrak{S}_{2^n} & \xleftarrow{\mu:=\mu_{2^{n-1},2^{n-1}}} & \left(F(\mathbb{R}^q,2^{n-1})/\mathfrak{S}_{2^{n-1}}\right)^2 \\ \uparrow{\scriptstyle i(q,n)} & & \uparrow{\scriptstyle i(q,n-1)^2} \\ M(q,n) & \xleftarrow{\varphi:=\varphi_{n-1}} & M(q,n-1)^2. \end{array}$$

This diagram induces the following commutative diagram in cohomology for which Hưng claimed that **each of the rows is exact**:[1]

$$
\begin{array}{ccccccccc}
0 & \longrightarrow & \ker(\mu^*) & \longrightarrow & H^*(F(\mathbb{R}^q, 2^n)/\mathfrak{S}_{2^n}) & \xrightarrow{\mu^*} & H^*((F(\mathbb{R}^q, 2^{n-1})/\mathfrak{S}_{2^{n-1}})^2) & \longrightarrow & 0 \\
 & & \Big\downarrow{\scriptstyle i^*(q,n)|_{\ker(\mu^*)}} & & \Big\downarrow{\scriptstyle i^*(q,n)} & & \Big\downarrow{\scriptstyle i^*(q,n-1)^2} & & \\
0 & \longrightarrow & \ker(\varphi^*) & \longrightarrow & H^*(M(q,n)) & \xrightarrow{\varphi^*} & H^*(M(q,n-1)^2) & \longrightarrow & 0
\end{array}
\tag{4.4}
$$

From induction hypothesis we have that $i^*(q,n-1)^2 = i^*(q,n-1) \otimes i^*(q,n-1)$ is a monomorphism. Then from 5-lemma in order to conclude the induction, and consequently prove [64, Thm. 3.1], it suffices to prove that

$$i^*(q,n)|_{\ker(\mu^*)} \colon \ker(\mu^*) \longrightarrow \ker(\varphi^*)$$

is a monomorphism.

At this point we already crossed path with the first critical point in the proof (indicated by a footnote). The following claim explains the nature of the problem that appears in the diagram (4.4).

Claim 4.4 For any integer $n \geq 2$, the map

$$\varphi^* \colon H^*(M(q,n)) \longrightarrow H^*(M(q,n-1)^2)$$

in (4.4) is not surjective.

Proof In the proof of the claim we use the notation of Hưng and keep in mind that

$$M(q,n) = \widetilde{M}(q,n)/\mathfrak{S}_{2^n,2} = \mathrm{Pe}(\mathbb{R}^q, 2^n)/\mathcal{S}_{2^n}.$$

The cohomology of this space is described in Chap. 3. For a reader convenience we repeat some of the arguments already presented.

From Definition 2.1 we have that

$$M(q,n) = \widetilde{M}(q,n)/\mathfrak{S}_{2^n,2} \cong (M(q,n-1) \times M(q,n-1)) \times_{\mathbb{Z}_2} S^{q-1}.$$

Since the action of $\mathbb{Z}_2$ on the sphere S^{q-1} is free the projection on the last coordinate induces the following fiber bundle

$$
\begin{aligned}
&M(q,n-1) \times M(q,n-1) \\
&\qquad\qquad \longrightarrow (M(q,n-1) \times M(q,n-1)) \times_{\mathbb{Z}_2} S^{q-1} \longrightarrow \mathbb{R}\mathrm{P}^{q-1},
\end{aligned}
$$

where the map $\varphi = \varphi_{n-1}$ is the fiber embedding.

[1]The first critical point that is explained in Claim 4.4.

The Serre spectral sequence associated to this fibration has the E_2-term given by

$$E_2^{r,s} = H^r(\mathbb{R}\mathrm{P}^{q-1}; \mathcal{H}^s(M(q,n-1) \times M(q,n-1))).$$

As we have seen in Corollary 3.5 this spectral sequence collapses at the E_2-term, that is $E_2^{r,s} \cong E_\infty^{r,s}$. In particular, for an arbitrary integer $k \geq 0$ this means that the map φ^* factors as follows:

$$\begin{aligned} H^k(M(q,n)) &\cong \bigoplus_{r+s=k} E_2^{r,s} \\ &\longrightarrow E_2^{0,k} \cong H^k(M(q,n-1) \times M(q,n-1))^{\pi_1(\mathbb{R}\mathrm{P}^{q-1})} \\ &\longrightarrow H^k(M(q,n-1) \times M(q,n-1)). \end{aligned}$$

Here the first map is the projection and the second map is the inclusion. It is important to recall that $\pi_1(\mathbb{R}\mathrm{P}^{q-1})$ acts on $H^s(M(q,n-1) \times M(q,n-1))$ by interchanging the factors in the product. Thus, while the first map—the projection—is surjective, the second map is not surjective in all positive dimensions where $H^k(M(q,n-1) \times M(q,n-1)) \neq 0$. □

Thus already at this point the proof of [64, Thm. 3.1] has the first problem. Nevertheless, we continue to outline next steps of the proof that now concentrates on proving that

$$i^*(q,n)|_{\ker(\mu^*)} \colon \ker(\mu^*) \longrightarrow \ker(\varphi^*)$$

is a monomorphism. The complexity of the proof suggests that we first explain the strategy that was used by Hưng and then study particular details. Consider the commutative diagram (4.5) on the next page, which is an enrichment of the diagram (4.4) that we have already considered. The proof of the injectivity of the map $i^*(q,n)|_{\ker(\mu^*)}$ presented by Hưng consists of several steps that we now list:

(A) Description of $\ker\left(\mathrm{res}^{\mathfrak{S}_{2^n}}_{\mathfrak{S}^2_{2^{n-1}}}\right)$ in terms of the dual Nakamura elements.

(B) Description of $\ker(\mu^*)$ via the surjectivity of the map

$$i^*(F,q)| = i^*(F,q)|_{\ker\left(\mathrm{res}^{\mathfrak{S}_{2^n}}_{\mathfrak{S}^2_{2^{n-1}}}\right)} \colon \ker\left(\mathrm{res}^{\mathfrak{S}_{2^n}}_{\mathfrak{S}^2_{2^{n-1}}}\right) \longrightarrow \ker(\mu^*).$$

(C) Description of the image

$$(i^*(M,q) \circ i^*(\infty,n))\left(\ker\left(\mathrm{res}^{\mathfrak{S}_{2^n}}_{\mathfrak{S}^2_{2^{n-1}}}\right)\right) \subseteq \ker(\varphi^*).$$

(D) A proof that

$$\ker(\mu^*) \cong (i^*(M,q) \circ i^*(\infty,n))\left(\ker\left(\mathrm{res}^{\mathfrak{S}_{2^n}}_{\mathfrak{S}^2_{2^{n-1}}}\right)\right)$$

as $\mathbb{F}_2$-vector spaces.

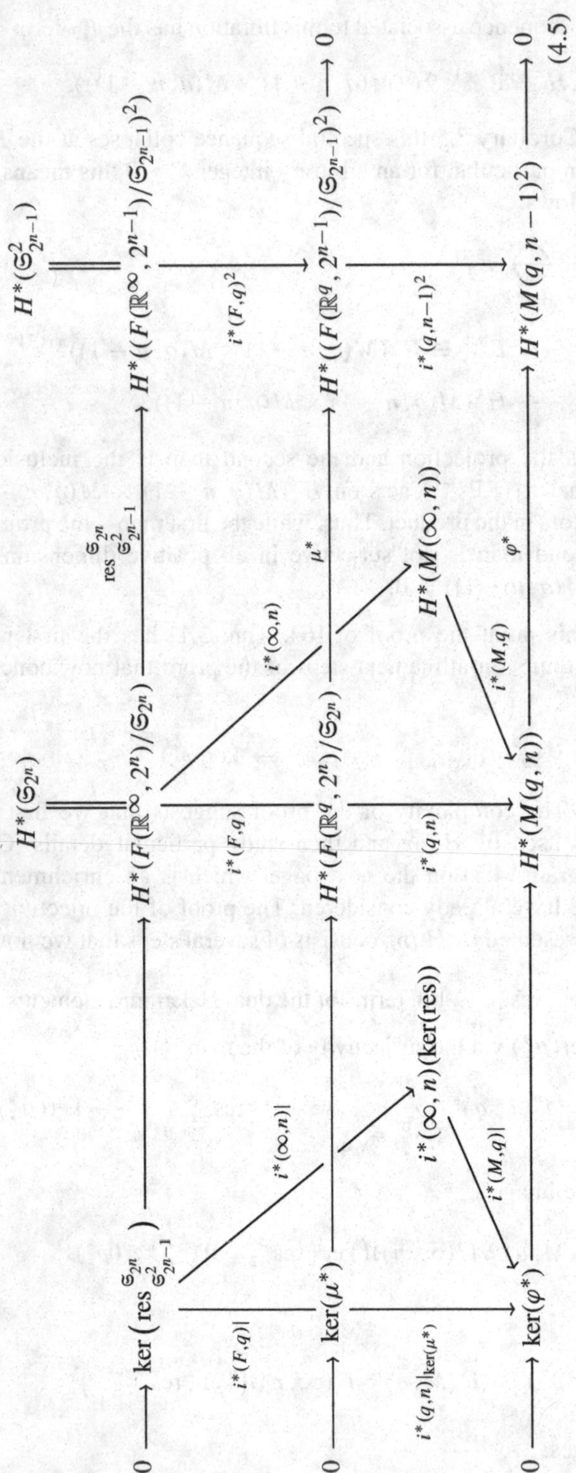

(4.5)

These claims along with commutativity of the diagram (4.5) imply that the map

$$i^*(q,n)|_{\ker(\mu^*)}\colon \ker(\mu^*) \longrightarrow (i^*(M,q)\circ i^*(\infty,n))\big(\ker\big(\mathrm{res}^{\mathfrak{S}_{2^n}}_{\mathfrak{S}^2_{2^{n-1}}}\big)\big)$$

is an isomorphism, and consequently

$$i^*(q,n)|_{\ker(\mu^*)}\colon \ker(\mu^*) \longrightarrow \ker(\varphi^*)$$

is a monomorphism. Thus, establishing claims we listed above would complete the proof of [64, Thm. 3.1]. Now we discuss these steps separately exhibiting the second critical point of the proof.

(A) and **(B)** For these steps multiple results of May [77], Nakaoka [87], Nakamura [85], and Hưng [62] are recalled and will be used in the proof. For the reader's convenience we collect the relevant facts as presented in [64, Sec. 3].

First, the homology $H_*(F(\mathbb{R}^q,\infty)/\mathfrak{S}_\infty)$ can be identified with a Hopf subalgebra of the homology $H_*(F(\mathbb{R}^\infty,\infty)/\mathfrak{S}_\infty) = H_*(\mathfrak{S}_\infty)$, [76, Sec. 5]. Furthermore, $H_*(F(\mathbb{R}^q,\infty)/\mathfrak{S}_\infty)$ is equipped with multiplicity in such a way that

$${}_mH_*(F(\mathbb{R}^q,\infty)/\mathfrak{S}_\infty) = H_*(F(\mathbb{R}^q,m)/\mathfrak{S}_m, F(\mathbb{R}^q,m-1)/\mathfrak{S}_{m-1}),$$

consult [62, Sec. 2]. In general, an algebra A equipped with multiplicity has a decomposition $A = \bigoplus_{n\geq 0} {}_nA$, and we define its filtration by multiplicities with $A(m) := \bigoplus_{0\leq n\leq m} {}_nA$. For more detailed definitions see [85, p. 96]. In our concrete situation we have that

$$H_*(F(\mathbb{R}^q,\infty)/\mathfrak{S}_\infty)(m) = H_*(F(\mathbb{R}^q,m)/\mathfrak{S}_m).$$

Let us further on, for integers $k_0,\dots,k_{n-1}\geq 0$, denote by

$$N_{k_0,\dots,k_{n-1}} \in H_*(F(\mathbb{R}^q,\infty)/\mathfrak{S}_\infty)$$

the so called Nakamura element of multiplicity 2^n, as introduced in [62, Sec. 2]. Now we can quote the following theorem from [64, Thm. 3.4].

3.4. THEOREM (Nakamura [85], May [77], Huỳnh Múi [83])

(i) *Let $q>0$ and*

$$J^+(q) = \Big\{K = (k_0,\dots,k_{n-1}) : n\geq 1, k_0\geq 1, k_1,\dots,k_{n-1}\geq 0, \sum_{i=0}^{n-1} k_i \leq q-1\Big\}.$$

Then

$$H_*(F(\mathbb{R}^q,\infty)/\mathfrak{S}_\infty) = \mathbb{F}_2[N_K : K\in J^+(q)]$$

as algebras with multiplicities. So we have for every $0 \leq n \leq \infty$ that

$$H_*(F(\mathbb{R}^q, m)/\mathfrak{S}_m) = \mathbb{F}_2[N_K : K \in J^+(q)](m).$$

In other words $H_*(F(\mathbb{R}^q, m)/\mathfrak{S}_m)$ has the $\mathbb{F}_2$-basis consisting of all monomials in $\mathbb{F}_2[N_K : K \in J^+(q)]$ of multiplicities $\leq m$. This is called Nakamura basis.

(ii) *The homomorphism*

$$i_*(F, q)\colon H_*(F(\mathbb{R}^q, \infty)/\mathfrak{S}_\infty) \longrightarrow H_*(F(\mathbb{R}^\infty, \infty)/\mathfrak{S}_\infty)$$

induced by the canonical embeddings $F(\mathbb{R}^q, m) \subset F(\mathbb{R}^\infty, m)$, $0 \leq m < \infty$, is an injection. It sends N_K to the element denoted by the same notation N_K for $K \in J^+(q)$.

Now a part of the commutative diagram (4.5) is considered:

$$\begin{array}{ccc} H^*(F(\mathbb{R}^\infty, 2^n)/\mathfrak{S}_{2^n}) & \xrightarrow{\operatorname{res}^{\mathfrak{S}_{2^n}}_{\mathfrak{S}^2_{2^{n-1}}}} & H^*((F(\mathbb{R}^\infty, 2^{n-1})/\mathfrak{S}_{2^{n-1}})^2) \\ \Big\downarrow{\scriptstyle i^*(F,q)} & & \Big\downarrow{\scriptstyle i^*(F,q)^2} \\ H^*(F(\mathbb{R}^q, 2^n)/\mathfrak{S}_{2^n}) & \xrightarrow{\mu^*} & H^*((F(\mathbb{R}^q, 2^{n-1})/\mathfrak{S}_{2^{n-1}})^2), \end{array}$$

where the maps $i^*(F, q)$ are induced by the inclusion $\mathbb{R}^q \longrightarrow \mathbb{R}^\infty$. Observe that

$$H^*(F(\mathbb{R}^\infty, 2^n)/\mathfrak{S}_{2^n}) \cong H^*(\mathfrak{S}_{2^n})$$

and

$$H^*(F(\mathbb{R}^\infty, 2^{n-1})/\mathfrak{S}_{2^{n-1}})^2 \cong H^*(\mathfrak{S}_{2^{n-1}} \times \mathfrak{S}_{2^{n-1}}).$$

It is now claimed that based of [64, Thm. 3.4], which we quoted in full, and the definition of the algebra structure on $H_*(F(\mathbb{R}^q, \infty)/\mathfrak{S}_\infty)$, see [63, Sec. 2], the following equality holds:

$$\ker\left(\operatorname{res}^{\mathfrak{S}_{2^n}}_{\mathfrak{S}^2_{2^{n-1}}}\right) = \operatorname{span}\{N^*_{k_0,\ldots,k_{n-1}} : k_0 \geq 1\},$$

where $N^*_{k_0,\ldots,k_{n-1}}$ is the dual of the Nakamura element $N_{k_0,\ldots,k_{n-1}}$. Furthermore

$$\begin{aligned} \ker(\mu^*) &= \operatorname{span}\{N^*_{k_0,\ldots,k_{n-1}} : (k_0, \ldots, k_{n-1}) \in J^+(q)\} \\ &= i^*(F, q)\big(\operatorname{span}\{N^*_{k_0,\ldots,k_{n-1}} : k_0 \geq 1\}\big) \\ &= i^*(F, q)\big(\ker\left(\operatorname{res}^{\mathfrak{S}_{2^n}}_{\mathfrak{S}^2_{2^{n-1}}}\right)\big). \end{aligned}$$

In particular, the map $i^*(F,q)|_{\ker\left(\mathrm{res}^{\mathfrak{S}_{2^n}}_{\mathfrak{S}^2_{2^{n-1}}}\right)}$ is surjective. With this parts (A) and (B) of Hưng's injectivity proof are concluded.

(C) In [64, Prop. 3.5] the image $(i^*(M,q)\circ i^*(\infty,n))\big(\ker\big(\mathrm{res}^{\mathfrak{S}_{2^n}}_{\mathfrak{S}^2_{2^{n-1}}}\big)\big)$ was described. We present this proposition with the paragraph that precedes it as in the original.

On the other, let $\rho_{2^n}\colon \mathfrak{S}_{2^n} \longrightarrow \mathrm{O}(2^n)$ denote the natural representation of the symmetric group $\mathfrak{S}_{2^n}$ in the orthogonal group $\mathrm{O}(2^n)$. As it is well known:

$$H^*(\mathrm{O}(2^n)) = \mathbb{Z}_2[W_1,\ldots,W_{2^n}],$$

where W_i denotes the ith universal Stiefel–Whitney class (of dimension i). We define $(2^n - 2^s)$th Stiefel–Whitney class of ρ_{2^n} by putting

$$W_{n,s} = \rho^*_{2^n} W_{2^n-2^s}, \qquad 0 \le s < n.$$

Further, we set

$$\overline{Q}_{n,s} = \mathrm{Res}(\mathfrak{S}_{2^n,2},\mathfrak{S}_{2^n})(W_{n,s}) \in H^*(\mathfrak{S}_{2^n,2}), \qquad 0 \le s < n.$$

3.5. PROPOSITION. *Let*

$$\begin{aligned} i(M,q)&\colon M(q,n) \longrightarrow M(\infty,n),\\ i(\infty,n)&\colon M(\infty,n) \longrightarrow F(\mathbb{R}^\infty,2^n)/\mathfrak{S}_{2^n} \end{aligned}$$

be well-known embeddings. Then we have

$$(i^*(M,q)\circ i^*(\infty,n))\big(\ker\big(\mathrm{res}^{\mathfrak{S}_{2^n}}_{\mathfrak{S}^2_{2^{n-1}}}\big)\big) = \overline{Q}_{n,0}\,\mathbb{F}_2[\overline{Q}_{n,0},\ldots,\overline{Q}_{n,n-1}]/I(\overline{Q},q).$$

Here $I(\overline{Q},q)$ denotes the ideal of $\overline{Q}_{n,0}\,\mathbb{F}_2[\overline{Q}_{n,0},\ldots,\overline{Q}_{n,n-1}]$ generated by monomials of degree q.

The proof given by Hưng proceeds as follows. According to Lemma 3.14, the decomposition (3.27), and the fact that $H^*(\mathfrak{S}_{2^n};\mathbb{F}_2)$ is detected by the subgroups E^n and $\mathfrak{S}^2_{2^{n-1}}$ we have that

$$\ker\big(\mathrm{res}^{\mathfrak{S}_{2^n}}_{\mathfrak{S}^2_{2^{n-1}}}\big) = \langle W_{n,0}\rangle = W_{n,0}\,\mathbb{F}_2[W_{n,0},\ldots,W_{n,n-1}].$$

Consequently, using the notation dictionary $i^*(\infty,n) = \mathrm{res}^{\mathfrak{S}_{2^n}}_{S_{2^n}}$, we get

$$\begin{aligned} i^*(\infty,n)\big(\ker\big(\mathrm{res}^{\mathfrak{S}_{2^n}}_{\mathfrak{S}^2_{2^{n-1}}}\big)\big) &= i^*(\infty,n)(\langle W_{n,0}\rangle)\\ &= i^*(\infty,n)\big(W_{n,0}\,\mathbb{F}_2[W_{n,0},\ldots,W_{n,n-1}]\big)\\ &= \overline{Q}_{n,0}\,\mathbb{F}_2[\overline{Q}_{n,0},\ldots,\overline{Q}_{n,n-1}]. \end{aligned}$$

On the other hand from Lemma 3.13 we have that

$$i^*(\infty, n)\big(\langle W_{n,0}\rangle\big) \subseteq \mathbb{F}_2[\overline{V}_{n,1}, \dots, \overline{V}_{n,n}].$$

Thus,

$$i^*(\infty, n)\big(\ker\big(\operatorname{res}^{\mathfrak{S}_{2n}}_{\mathfrak{S}^2_{2^{n-1}}}\big)\big) = \overline{Q}_{n,0}\,\mathbb{F}_2[\overline{Q}_{n,0}, \dots, \overline{Q}_{n,n-1}] \subseteq \mathbb{F}_2[\overline{V}_{n,1}, \dots, \overline{V}_{n,n}].$$

Now from Theorem 3.11 we have that

$$i^*(M, q)(\mathbb{F}_2[\overline{V}_{n,1}, \dots, \overline{V}_{n,n}]) \cong \mathbb{F}_2[\overline{V}_{n,1}, \dots, \overline{V}_{n,n}]/\langle \overline{V}^q_{n,1}, \dots, \overline{V}^q_{n,n}\rangle.$$

Therefore,

$$(i^*(M, q) \circ i^*(\infty, n))\big(\ker\big(\operatorname{res}^{\mathfrak{S}_{2n}}_{\mathfrak{S}^2_{2^{n-1}}}\big)\big) \cong \overline{Q}_{n,0}\,\mathbb{F}_2[\overline{Q}_{n,0}, \dots, \overline{Q}_{n,n-1}]/I(\overline{Q}, q)$$

where ideal $I(\overline{Q}, q)$ is given by

$$I(\overline{Q}, q) = \overline{Q}_{n,0}\,\mathbb{F}_2[\overline{Q}_{n,0}, \dots, \overline{Q}_{n,n-1}] \cap \langle \overline{V}^q_{n,1}, \dots, \overline{V}^q_{n,n}\rangle. \tag{4.6}$$

In order to complete the step (C) in the proof of [64, Prop. 3.5] it remains to show that the ideal $I(\overline{Q}, q)$ is the ideal generated by the monomials of degree q. The necessary argument for this was given in [64, Lem. 3.19] by using the restriction homomorphism $\operatorname{res}^{S_{2^n}}_{\mathcal{E}_n}$ and considering the corresponding claim in $H^*(\mathcal{E}_n)$. We give the original formulation without introducing new variables.

3.19. LEMMA. *Let*

$$\mathrm{pr}\colon \mathbb{F}_2[V_{n,1}, \dots, V_{n,n}] \longrightarrow \mathbb{F}_2[V_{n,1}, \dots, V_{n,n}]/\langle V^q_{n,1}, \dots, V^q_{n,n}\rangle$$

be the projection. Then, for the subring $\mathbb{F}_2[Q_{n,0}, \dots, Q_{n,n-1}]$ *of* $\mathbb{F}_2[V_{n,1}, \dots, V_{n,n}]$, *we have*

$$\mathrm{pr}\big(\mathbb{F}_2[Q_{n,0}, \dots, Q_{n,n-1}]\big) = \mathbb{F}_2[Q_{n,0}, \dots, Q_{n,n-1}]/I(Q, q).$$

Here $I(Q, q)$ *denotes the ideal of* $\mathbb{F}_2[Q_{n,0}, \dots, Q_{n,n-1}]$ *generated by monomials of degree* q.[2]

It will turn out that the ideal $I(Q, q)$ is *not* the ideal of $\mathbb{F}_2[Q_{n,0}, \dots, Q_{n,n-1}]$ generated by monomials of degree q, implying that the claim of [64, Lem. 3.19] does not stand.

First we equivalently transform the statement of the lemma. Recall that the classes $Q_{n,0}, \dots, Q_{n,n-1}$ are Dickson invariants and therefore $\mathrm{GL}_n(\mathbb{F}_2)$-invariants,

[2] The second critical point that is explained in Claim 4.5.

while $V_{n,1}, \ldots, V_{n,n}$ are $\mathrm{L}_n(\mathbb{F}_2)$-invariants. Furthermore, from (3.30) we have that

$$Q_{n,r} = (\chi_n Q_{n-1,r})\, V_{n,n} + (\chi_n Q_{n-1,r-1})^2 \tag{4.7}$$

where $\chi_n \in \mathrm{GL}_n(\mathbb{F}_2)$ can be interpreted as the variable change in $\mathbb{F}_2[y_1, \ldots, y_n]$ given by $y_i \longmapsto y_{n-i+1}$ for $1 \leq i \leq n$. Applying χ_n to the equality (4.7) yields

$$\chi_n Q_{n,r} = (\chi_n^2 Q_{n-1,r})\, \chi_n V_{n,n} + (\chi_n^2 Q_{n-1,r-1})^2.$$

Since $\chi_n^2 = \mathrm{id}$ and $Q_{n,r}$ is a $\mathrm{GL}_n(\mathbb{F}_2)$-invariant we get the following recurrent relation

$$Q_{n,r} = Q_{n-1,r}\,(\chi_n V_{n,n}) + (Q_{n-1,r-1})^2.$$

For simplicity the following notation was introduced $V_r := \chi_n V_{n,r}$ for all $1 \leq r \leq n$. Then the projection map

$$\mathrm{pr}\colon \mathbb{F}_2[V_{n,1}, \ldots, V_{n,n}] \longrightarrow \mathbb{F}_2[V_{n,1}, \ldots, V_{n,n}]/\langle V_{n,1}^q, \ldots, V_{n,n}^q\rangle$$

can written by

$$\mathrm{pr}\colon \mathbb{F}_2[V_1, \ldots, V_n] \longrightarrow \mathbb{F}_2[V_1, \ldots, V_n]/\langle V_1^q, \ldots, V_n^q\rangle.$$

Thus [64, Lem. 3.19], equivalently, states that

$$\mathrm{pr}\big(\mathbb{F}_2[Q_{n,0}, \ldots, Q_{n,n-1}]\big) = \mathbb{F}_2[Q_{n,0}, \ldots, Q_{n,n-1}]/I(Q,q),$$

where $I(Q,q)$ is the ideal of $\mathbb{F}_2[Q_{n,0}, \ldots, Q_{n,n-1}]$ generated by monomials of degree q, and

$$Q_{n,r} = Q_{n-1,r}\, V_n + (Q_{n-1,r-1})^2, \tag{4.8}$$

where we assume that $Q_{k,k} = 1$ and $Q_{k,-1} = 0$ for any integer $k \geq 1$. In particular, from the second equality in (3.32) we get that

$$Q_{n,0} = V_1 \cdots V_n. \tag{4.9}$$

Now we explain the problem that occurs in the description of the ideal $I(Q,q)$.

Claim 4.5 The ideal $I(Q,q)$, defined in (4.6), is not in general the ideal of the ring $\mathbb{F}_2[Q_{n,0}, \ldots, Q_{n,n-1}]$ generated by the monomials of degree q. For example, this fails for $n = 2$ and $q = 3$ or $q = 4$.

Proof The proof is given by exhibiting several counterexamples in the case when $n = 2$. From equalities (4.8) and (4.9) we get that

$$Q_{2,0} = V_1 V_2 \qquad \text{and} \qquad Q_{2,1} = V_2 + V_1^2.$$

Now we discuss different values of q.

(1) Let $q = 3$. Then by a direct computation in the ring $\mathbb{F}_2[V_1, V_2]$ we have that monomials

$$Q_{2,0}^3 = V_1^3 V_2^3, \quad Q_{2,0}^2 Q_{2,1} = V_1^2 V_2^3 + V_1^4 V_2^2, \quad Q_{2,0} Q_{2,1}^2 = V_1 V_2^3 + V_1^5 V_2$$

belong to the ideal $I(Q, 3)$. Furthermore, since $Q_{2,0}^2 = V_1^2 V_2^2 \notin I(Q, 3)$ and

$$Q_{2,1}^3 = V_2^3 + V_2^2 V_1^2 + V_2 V_1^4 + V_1^6 \notin I(Q, 3)$$

we obtain that the monomial

$$Q_{2,0}^2 + Q_{2,1}^3 = V_2^3 + V_2 V_1^4 + V_1^6 \in I(Q, 3).$$

Thus, $Q_{2,1}^3 \notin I(Q, 3)$ and $Q_{2,0}^2 + Q_{2,1}^3 \in I(Q, 3)$ giving us the first counterexample to the description of the ideal $I(Q, q)$.

(2) Let $q = 4$. Then working in the ring $\mathbb{F}_2[V_1, V_2]$ we get that the monomials

$$Q_{2,0}^4 = V_1^4 V_2^4, \qquad Q_{2,0}^3 Q_{2,1} = V_1^3 V_2^4 + V_1^5 V_2^3,$$
$$Q_{2,0}^2 Q_{2,1}^2 = V_1^2 V_2^4 + V_1^6 V_2^2, \qquad Q_{2,1}^4 = V_1^8 + V_2^4$$

are in the ideal $I(Q, 4)$, while

$$Q_{2,0} Q_{2,1}^3 = V_1 V_2^4 + V_1^3 V_2^3 + V_1^5 V_2^2 + V_1^7 V_2 \notin I(Q, 4).$$

Thus we obtained yet another evidence that the description of the ideal $I(Q, q)$ is incorrect.

This concludes the proof of the claim and opens a question of correct description of the ideal $I(Q, q)$. □

We explained an essential gap in the proof of the step (C): "description of the image $(i^*(M, q) \circ i^*(\infty, n))\big(\ker\big(\operatorname{res}^{\mathfrak{S}_{2^n}}_{\mathfrak{S}^2_{2^{n-1}}}\big)\big)$." This automatically invalidates the proof of the next step (D). Thus based on two gaps explained in Claims 4.4 and 4.5 we have shown that the proof of [64, Thm. 3.1] is incorrect. Moreover, we do not see how the approach taken by Hưng can be easily repaired.

Remark 4.6 The presented gaps also invalidate results of [64, Sec. 4]. In particular, counterexamples given in the proof of Claim 4.5 are also counter examples of the equality [64, (4.7)] that we copy as in the original:

$$H^*(F(\mathbb{R}^q, 2^n)/\mathfrak{S}_{2^n}) = i^*(F, q)R \oplus i^*(F, q)(\langle W_{n,0}\rangle). \tag{4.7}$$

4.2 Proof of the Injectivity Theorem

In this section we prove that the map

$$\rho^*_{d,2^m}\colon H^*(\mathrm{F}(\mathbb{R}^d, 2^m)/\mathfrak{S}_{2^m}; \mathbb{F}_2) \longrightarrow H^*(\mathrm{Pe}(\mathbb{R}^d, 2^m)/\mathcal{S}_{2^m}; \mathbb{F}_2)$$

is a monomorphism for all $d \geq 2$ and all $m \geq 0$. For the case when $d = \infty$, in Remark 4.3, we already explained why $\rho^*_{\infty,2^m}$ is a monomorphism.

The proof we present is by induction on m. For $m = 0$ we have that

$$\mathrm{Pe}(\mathbb{R}^d, 2^0)/\mathcal{S}_{2^0} = \mathrm{Pe}(\mathbb{R}^d, 2^0) = \{\mathrm{pt}\} \qquad \text{and} \qquad \mathrm{F}(\mathbb{R}^d, 2^0)/\mathfrak{S}_{2^0} = \mathrm{F}(\mathbb{R}^d, 2^0) = \mathbb{R}^d,$$

where $\mathrm{ecy}_{d,2^0}\colon \mathrm{Pe}(\mathbb{R}^d, 2^0) \longrightarrow \mathrm{F}(\mathbb{R}^d, 2^0)$ is given my $\mathrm{pt} \longmapsto 0 \in \mathbb{R}^d$. Thus, $\rho^*_{d,2^0}$ is obviously an isomorphism and consequently a monomorphism. Let $m \geq 1$, and let us assume that

$$\rho^*_{d,2^{m-1}}\colon H^*(\mathrm{F}(\mathbb{R}^d, 2^{m-1})/\mathfrak{S}_{2^{m-1}}; \mathbb{F}_2) \longrightarrow H^*(\mathrm{Pe}(\mathbb{R}^d, 2^{m-1})/\mathcal{S}_{2^{m-1}}; \mathbb{F}_2)$$

is a monomorphism. From Corollary 3.8 we have that the map:

$$\begin{aligned}(\rho_{d,2^{m-1}} \times \rho_{d,2^{m-1}}) \times_{\mathbb{Z}_2} \mathrm{id}\colon&\\ &(\mathrm{Pe}(\mathbb{R}^d, 2^{m-1})/\mathcal{S}_{2^{m-1}} \times \mathrm{Pe}(\mathbb{R}^d, 2^{m-1})/\mathcal{S}_{2^{m-1}}) \times_{\mathbb{Z}_2} S^{d-1}\\ &\longrightarrow (\mathrm{F}(\mathbb{R}^d, 2^{m-1})/\mathfrak{S}_{2^{m-1}} \times \mathrm{F}(\mathbb{R}^d, 2^{m-1})/\mathfrak{S}_{2^{m-1}}) \times_{\mathbb{Z}_2} S^{d-1}\end{aligned}$$

induces a monomorphism $((\rho_{d,2^{m-1}} \times \rho_{d,2^{m-1}}) \times_{\mathbb{Z}_2} \mathrm{id})^*$ in cohomology. Since from Definition 2.1 we know that

$$\mathrm{Pe}(\mathbb{R}^d, 2^m)/\mathcal{S}_{2^m} = (\mathrm{Pe}(\mathbb{R}^d, 2^{m-1})/\mathcal{S}_{2^{m-1}} \times \mathrm{Pe}(\mathbb{R}^d, 2^{m-1})/\mathcal{S}_{2^{m-1}}) \times_{\mathbb{Z}_2} S^{d-1}$$

we have obtained the monomorphism $((\rho_{d,2^{m-1}} \times \rho_{d,2^{m-1}}) \times_{\mathbb{Z}_2} \mathrm{id})^*$ in cohomology:

$$\begin{aligned}H^*((\mathrm{F}(\mathbb{R}^d, 2^{m-1})/\mathfrak{S}_{2^{m-1}} \times \mathrm{F}(\mathbb{R}^d, 2^{m-1})/\mathfrak{S}_{2^{m-1}}) \times_{\mathbb{Z}_2} S^{d-1}; \mathbb{F}_2)&\\ \longrightarrow H^*(\mathrm{Pe}(\mathbb{R}^d, 2^m)/\mathcal{S}_{2^m}; \mathbb{F}_2).&\end{aligned}$$

Next, consider the following diagram of spaces

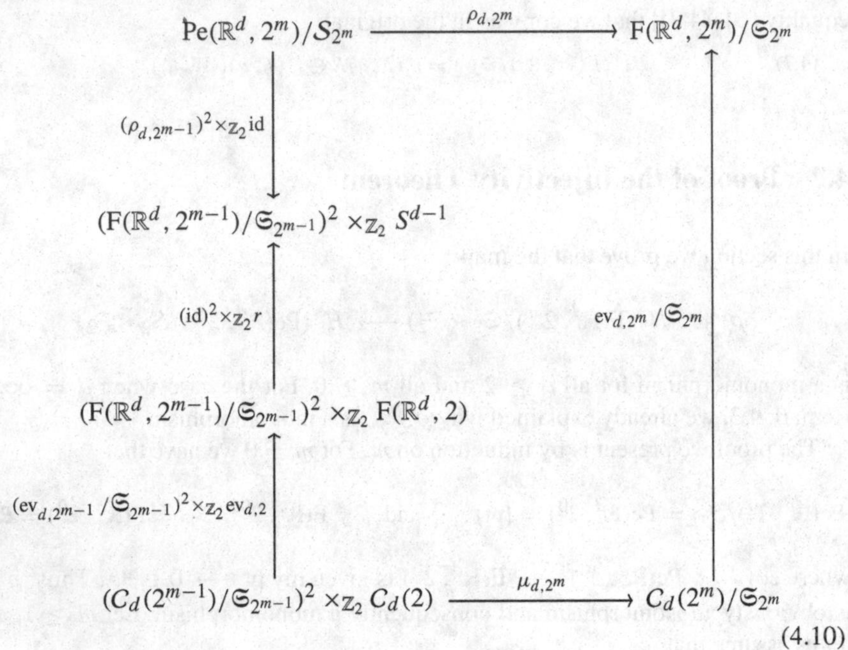

(4.10)

which "commutes up to a homotopy."

Here

- $r\colon \mathrm{F}(\mathbb{R}^d, 2) \longrightarrow S^{d-1}$ is the deformation retraction $r(x_1, x_2) := \frac{x_1 - x_2}{\|x_1 - x_2\|}$,
- C_d is the little d-cubes operad, consult Definition 7.8,
- $\mathrm{ev}_{d,n}\colon C_d(n) \longrightarrow \mathrm{F}(\mathbb{R}^d, n)$ is the evaluation map introduced in (7.1) which is an $\mathfrak{S}_n$-equivariant homotopy equivalence, as stated in Lemma 7.9, and
- $\mu_{d,2^m}$ is induced by the structural map of the little d-cubes operad:

$$\mu\colon (C_d(2^{m-1}) \times C_d(2^{m-1})) \times C_d(2) \longrightarrow C_d(2^m).$$

(See Definition 7.8.)

What we mean by "commutes up to a homotopy" here is that the diagram (4.10), after substituting the maps $(\mathrm{id})^2 \times_{\mathbb{Z}_2} r$ and $(\mathrm{ev}_{d,2^{m-1}}/\mathfrak{S}_{2^{m-1}})^2 \times_{\mathbb{Z}_2} \mathrm{ev}_{d,2}$ with its homotopy inverses, becomes commutative up to a homotopy. Since we know that the maps $(\mathrm{id})^2 \times_{\mathbb{Z}_2} r$, $(\mathrm{ev}_{d,2^{m-1}}/\mathfrak{S}_{2^{m-1}})^2 \times_{\mathbb{Z}_2} \mathrm{ev}_{d,2}$ and $\mathrm{ev}_{d,2^m}/\mathfrak{S}_{2^m}$ are homotopy equivalences, and $((\rho_{d,2^{m-1}} \times \rho_{d,2^{m-1}}) \times_{\mathbb{Z}_2} \mathrm{id})^*$ is an injection by induction hypothesis, we obtained the following claim.

Lemma 4.7 *If the homomorphism*

$$(\mu_{d,2^m})^*\colon H^*(C_d(2^m)/\mathfrak{S}_{2^m};\mathbb{F}_2)$$
$$\longrightarrow H^*((C_d(2^{m-1})/\mathfrak{S}_{2^{m-1}}\times C_d(2^{m-1})/\mathfrak{S}_{2^{m-1}})\times_{\mathbb{Z}_2} C_d(2);\mathbb{F}_2)$$

is a monomorphism, then the homomorphism

$$\rho^*_{d,2^m}\colon H^*(\mathrm{F}(\mathbb{R}^d,2^m)/\mathfrak{S}_{2^m};\mathbb{F}_2)\longrightarrow H^*(\mathrm{Pe}(\mathbb{R}^d,2^m)/S_{2^m};\mathbb{F}_2)$$

is also a monomorphism.

Hence, the induction step, in the proof of Theorem 4.1, would follow from the proof of injectivity of the homomorphism $(\mu_{d,2^m})^*$. To conclude the induction step, and consequently complete the proof of Theorem 4.1, we show the following dual theorem.

Theorem 4.8 *Let $d\geq 2$ and $m\geq 1$ be integers. The homomorphism*

$$(\mu_{d,2^m})_*\colon H_*((C_d(2^{m-1})/\mathfrak{S}_{2^{m-1}}\times C_d(2^{m-1})/\mathfrak{S}_{2^{m-1}})\times_{\mathbb{Z}_2} C_d(2);\mathbb{F}_2)$$
$$\longrightarrow H_*(C_d(2^m)/\mathfrak{S}_{2^m};\mathbb{F}_2) \qquad (4.11)$$

is an epimorphism.

For the proof of the theorem we use results of many authors that we first review in generality we need, and then in Sect. 4.2.2 we give the proof of Theorem 4.8. This will finalize the proof of Theorem 4.1.

4.2.1 Prerequisites

Let X be a path-connected space in $\mathrm{Top}_{\mathrm{pt}}$. The free C_d-space generated by X is defined as the quotient space

$$C_d(X):=\Big(\coprod_{m\geq 0} C_d(m)\times_{\mathfrak{S}_m} X^m\Big)/{\approx},$$

where for $(\mathbf{c}_1,\dots,\mathbf{c}_m)\in C_d(m)$ and $(x_1,\dots,x_{m-1},x_m)\in X^m$, with $x_m=\mathrm{pt}$ the base point, we define

$$((\mathbf{c}_1,\dots,\mathbf{c}_{m-1},\mathbf{c}_m),(x_1,\dots,x_{m-1},x_m))\approx((\mathbf{c}_1,\dots,\mathbf{c}_{m-1}),(x_1,\dots,x_{m-1})).$$

(For more details consult [76, Cons. 2.4] or Sect. 7.5.) The space $C_d(X)$ is equipped with the natural filtration given by the number of cubes, that is for $k \geq 0$ we define

$$\begin{aligned} \mathrm{F}_k C_d(X) := \mathrm{im}\Big(&\coprod_{0\leq m\leq k} C_d(m) \times_{\mathfrak{S}_m} X^m \\ &\longrightarrow \coprod_{m\geq 0} C_d(m) \times_{\mathfrak{S}_m} X^m \longrightarrow \Big(\coprod_{m\geq 0} C_d(m) \times_{\mathfrak{S}_m} X^m\Big)/\approx\Big), \end{aligned}$$

where the first map is the obvious inclusion and the second map is the identification map. Thus, we obtained the filtration of $C_d(X)$:

$$\varnothing = \mathrm{F}_{-1}C_d(X) \subseteq \mathrm{F}_0 C_d(X) \subseteq \mathrm{F}_1 C_d(X) \subseteq \cdots \subseteq \mathrm{F}_{k-1}C_d(X) \subseteq \mathrm{F}_k C_d(X) \subseteq \cdots,$$

where each pair of spaces $(\mathrm{F}_k C_d(X), \mathrm{F}_{k-1}C_d(X))$ is an NDR-pair; see [76, Prop. 2.6]. Further on we denote successive quotients by

$$\mathrm{D}_k C_d(X) := \mathrm{F}_k C_d(X)/\mathrm{F}_{k-1}C_d(X).$$

Recall that we have set $\mathrm{F}_{-1}C_d(X) = \varnothing$. Next, we introduce the real k-dimensional vector bundle $\xi_{d,k}$ over the quotient space $C_d(k)/\mathfrak{S}_k$ by

$$\mathbb{R}^k \longrightarrow C_d(k) \times_{\mathfrak{S}_k} \mathbb{R}^k \longrightarrow C_d(k)/\mathfrak{S}_k, \tag{4.12}$$

where $\mathbb{R}^k$ is assumed to be the real $\mathfrak{S}_k$-representation with the action given by permutation of the coordinates. The facts we are going to use—in the case when the space X is a sphere—are collected in the following theorem, see [20, 39, Thm. A-C] and [40, Thm. 2.6].

Theorem 4.9 *Let $L \geq 1$ and $N \geq 1$ be integers.*

(1) *The space $\mathrm{D}_k C_d(S^L)$ is homeomorphic to the Thom space of the vector bundle $\xi_{d,k}^{\oplus L}$, that is $\mathrm{D}_k C_d(S^L) \approx \mathrm{Th}(\xi_{d,k}^{\oplus L})$. Consequently, for every $i \geq 0$ there is the Thom isomorphism*

$$\begin{aligned} \widetilde{H}_{i+Lk}(\mathrm{D}_k C_d(S^L); \mathbb{F}_2) &\cong \widetilde{H}_{i+Lk}(\mathrm{Th}(\xi_{d,k}^{\oplus L})) \\ &\cong H_i(C_d(k)/\mathfrak{S}_k; \mathbb{F}_2) \cong H_i(\mathrm{F}(\mathbb{R}^d, k)/\mathfrak{S}_k; \mathbb{F}_2). \end{aligned}$$

(2) *For N large enough there is a homotopy equivalence*

$$\Sigma^N\big(\mathrm{F}_k C_d(S^L)\big) \simeq \Sigma^N\big(\mathrm{D}_k C_d(S^L) \vee \mathrm{F}_{k-1}C_d(S^L)\big).$$

Here $\Sigma(X)$ denotes the suspension of the spaces X.

The Approximation theorem of May, Theorem 7.10, applied to the sphere S^L, where $L \geq 1$ is an integer, yields the weak homotopy equivalence

$$\alpha_d \colon C_d(S^L) \longrightarrow \Omega^d \Sigma^d S^L.$$

In particular, α_d induces the isomorphism in homology with $\mathbb{F}_2$ coefficients:

$$(\alpha_d)_* \colon H_*(C_d(S^L); \mathbb{F}_2) \longrightarrow H_*(\Omega^d \Sigma^d S^L; \mathbb{F}_2).$$

(The results in [33, Sec. 3] describe further properties of the homology isomorphism $(\alpha_d)_*$.) The homology of the iterated loop space $\Omega^d \Sigma^d S^L = \Omega^d S^{d+L}$ with $\mathbb{F}_2$ coefficients was described as a Pontryagin ring by Araki and Kudo in their seminal paper [70, Thm. 7.1] using, what we call now, Araki–Kudo–Dyer–Lashof homology operations; see Sect. 7.6 for more details.

Theorem 4.10 *Let $d \geq 1$ and $L \geq 1$ be integers. The homology $H_*(\Omega^d \Sigma^d S^L; \mathbb{F}_2)$, as a Pontryagin ring, is a polynomial algebra generated by a generator u_L and all $Q_{i_1} Q_{i_2} \cdots Q_{i_s} u_L$ where $1 \leq i_1 \leq i_2 \leq \cdots \leq i_s \leq d-1$. Furthermore, $\deg(u_L) = L$ and*

$$\deg(Q_{i_1} Q_{i_2} \cdots Q_{i_s} u_L) = i_1 + 2i_2 + 4i_3 + \cdots + 2^{s-1} i_s + 2^s L.$$

We use notation

$$\begin{aligned} &H_*(\Omega^d \Sigma^d S^L; \mathbb{F}_2) \\ &\quad \cong \mathbb{F}_2\big[\{u_L\} \cup \{Q_{i_1} Q_{i_2} \cdots Q_{i_s} u_L \ : s \geq 1, 1 \leq i_1 \leq i_2 \leq \cdots \leq i_s \leq d-1\}\big]. \end{aligned}$$

Remark 4.11 The generators of the homology $H_*(\Omega^d \Sigma^d S^L; \mathbb{F}_2)$ have a concrete geometric description. Consider the map $e_d \colon S^L \longrightarrow \Omega^d \Sigma^d S^L$ that corresponds to the identity map $\mathrm{id} \colon \Sigma^d S^L \longrightarrow \Sigma^d S^L$ along the adjunction $[A, \Omega^d B]_{\mathrm{pt}} \longleftrightarrow [\Sigma^d A, B]_{\mathrm{pt}}$. Here $[\cdot, \cdot]_{\mathrm{pt}}$ denotes the set of all homotopy classes of pointed maps between the pointed spaces. After taking d-fold suspension, or the smash product with the sphere S^d, we get the following commutative triangle

$$\begin{array}{ccc} \Sigma^d S^L = S^d \wedge S^L & \xrightarrow{\ \mathrm{id} \wedge e_d\ } & \Sigma^d \Omega^d \Sigma^d S^L = S^d \wedge (\Omega^d \Sigma^d S^L) \\ & \searrow^{\mathrm{id}} & \downarrow ev_d \\ & & \Sigma^d S^L = S^d \wedge S^L, \end{array}$$

where $ev_d \colon \Sigma^d \Omega^d X \longrightarrow X$ is the evaluation map. Thus, the induced map in homology

$$(e_d)_* \colon H_*(S^L; \mathbb{F}_2) \longrightarrow H_*(\Omega^d \Sigma^d S^L; \mathbb{F}_2)$$

is a monomorphism. The generator u_L, the so called fundamental class of $\Omega^d\Sigma^d S^L$, is the image of the generator of $H_L(S^L;\mathbb{F}_2)$ along $(e_d)_*$. Now the generators $Q_{i_1}Q_{i_2}\cdots Q_{i_s}u_L$ for $1\le i_1\le i_2\le\cdots\le i_s\le d-1$ are images of u_L under the sequence $Q_{i_1}Q_{i_2}\cdots Q_{i_s}$ of Araki–Kudo–Dyer–Lashof homology operations. Thus we have complete description of the homology $H_*(\Omega^d\Sigma^d S^L;\mathbb{F}_2)$.

Next we define a filtration of the polynomial algebra

$$\begin{aligned}&\mathbb{F}_2\big[\{u_L\}\cup\{Q_{i_1}Q_{i_2}\cdots Q_{i_s}u_L\ :\ s\ge 1,\ 1\le i_1\le i_2\le\cdots\le i_s\le d-1\}\big]\\ &\quad\cong H_*(\Omega^d\Sigma^d S^L;\mathbb{F}_2)\end{aligned}$$

by defining the weight function ω on its monomials as follows:

$$\omega(u_L)=1,\qquad \omega(Q_i u)=2\omega(u),\qquad \omega(v_1\cdot v_2)=\omega(v_1)+\omega(v_2),$$

where u is an algebra generator, and v_1, v_2 are monomials is algebra generators. In particular, the weight of algebra generators are always powers of two, that is

$$\omega(Q_{i_1}Q_{i_2}\cdots Q_{i_s}u_L)=2\omega(Q_{i_2}Q_{i_3}\cdots Q_{i_s}u_L)=\cdots=2^s.$$

Now, for $k\ge 0$, we set $\mathcal{F}_kH_*(\Omega^d\Sigma^d S^L;\mathbb{F}_2)$ to be the vector subspace of the polynomial ring $H_*(\Omega^d\Sigma^d S^L;\mathbb{F}_2)$ generated by all monomials with weight at most k. In this way we obtained a filtration of the polynomial ring $H_*(\Omega^d\Sigma^d S^L;\mathbb{F}_2)$ by vector spaces:

$$\begin{aligned}0=\mathcal{F}_0H_*(\Omega^d\Sigma^d S^L;\mathbb{F}_2)&\subseteq\mathcal{F}_1H_*(\Omega^d\Sigma^d S^L;\mathbb{F}_2)\\ &\subseteq\cdots\subseteq\mathcal{F}_{k-1}H_*(\Omega^d\Sigma^d S^L;\mathbb{F}_2)\subseteq\mathcal{F}_kH_*(\Omega^d\Sigma^d S^L;\mathbb{F}_2)\subseteq\cdots,\end{aligned}$$

the so called **weight filtration**. Furthermore we denote the sequence of quotients by

$$\mathcal{D}_kH_*(\Omega^d\Sigma^d S^L;\mathbb{F}_2):=\mathcal{F}_kH_*(\Omega^d\Sigma^d S^L;\mathbb{F}_2)/\mathcal{F}_{k-1}H_*(\Omega^d\Sigma^d S^L;\mathbb{F}_2).$$

Example 4.12 In order to illustrate the notion of the weight filtration in the following table we list all the generators of $H_*(\Omega^d\Sigma^d S^L;\mathbb{F}_2)$ with weight at most 4.

Thus, for example:

$$\begin{aligned}\mathcal{F}_1H_*(\Omega^d\Sigma^d S^L;\mathbb{F}_2)&=\text{the }\mathbb{F}_2\text{-vector space with a basis }\{u_L\},\\ \mathcal{F}_2H_*(\Omega^d\Sigma^d S^L;\mathbb{F}_2)&=\text{the }\mathbb{F}_2\text{-vector space with a basis }\{u_L,u_L^2,Q_iu_L\},\\ \mathcal{F}_3H_*(\Omega^d\Sigma^d S^L;\mathbb{F}_2)&=\text{the }\mathbb{F}_2\text{-vector space with a basis}\\ &\qquad\{u_L,u_L^2,Q_iu_L,u_L^3,u_LQ_iu_L\}.\end{aligned}$$

Monomial	Weight	Degree	
u_L	1	L	
u_L^2	2	$2L$	
$Q_i u_L$	2	$i+2L$	$1 \le i \le d-1$
u_L^3	3	$3L$	
$u_L Q_i u_L$	3	$i+3L$	$1 \le i \le d-1$
u_L^4	4	$4L$	
$u_L^2 Q_i u_L$	4	$i+4L$	$1 \le i \le d-1$
$Q_i u_L Q_j u_L$	4	$i+j+4L$	$1 \le i, j \le d-1$
$Q_i Q_j u_L$	4	$i+2j+4L$	$1 \le i \le j \le d-1$

Consequently,

$$\mathcal{D}_1 H_*(\Omega^d \Sigma^d S^L; \mathbb{F}_2) = \text{the } \mathbb{F}_2\text{-vector space with a basis } \{u_L\},$$
$$\mathcal{D}_2 H_*(\Omega^d \Sigma^d S^L; \mathbb{F}_2) = \text{the } \mathbb{F}_2\text{-vector space with a basis } \{u_L^2,\ Q_i u_L\},$$
$$\mathcal{D}_3 H_*(\Omega^d \Sigma^d S^L; \mathbb{F}_2) = \text{the } \mathbb{F}_2\text{-vector space with a basis } \{u_L^3,\ u_L Q_i u_L\},$$
$$\mathcal{D}_4 H_*(\Omega^d \Sigma^d S^L; \mathbb{F}_2) = \text{the } \mathbb{F}_2\text{-vector space with a basis } \{u_L^4,\ u_L^2 Q_i u_L,\ Q_i u_L Q_j u_L,\ Q_i Q_j u_L\}.$$

For more details calculation of this type see [20, Sec. 5].

The central property of the filtration we just defined is given in the following theorem, see [33, Cor. 3.3].

Theorem 4.13 *Let $d \ge 2$ and $L \ge 1$ be integers. For every $k \ge 0$ there is an isomorphism of graded vector spaces*

$$\alpha_{d,k} \colon H_*(\mathrm{F}_k C_d(S^L); \mathbb{F}_2) \longrightarrow \mathcal{F}_k H_*(\Omega^d \Sigma^d S^L; \mathbb{F}_2).$$

Moreover, for every $k \ge 0$ there is an isomorphism of vector spaces

$$\overline{\alpha}_{d,k} \colon \widetilde{H}_*(\mathrm{D}_k C_d(S^L); \mathbb{F}_2) \longrightarrow \mathcal{D}_k H_*(\Omega^d \Sigma^d S^L; \mathbb{F}_2),$$

such that the following diagram commutes

$$\begin{array}{ccc} H_*(\mathrm{F}_k C_d(S^L); \mathbb{F}_2) & \xrightarrow{\alpha_{d,k}} & \mathcal{F}_k H_*(\Omega^d \Sigma^d S^L; \mathbb{F}_2) \\ \downarrow & & \downarrow \\ \widetilde{H}_*(\mathrm{D}_k C_d(S^L); \mathbb{F}_2) & \xrightarrow{\overline{\alpha}_{d,k}} & \mathcal{D}_k H_*(\Omega^d \Sigma^d S^L; \mathbb{F}_2). \end{array}$$

Here the left vertical map is induced by the quotient map of topological spaces

$$\mathrm{F}_k C_d(S^L) \longrightarrow \mathrm{D}_k C_d(S^L) = \mathrm{F}_k C_d(S^L)/\mathrm{F}_{k-1} C_d(S^L),$$

while the right vertical map is the quotient map of vector spaces

$$\begin{aligned}\mathcal{F}_k H_*(\Omega^d \Sigma^d S^L; \mathbb{F}_2) &\longrightarrow \mathcal{D}_k H_*(\Omega^d \Sigma^d S^L; \mathbb{F}_2)\\ &= \mathcal{F}_k H_*(\Omega^d \Sigma^d S^L; \mathbb{F}_2)/\mathcal{F}_{k-1} H_*(\Omega^d \Sigma^d S^L; \mathbb{F}_2).\end{aligned}$$

The results presented in [33, Sec. 3 and Sec.4] imply that the isomorphisms $\alpha_{d,k}$ and $\overline{\alpha}_{d,k}$ are induced by the isomorphism $(\alpha_d)_*$.

Now, directly from Theorems 4.9 and 4.13, we get a description of homology of the unordered configuration space with $\mathbb{F}_2$ coefficients, see [20, Sec. 4.4].

Corollary 4.14 *Let $d \geq 2$, $k \geq 1$ and $L \geq 1$ be integers. There is an isomorphism of graded vector space*

$$H_{*-kL}(\mathrm{F}(\mathbb{R}^d, k)/\mathfrak{S}_k; \mathbb{F}_2) \cong \mathcal{D}_k H_*(\Omega^d \Sigma^d S^L; \mathbb{F}_2).$$

Alternatively, we can describe the homology of the unordered configuration space $\mathrm{F}(\mathbb{R}^d, k)/\mathfrak{S}_k$ with $\mathbb{F}_2$ coefficients in the following way.

Corollary 4.15 *Let $d \geq 2$, $k \geq 1$, $L \geq 1$ and $0 \leq i \leq (d-1)(k-1)$ be integers. The homology of the unordered configuration space*

$$H_i(\mathrm{F}(\mathbb{R}^d, k)/\mathfrak{S}_k; \mathbb{F}_2)$$

is isomorphic to the $\mathbb{F}_2$ vector space spanned by all monomials of degree $i + kL$ of the polynomial ring

$$\mathbb{F}_2\big[\{u_L\} \cup \{Q_{i_1} Q_{i_2} \cdots Q_{i_s} u_L \,:\, s \geq 1,\ 1 \leq i_1 \leq i_2 \leq \cdots \leq i_s \leq d-1\}\big]$$

whose weights are exactly k.

In order to illustrate the previous result we give a few simple examples of evaluation of the homology of the unordered configuration space.

Example 4.16 From Corollary 4.15 we have that the homology $H_*(\mathrm{F}(\mathbb{R}^d, k)/\mathfrak{S}_k; \mathbb{F}_2)$ of the unordered configuration space is

(1) for the case $d = 2$ and $k = 2$ spanned in

dimension 0 by the monomial: u_L^2,

dimension 1 by the monomial: $Q_1 u_L$,

(Thus, indeed $H_*(\mathrm{F}(\mathbb{R}^2, 2)/\mathfrak{S}_2; \mathbb{F}_2) \cong H_*(\mathbb{R}\mathrm{P}^1; \mathbb{F}_2)$ as expected.)

(2) for the case $d = 2$ and $k = 3$ spanned in

dimension 0 by the monomial: u_L^3
dimension 1 by the monomial: $u_L^3\,(Q_1u_L)$

(3) for the case $d = 2$ and $k = 4$ spanned in

dimension 0 by the monomial: u_L^4
dimension 1 by the monomial: $u_L^2\,(Q_1u_L)$
dimension 2 by the monomial: $(Q_1u_L)\,(Q_1u_L)$
dimension 3 by the monomial: $Q_1Q_1u_L$

(4) for the case $d = 3$ and $k = 2$ spanned in

dimension 0 by the monomial: u_L^2
dimension 1 by the monomial: Q_1u_L
dimension 1 by the monomial: Q_2u_L

(Again, we see that $H_*(\mathrm{F}(\mathbb{R}^3, 2)/\mathfrak{S}_2; \mathbb{F}_2) \cong H_*(\mathbb{RP}^2; \mathbb{F}_2)$ as expected.)

(5) for the case $d = 3$ and $k = 4$ spanned in

dimension 0 by the monomial: u_L^4
dimension 1 by the monomial: $u_L^2\,(Q_1u_L)$
dimension 2 by the monomial: $u_L^2\,(Q_2u_L)$ and $(Q_1u_L)\,(Q_1u_L)$
dimension 3 by the monomial: $(Q_1u_L)\,(Q_2u_L)$ and $Q_1Q_1u_L$
dimension 4 by the monomial: $(Q_2u_L)\,(Q_2u_L)$
dimension 5 by the monomial: $Q_1Q_2u_L$
dimension 6 by the monomial: $Q_2Q_2u_L$

Thus, we have that

$$H_i(\mathrm{F}(\mathbb{R}^3, 4)/\mathfrak{S}_4; \mathbb{F}_2) \cong \begin{cases} \mathbb{F}_2, & i = 0, 1, 4, 5, 6, \\ \mathbb{F}_2^{\oplus 2}, & i = 3, 4, \\ 0, & \text{otherwise.} \end{cases}$$

With this result we collected all necessary ingredients that we need for the proof of Theorem 4.8.

4.2.2 Proof of the Dual Epimorphism Theorem

Let the integers $d \geq 2$ and $m \geq 1$ be fixed. Now we prove that the following homomorphism, induced by a structural map of the little cubes operad C_d,

$$(\mu_{d,2^m})_*\colon H_*((C_d(2^{m-1})/\mathfrak{S}_{2^{m-1}} \times C_d(2^{m-1})/\mathfrak{S}_{2^{m-1}}) \times_{\mathbb{Z}_2} C_d(2); \mathbb{F}_2)$$
$$\longrightarrow H_*(C_d(2^m)/\mathfrak{S}_{2^m}; \mathbb{F}_2)$$

is an epimorphism.

Let $L \geq 1$ be an arbitrary integer, and consider the free C_d-space $C_d(S^L)$. From Theorem 4.10, using the weak homotopy equivalence

$$\alpha_d\colon C_d(S^L) \longrightarrow \Omega^d \Sigma^d S^L,$$

we get the following isomorphisms

$$H_*(C_d(S^L); \mathbb{F}_2) \cong H_*(\Omega^d \Sigma^d S^L; \mathbb{F}_2) \cong \bigoplus_{k \geq 0} \mathcal{D}_k H_*(\Omega^d \Sigma^d S^L; \mathbb{F}_2)$$
$$\cong \mathbb{F}_2\big[\{u_L\} \cup \{Q_{i_1} Q_{i_2} \cdots Q_{i_s} u_L \,:\, s \geq 1,\ 1 \leq i_1 \leq i_2 \leq \cdots \leq i_s \leq d-1\}\big].$$

Since $C_d(S^L)$ is a C_d-space, as explained in Lemma 7.6, there exists a map

$$\Theta_2\colon (C_d(S^L) \times C_d(S^L)) \times C_d(2) \longrightarrow C_d(S^L).$$

(In the following we will abuse notation and all maps induced by Θ_2 will be denoted in the same way.) On the level of space filtration, for integers $k_1 \geq 0$ and $k_2 \geq 0$, we have that Θ_2 induces a map

$$\Theta_2\colon (\mathrm{F}_{k_1} C_d(S^L) \times \mathrm{F}_{k_2} C_d(S^L)) \times C_d(2) \longrightarrow \mathrm{F}_{k_1+k_2} C_d(S^L).$$

Thus, for any integer $k \geq 0$ we have an induced map on quotient spaces

$$\Theta_2\colon (\mathrm{D}_k C_d(S^L) \times \mathrm{D}_k C_d(S^L)) \times C_d(2) \longrightarrow \mathrm{D}_{2k} C_d(S^L).$$

The symmetric group $\mathfrak{S}_2 \cong \mathbb{Z}_2$ acts naturally on: the space $C_d(2)$ of pairs of little d-cubes by interchanging the cubes, and on the product $\mathrm{D}_k C_d(S^L) \times \mathrm{D}_k C_d(S^L)$ by interchanging the factors. If the trivial action on the space $\mathrm{D}_{2k} C_d(S^L)$ is assumed, the equivariance property of the structural map of the little cubes operad implies that the last Θ_2 map is an $\mathfrak{S}_2$-equivariant map. Consequently, it induces maps such that the following diagram commutes

$$\begin{array}{ccc} (C_d(S^L) \times C_d(S^L)) \times_{\mathbb{Z}_2} C_d(2) & \xrightarrow{\Theta_2} & C_d(S^L) \\ \uparrow & & \uparrow \\ (\mathrm{F}_k C_d(S^L) \times \mathrm{F}_k C_d(S^L)) \times_{\mathbb{Z}_2} C_d(2) & \xrightarrow{\Theta_2} & \mathrm{F}_{2k} C_d(S^L) \\ \downarrow & & \downarrow \\ (\mathrm{D}_k C_d(S^L) \times \mathrm{D}_k C_d(S^L)) \times_{\mathbb{Z}_2} C_2(2) & \xrightarrow{\Theta_2} & \mathrm{D}_{2k} C_d(S^L). \end{array}$$

Here the vertical maps are either inclusions or quotient maps. Passing to homology we get the following commutative diagram:

$$\begin{array}{ccc} H_*((C_d(S^L) \times C_d(S^L)) \times_{\mathbb{Z}_2} C_d(2); \mathbb{F}_2) & \xrightarrow{(\Theta_2)_*} & H_*(C_d(S^L); \mathbb{F}_2) \\ \uparrow & & \uparrow \\ H_*((\mathrm{F}_k C_d(S^L) \times \mathrm{F}_k C_d(S^L)) \times_{\mathbb{Z}_2} C_d(2); \mathbb{F}_2) & \xrightarrow{(\Theta_2)_*} & H_*(\mathrm{F}_{2k} C_d(S^L); \mathbb{F}_2) \\ \downarrow & & \downarrow \\ H_*((\mathrm{D}_k C_d(S^L) \times \mathrm{D}_k C_d(S^L)) \times_{\mathbb{Z}_2} C_2(2); \mathbb{F}_2) & \xrightarrow{(\Theta_2)_*} & H_*(\mathrm{D}_{2k} C_d(S^L); \mathbb{F}_2). \end{array} \tag{4.13}$$

In the case when $k = 2^{m-1}$ the homomorphism

$$(\Theta_2)_*\colon H_*((\mathrm{D}_k C_d(S^L) \times \mathrm{D}_k C_d(S^L)) \times_{\mathbb{Z}_2} C_2(2); \mathbb{F}_2) \longrightarrow H_*(\mathrm{D}_{2k} C_d(S^L); \mathbb{F}_2) \tag{4.14}$$

coincides with the map $(\mu_{d,2^m})_*$ from (4.11)—after appropriate dimension shift by $-2kL$. Therefore in order to complete the proof of Theorem 4.8 it suffices to prove that the homomorphism $(\Theta_2)_*$, from (4.14), is surjective.

The map $H_*(\mathrm{F}_k C_d(S^L); \mathbb{F}_2) \longrightarrow H_*(\mathrm{D}_k C_d(S^L); \mathbb{F}_2)$ is surjective for every k. Therefore, using Corollary 3.9 and commutativity of the diagram (4.13), we should first consider the homomorphism

$$(\Theta_2)_*\colon H_*((\mathrm{F}_k C_d(S^L) \times \mathrm{F}_k C_d(S^L)) \times_{\mathbb{Z}_2} C_2(2); \mathbb{F}_2) \longrightarrow H_*(\mathrm{F}_{2k} C_d(S^L); \mathbb{F}_2). \tag{4.15}$$

For $d \geq 2$, let

$$\mathcal{I}_d := \{(i_1, \ldots, i_s) : s \geq 1,\ 1 \leq i_1 \leq \cdots \leq i_s \leq d-1\}$$

be the set of all admissible iterations. Then, every element $I \in \mathcal{I}_d$ defines the iterated Araki–Kudo–Dyer–Lashof homology operation

$$Q_I := Q_{i_1} Q_{i_2} \cdots Q_{i_s}.$$

Hence, as we have seen in Theorem 4.13, the homology $H_*(\mathrm{F}_k C_d(S^L); \mathbb{F}_2)$ can be identified with the vector space spanned by all monomials in, variables from the set $\{Q_I u_L : I \in \mathcal{I}_d\} \cup \{u_L\}$ with weight at most k. Furthermore, from definition of Araki–Kudo–Dyer–Lashof homology operations given in Sect. 7.6, it follows that for every $Q_I u_L$, $I \in \mathcal{I}_d$, of weight exactly 2^m, there exists $i \geq 1$ and an element $Q_J u_L$ or u_L, $J \in \mathcal{I}_d$, of weight exactly 2^{m-1} such that

$$Q_I u_L = (\Theta_2)_*((Q_J u_L \otimes Q_J u_L) \otimes_{\mathbb{Z}_2} f_i). \tag{4.16}$$

It is important to notice that this **does not** mean that the homomorphism $(\Theta_2)_*$ from (4.15) is surjective.

Now we specialize to the case $k = 2^{m-1}$ and analyze the map (4.14):

$$(\Theta_2)_* \colon H_*((\mathrm{D}_{2^{m-1}} C_d(S^L) \times \mathrm{D}_{2^{m-1}} C_d(S^L)) \times_{\mathbb{Z}_2} C_d(2); \mathbb{F}_2) \longrightarrow H_*(\mathrm{D}_{2^m} C_d(S^L); \mathbb{F}_2)$$

and prove that it is surjective. The vector space $H_*(\mathrm{D}_{2^m} C_d(S^L); \mathbb{F}_2)$ is a quotient of $H_*(\mathrm{F}_{2^m} C_d(S^L); \mathbb{F}_2)$. It can be identified with the vector space spanned by all monomials of weight exactly 2^m in variables $\{u_L\} \cup \{Q_I u_L : I \in \mathcal{I}_d\}$. From (4.16) we that all elements of the form $Q_I u_L$ of weight 2^m (variables) are in the image of $(\Theta_2)_*$. It remains to show that all other monomials are also in $\mathrm{im}((\Theta_2)_*)$.

Let $Q_{I_1} u_L \cdots Q_{I_t} u_L$ with $t \geq 2$, $I_1, \ldots I_t \in \mathcal{I}_d \cup \{0\}$ and $Q_0 u_L := u_L$, be a typical element of the basis of $H_*(\mathrm{D}_{2^m} C_d(S^L); \mathbb{F}_2)$ which is not a variable. Then,

$$2^m = \omega(Q_{I_1} u_L \cdots Q_{I_t} u_L) = \omega(Q_{I_1} u_L) + \cdots + \omega(Q_{I_t} u_L) = 2^{b_1} + \cdots + 2^{b_t},$$

where

$$\omega(Q_{I_1} u_L) = 2^{b_1},\ \omega(Q_{I_2} u_L) = 2^{b_2},\ \ldots,\ \omega(Q_{I_t} u_L) = 2^{b_t}.$$

Without loss of generality, we can assume that there exists an integer $1 \leq \ell \leq t-1$ such that

$$\omega(Q_{I_1} u_L \cdots Q_{I_\ell} u_L) = 2^{m-1} \quad \text{and} \quad \omega(Q_{I_{\ell+1}} u_L \cdots Q_{I_t} u_L) = 2^{m-1}.$$

Consequently,

$$Q_{I_1} u_L \cdots Q_{I_\ell} u_L \in H_*(\mathrm{D}_{2^{m-1}} C_d(S^L); \mathbb{F}_2)$$

and

$$Q_{I_{\ell+1}}u_L \cdots Q_{I_t}u_L \in H_*(\mathrm{D}_{2^{m-1}}C_d(S^L); \mathbb{F}_2).$$

The product structure on the homology of C_d-space $C_d(S^L)$ is introduced by an arbitrary element of $(\mathbf{c}_1, \mathbf{c}_2) \in C_d(2)$ via the map

$$\Theta_2(\mathbf{c}_1, \mathbf{c}_2)\colon C_d(S^L) \times C_d(S^L) \longrightarrow C_d(S^L),$$

see [76, Lem. 1.9(i)]. The invariance of the action of the little cubes operad with respected to the actions of related symmetric groups implies that the product map $\Theta_2(\mathbf{c}_1, \mathbf{c}_2)$ factors thought the induced map

$$\Theta_2\colon (C_d(S^L) \times C_d(S^L)) \times_{\mathbb{Z}_2} C_d(2) \longrightarrow C_d(S^L).$$

Hence, on the level of homology with $\mathbb{F}_2$ coefficients we have that

$$\begin{aligned} Q_{I_1}u_L \dots Q_{I_t}u_L &= (Q_{I_1}u_L \cdots Q_{I_\ell}u_L) \cdot (Q_{I_{\ell+1}}u_L \cdots Q_{I_t}u_L) \\ &= (\Theta_2)_*\big((Q_{I_1}u_L \cdots Q_{I_\ell}u_L) \otimes (Q_{I_{\ell+1}}u_L \cdots Q_{I_t}u_L) \otimes_{\mathbb{Z}_2} h_0\big), \end{aligned}$$

where h_0 is the generator of $H_0(\mathbb{RP}^{d-1}; \mathcal{M})$, as explained in Sect. 10.5. Thus, we have proved that all additive generators of $H_*(\mathrm{D}_{2^m}C_d(S^L); \mathbb{F}_2)$ are in the image of $(\Theta_2)_*$ implying that the map $(\Theta_2)_*$ in (4.14) is surjective. This concludes the proof of Theorem 4.8 and consequently we have proved Theorem 4.1.

4.3 An Unexpected Corollary

In this section we take a small detour from the objectives of this book and present an unexpected corollary which is an artefact of the proof of Theorem 4.8. This result shares the spirit with the classical result of Michael Atiyah [10] about the complex representation ring of the symmetric group $R(\mathfrak{S}_k) \cong K^*_{\mathfrak{S}_k}(\mathrm{pt})$ and the modern breakthrough of Chad Giusti et al. [55, Thm. 1.2] in describing the cohomology of the symmetric groups collected disjointly together $H^*(\coprod_{k\geq 1} \mathrm{B}\mathfrak{S}_k; \mathbb{F}_2)$.

4.3.1 Motivation

In 1966 Atiyah studied the following direct sum of abelian groups

$$R_* := \bigoplus_{k\geq 0} \hom(R(\mathfrak{S}_k), \mathbb{Z}) \cong \bigoplus_{k\geq 0} \hom(K^*_{\mathfrak{S}_k}(\mathrm{pt}), \mathbb{Z})$$

which was additionally equipped with the structure of a commutative ring via the product induced by the inclusion maps $\mathfrak{S}_r \times \mathfrak{S}_s \longrightarrow \mathfrak{S}_{r+s}$. For details and original presentation see [10, p. 169]. He showed [10, Cor. 1.3] that the ring R_* is isomorphic to the polynomial ring

$$\mathbb{Z}[\sigma^1, \ldots, \sigma^k, \ldots]$$

on generators $\sigma^k \colon R(\mathfrak{S}_k) \longrightarrow \mathbb{Z}$, for $k \geq 1$, given by the dimension of the fixed $\mathfrak{S}_k$-submodule.

The study of the homology and cohomology of symmetric groups has a rich and exciting history with many spectacular breakthroughs and crucial applications in different areas of mathematics. The major contributions go back to the extraordinary work of Nakaoka [86–88], Daniel Quillen [89, 90], Quillen and Boris Venkov [91], Huỳnh Mùi [82], Benjamin Mann [73], Adem et al. [1], Adem and Milgram [2], all the way towards the modern results by Mark Feshbach [51] and Giusti et al. [55].

A detailed understanding of the cohomology of a particular symmetric group is still a challenging problem in general. On the other hand, Giusti, Salvatore and Sinha [55], considering all symmetric groups together, like Atiyah did, gave a compact description of the cohomology $H^*\big(\coprod_{k\geq 1} \mathrm{B}\mathfrak{S}_k; \mathbb{F}_2\big)$ in the language of Hopf rings where the operations are naturally introduced.

In more details, a Hopf ring is a five-tuple $(V, \odot, \cdot, \Delta, S)$ of the vectors space V, two multiplications $\odot$ and $\cdot$, one comultiplication Δ, and an antipode S such that:

1. $(V, \odot, \Delta, S)$ is a Hopf algebra,
2. $(V, \cdot, \Delta)$ is a bialgebra, and
3. $u \cdot (v \odot w) = \sum_{\Delta u = \sum u' \otimes u''} (u' \cdot v) \odot (u'' \cdot w)$ for all $u, v, w \in V$.

Next, the cohomology $H^*\big(\coprod_{k\geq 1} \mathrm{B}\mathfrak{S}_k; \mathbb{F}_2\big)$ can be equipped with a structure of a Hopf ring where:

- the first product $\odot$ is the so call transfer product, for a definition consult [55, Sec. 3],
- the second product $\cdot$ is the cup product, extended to be zero on the classes coming from different disjoint components,
- the coproduct Δ on is the dual to the standard Pontryagin product on the homology of $H_*\big(\coprod_{k\geq 1} \mathrm{B}\mathfrak{S}_k; \mathbb{F}_2\big)$, that is the coproduct induced by the natural inclusions of the symmetric groups $\mathfrak{S}_r \times \mathfrak{S}_s \longrightarrow \mathfrak{S}_{r+s}$, and finally
- the anipode S is just the identity map.

Now the main result of Giusti et al. [55, Thm. 1.2] is as follows.

Theorem 4.17 *The cohomology $H^*\big(\coprod_{k\geq 1} \mathrm{B}\mathfrak{S}_k; \mathbb{F}_2\big)$, as a Hopf ring, is generated by the unit classes on each component and classes $u_{\ell,n} \in H^*(\mathrm{B}\mathfrak{S}_{n2^\ell}; \mathbb{F}_2)$. The coproduct is generated by the formula*

$$\Delta u_{\ell,n} = \sum_{r+s=n} u_{\ell,r} \otimes u_{\ell,s},$$

while the relations between the transfer products on generators are given by

$$u_{\ell,r} \odot u_{\ell,s} = \binom{r+s}{r} u_{\ell,r+s}.$$

The second (cup-)product $\cdot$ between the generators from different components is zero, and there are no further relations between products of the generators. In addition, the antipode is the identity.

4.3.2 Corollary

Motivated by the work of Atiyah and Giusti, Salvatore and Sinha, for a fixed integer $d \geq 2$ or $d = \infty$, we describe the $\mathbb{F}_2$ homology of the space of all finite subsets of $\mathbb{R}^d$ with additional base point, that is $H_*\big(\coprod_{k\geq 0} \mathrm{F}(\mathbb{R}^d, k)/\mathfrak{S}_k; \mathbb{F}_2\big)$. Here, $\mathrm{F}(\mathbb{R}^d, 0)$ and $\mathrm{F}(\mathbb{R}^d, 0)/\mathfrak{S}_0$ stand for base points.

In addition to the additive structure on the homology, which can be already read from Corollary 4.15, we can also identify the ring structure with the respect to a naturally induced multiplication. For a clear definition of the product structure on this homology we use the little d-cubes operad isomorphic model and set

$$\mathcal{T}_d := H_*\Big(\coprod_{k\geq 0} C_d(k)/\mathfrak{S}_k; \mathbb{F}_2\Big) \cong H_*\Big(\coprod_{k\geq 0} \mathrm{F}(\mathbb{R}^d, k)/\mathfrak{S}_k; \mathbb{F}_2\Big).$$

Recall that by Definition 7.1 the space $C_d(0)$ is just a point and therefore coincides with the $\mathrm{F}(\mathbb{R}^d, 0)$. Now, the product structure on $\mathcal{T}_d$, we care about, is induced by the structural maps of the little d-cubes operad

$$\mu_{C_d}(\mathbf{c}_1, \mathbf{c}_2)\colon C_d(r) \times C_d(s) \longrightarrow C_d(r+s), \tag{4.17}$$

where $(\mathbf{c}_1, \mathbf{c}_2) \in C_d(2)$ is assumed to be an arbitrary but fixed pair of little d-cubes.

Theorem 4.18 *Let $d \geq 2$ be an integer or $d = \infty$. Then $\mathcal{T}_d$ is isomorphic to the polynomial ring*

$$\mathcal{R}_d := \mathbb{F}_2\big[\{u_0\} \cup \{Q_{i_1} Q_{i_2} \cdots Q_{i_s} u_0 \,:\, s \geq 1,\ 1 \leq i_1 \leq i_2 \leq \cdots \leq i_s \leq d-1\}\big]$$

on generators

$$u_0 \in H_0(C_d(1)/\mathfrak{S}_1; \mathbb{F}_2) \qquad \text{and} \qquad Q_{i_1} Q_{i_2} \cdots Q_{i_s} u_0 \in H_i(C_d(2^s)/\mathfrak{S}_{2^s}; \mathbb{F}_2)$$

where $s \geq 1$, $1 \leq i_1 \leq i_2 \leq \cdots \leq i_s \leq d-1$, and $i = i_1+2i_2+2^2i_3+\cdots+2^{s-1}i_s$. The unit of the ring $\mathcal{R}_d$ is the generator of the 0-homology of the base point, that is $1 \in H_0(C_d(0)/\mathfrak{S}_0; \mathbb{F}_2)$.

Proof The ring $\mathcal{T}_d$ has two gradings, one with respect to the number of cubes (points) and the second one with respect to the homology degree:

$$\mathcal{T}_d = H_*\Big(\coprod_{k\geq 0} C_d(k)/\mathfrak{S}_k; \mathbb{F}_2\Big) \cong \bigoplus_{k\geq 0} H_*(C_d(k)/\mathfrak{S}_k; \mathbb{F}_2)$$

$$\cong \bigoplus_{k\geq 0}\bigoplus_{i\geq 0} H_i(C_d(k)/\mathfrak{S}_k; \mathbb{F}_2).$$

Indeed the product, induced by the map (4.17), is the composition homomorphism:

$$\begin{array}{ccc} H_i(C_d(r)/\mathfrak{S}_r; \mathbb{F}_2) \otimes H_j(C_d(s)/\mathfrak{S}_k; \mathbb{F}_2) & \xrightarrow{\times} & H_{i+j}(C_d(r)/\mathfrak{S}_r \times C_d(s)/\mathfrak{S}_k; \mathbb{F}_2) \\ & \searrow & \downarrow (\mu_{C_d}(c_1,c_2))_* \\ & & H_{i+j}(C_d(r+s)/\mathfrak{S}_{r+s}; \mathbb{F}_2). \end{array}$$

The horizontal map is the homology cross product which is an isomorphism in our situation; consult for example [23, Thm. VI.1.6].

In order to simplify notation for $d \geq 2$ we set

$$\mathcal{I}_d := \{(i_1, \ldots, i_s) : s \geq 1,\ 1 \leq i_1 \leq \cdots \leq i_s \leq d-1\}.$$

Let $I = (i_1, \ldots, i_s) \in \mathcal{I}_d$, then we set the length of I to be $|I| := s$. Like in the previous section, we denote the iterated Araki–Kudo–Dyer–Lashof homology operation $Q_{i_1} Q_{i_2} \cdots Q_{i_s}$ applied on u_L by $Q_I u_L$ where $I = (i_1, \ldots, i_s) \in \mathcal{I}_d$. Furthermore, we write $\mathcal{Q}_I u_0 := \mathcal{Q}_{i_1}\mathcal{Q}_{i_2}\cdots\mathcal{Q}_{i_s} u_0$ for a typical generator of the polynomial ring $\mathcal{R}$ which differs from the generator u_0. Recall, that $\deg(u_L) = L$, and

$$\deg(Q_I u_l) = \deg(Q_{i_1} Q_{i_2}\cdots Q_{i_s} u_L) = i_1 + 2i_2 + 4i_3 + \cdots + 2^{s-1}i_s + 2^s L.$$

On the other hand, we have set that

$$\deg(\mathcal{Q}_I u_0) = \deg(\mathcal{Q}_{i_1}\mathcal{Q}_{i_2}\cdots\mathcal{Q}_{i_s} u_0) = i_1 + 2i_2 + 4i_3 + \cdots + 2^{s-1}i_s,$$

with $\deg(u_0) = 0$. In other words, if L could be zero the rings $\mathcal{T}_d$ and $\mathcal{R}_d$ would coincide precisely.

From Corollary 4.15 we know that the homology $H_i(C_d(k)/\mathfrak{S}_k;\mathbb{F}_2)$, for $0 \le i \le (d-1)(k-1)$, is the $\mathbb{F}_2$ vector space spanned by all monomials of degree $i+kL$ of the polynomial ring $\mathbb{F}_2[\{u_L\} \cup \{Q_I u_L : I \in \mathcal{I}_d\}]$ whose weights are exactly k. It is important to keep in mind that the base elements of this vector space are indexed by, but are not equal to, the iterated Araki–Kudo–Dyer–Lashof homology operations applied to the class u_L. Furthermore, recall that for $I = (i_1, \dots, i_s) \in \mathcal{I}_d$ the weight of $Q_I u_L$ is 2^s, while the weight of u_L is 1. Then we the ring $\mathcal{T}_d$ can be decomposed into a direct sum of vector spaces as follows:

$$
\begin{aligned}
\mathcal{T}_d = H_*\Big(\coprod_{k\ge 0} C_d(k)/\mathfrak{S}_k;\mathbb{F}_2\Big) &\cong \bigoplus_{k\ge 0} H_*(C_d(k)/\mathfrak{S}_k;\mathbb{F}_2)\\
&\cong \bigoplus_{k\ge 0}\bigoplus_{i\ge 0} H_i(C_d(k)/\mathfrak{S}_k;\mathbb{F}_2)\\
&\cong \bigoplus_{k\ge 0}\bigoplus_{i\ge 0}\Big\langle (Q_{I_1}u_L)\cdots(Q_{I_t}u_L)\,u_L^a \;:\; \begin{matrix}\omega((Q_{I_1}u_L)\cdots(Q_{I_t}u_L)\,u_L^a) = k\\ \deg((Q_{I_1}u_L)\cdots(Q_{I_t}u_L)\,u_L^a) = i+kL\end{matrix}\Big\rangle\\
&= \bigoplus_{k\ge 0}\bigoplus_{i\ge 0}\Big\langle (Q_{I_1}u_L)\cdots(Q_{I_t}u_L)\,u_L^a \;:\; \begin{matrix}2^{s_1}+\dots+2^{s_t}+a = k\\ \iota(I_1,\dots,I_t)+(2^{s_1}+\dots+2^{s_t}+a)L = i+kL\end{matrix}\Big\rangle\\
&= \bigoplus_{k\ge 0}\bigoplus_{i\ge 0}\Big\langle (Q_{I_1}u_L)\cdots(Q_{I_t}u_L)\,u_L^a \;:\; \begin{matrix}2^{s_1}+\dots+2^{s_t}+a = k\\ \iota(I_1,\dots,I_t) = i\end{matrix}\Big\rangle.
\end{aligned}
$$

Here we took $I_1 = (i_{11}, i_{12}, \dots, i_{1s_1}), \dots, I_t = (i_{t1}, i_{t2}, \dots, i_{ts_t})$ and denoted by $\iota(I_1,\dots,I_t)$ the sum $(i_{11}+\dots+2^{s_1-1}i_{1s_1})+\dots+(i_{t1}+\dots+2^{s_t-1}i_{ts_t})$.

On the other hand, in a similar way as above, the polynomial ring $\mathcal{R}_d$ can be decomposed into a direct sum of vector subspaces as follows:

$$
\begin{aligned}
\mathcal{R}_d &:= \mathbb{F}_2[\{u_0\}\cup\{Q_I u_0 : I \in \mathcal{I}_d\}]\\
&\cong \bigoplus_{k\ge 0}\bigoplus_{i\ge 0}\Big\langle (Q_{I_1}u_0)\cdots(Q_{I_t}u_0)\,u_0^a \;:\; \begin{matrix}2^{s_1}+\dots+2^{s_t}+a = k\\ \iota(I_1,\dots,I_t) = i\end{matrix}\Big\rangle.
\end{aligned}
$$

Thus, the bijective correspondence $Q_I u_L \longleftrightarrow Q_I u_0$, for $I \in \mathcal{I}_d$, and $u_L \longleftrightarrow u_0$ induces bijection between the (monomial) basis of $\mathcal{T}_d$ and $\mathcal{R}_d$, which further on induces an isomorphism of $\mathcal{T}_d$ and $\mathcal{R}_d$ considered as $\mathbb{F}_2$ vector spaces.

It remains to show that the induced isomorphism on the underlying vector space structures of $\mathcal{T}_d$ and $\mathcal{R}_d$ also respects the corresponding product structures. This fact follows from the commutativity of the following diagram:

$$\begin{array}{ccc}
H_i(C_d(r)/\mathfrak{S}_r)\otimes H_j(C_d(s)/\mathfrak{S}_s) & \xleftarrow{\cong} & \widetilde{H}_{i+rL}(\mathrm{D}_rC_d(S^L))\otimes \widetilde{H}_{j+sL}(\mathrm{D}_sC_d(S^L)) \\
\Big\downarrow = & & \Big\downarrow \cong \\
H_i(C_d(r)/\mathfrak{S}_r)\otimes H_j(C_d(s)/\mathfrak{S}_s) & \xleftarrow{\cong} & \widetilde{H}_{i+rL}(\mathrm{Th}(\xi_{d,r}^{\oplus L}))\otimes \widetilde{H}_{j+sL}(\mathrm{Th}(\xi_{d,s}^{\oplus L})) \\
\Big\downarrow \cong & & \Big\downarrow \cong \\
H_{i+j}(C_d(r)/\mathfrak{S}_r\times C_d(s)/\mathfrak{S}_s) & \xleftarrow{\cong} & \widetilde{H}_{i+j+(r+s)L}(\mathrm{Th}(\xi_{d,r}^{\oplus L})\wedge \mathrm{Th}(\xi_{d,s}^{\oplus L})) \\
\Big\downarrow = & & \Big\downarrow \cong \\
H_{i+j}(C_d(r)/\mathfrak{S}_r\times C_d(s)/\mathfrak{S}_s) & \xleftarrow{\cong} & \widetilde{H}_{i+j+(r+s)L}(\mathrm{Th}(\xi_{d,r}^{\oplus L}\times \xi_{d,s}^{\oplus L})) \\
\Big\downarrow (\mu_{C_d}(\mathbf{c}_1,\mathbf{c}_2))_* & & \Big\downarrow (\mu_{C_d}(\mathbf{c}_1,\mathbf{c}_2))_* \\
H_{i+j}(C_d(r+s)/\mathfrak{S}_{r+s}) & \xleftarrow{\cong} & \widetilde{H}_{i+j+(r+s)L}(\mathrm{Th}(\xi_{d,r+s}^{\oplus L})) \\
\Big\downarrow = & & \Big\downarrow \cong \\
H_{i+j}(C_d(r+s)/\mathfrak{S}_{r+s}) & \xleftarrow{\cong} & \widetilde{H}_{i+j+(r+s)L}(\mathrm{D}_{r+s}C_d(S^L))
\end{array} \tag{4.18}$$

where the coefficients in all homologies are assumed to be in the field $\mathbb{F}_2$. (Recall, here $\mathrm{Th}(\xi)$ stands for the Thom space of the vector bundle ξ.) Thus, we have proved that $\mathcal{T}_d$ and $\mathcal{R}_d$ are isomorphic as rings. □

Remark 4.19 It is important to observe that alternatively we could see the disjoint union $\coprod_{k\geq 0}\mathrm{F}(\mathbb{R}^d,k)/\mathfrak{S}_k$ as the free C_d-space $C_d(S^d)$ generated by the sphere S^0. Then the homology $H_*(\coprod_{k\geq 0}\mathrm{F}(\mathbb{R}^d,k)/\mathfrak{S}_k;\mathbb{F}_p)$, for p a prime, was described already in 1976 by Cohen [33, Thm. 3.1] using the framework of an appropriate class of Hopf algebras.

Part II
Applications to the (Non-)Existence of Regular and Skew Embeddings

Chapter 5
On Highly Regular Embeddings: Revised

The study of highly regular embeddings has a long history, which starts in 1957 with the work of Borsuk [22] on the existence k-regular embeddings. Boltjanskiĭ et al. [21] in 1963 gave the first lower bounds for the existence of $2k$-regular embeddings. In the 1970s and 1980s the accumulated knowledge about the topology of the unordered configuration space allowed for further progress, which was made by Cohen and Handel [36], Chisholm [29], Handel [57], and Handel and Segal [59]. In the 1990s the existence of k-regular embeddings was related to the notion of k-neighbourly submanifolds; this was considered by Vassiliev [100] and Handel [58]. The result of Chisholm [29, Theorem 2] is the strongest result from that time.

In the first decade of twenty-first century the notion of ℓ-skew embeddings was introduced and studied by Ghomi and Tabachnikov [54]. Furthermore, combining the notions of k-regular embeddings and of ℓ-skew embeddings, in 2006 Stojenović [96] defined the notion of k-regular-ℓ-skew embeddings and gave the first bounds for their existence.

The second decade of the new century also brought the first non-elementary constructions of k-regular embedding. In 2019, Jarosław Buczyński et al. [26], using advanced methods of algebraic geometry, constructed k-regular embeddings of finite dimensional real and complex vector spaces.

In this chapter we revise the results of Blagojević, Lück and Ziegler presented in the paper [15]. The proofs of [15, Thm. 2.1, Thm. 3.1, Thm. 4.1] relied essentially on a result of Hưng [64, (4.7)] that turned out to be incorrect, see Remark 4.6. Specifically, [15, Lem. 2.15] is not true when d is not a power of 2.

P. V. M. Blagojević et al., *Equivariant Cohomology of Configuration Spaces Mod 2*, Lecture Notes in Mathematics 2282, https://doi.org/10.1007/978-3-030-84138-6_5

5.1 *k*-Regular Embeddings

In this section we correct [15, Sec. 2]. In the following all topological spaces are Hausdorff spaces and all maps are assumed to be continuous.

Definition 5.1 Let $k \geq 1$ and $N \geq 1$ be integers, and let X be a topological space. A continuous map $f\colon X \longrightarrow \mathbb{R}^N$ is a ***k*-regular embedding** if for every $(x_1, \ldots, x_k) \in \mathrm{F}(X, k)$ the vectors $f(x_1), \ldots, f(x_k)$ are linearly independent. In particular, $0 \notin \mathrm{im}(f)$.

For $N \geq 1$ and $1 \leq k \leq N$ integers, the Stiefel manifold $V_k(\mathbb{R}^N)$ of all ordered k-frames is a (topological) subspace of the product $(\mathbb{R}^N)^k$ given by

$$V_k(\mathbb{R}^N) = \{(y_1, \ldots, y_k) \in (\mathbb{R}^N)^k : y_1, \ldots, y_k \text{ are linearly independent}\}.$$

The symmetric group $\mathfrak{S}_k$ acts freely on the Stiefel manifold by permuting the vectors in the frame. Now from Definition 5.1 we get a necessary condition for the existence of a k-regular embedding phrased in term of the existence of an equivariant map.

Lemma 5.2 *If there exists a k-regular embedding $X \longrightarrow \mathbb{R}^N$, then there exists an $\mathfrak{S}_k$-equivariant map*

$$\mathrm{F}(X, k) \longrightarrow V_k(\mathbb{R}^N). \tag{5.1}$$

In order to study the existence of an $\mathfrak{S}_k$-equivariant map (5.1) we use the criterion of Cohen and Handel [36, Prop. 2.1], see also [15, Lem. 2.11]. Let $\mathbb{R}^k$ be endowed with an $\mathfrak{S}_k$-action given by coordinate permutation. Then the subspace $W_k = \{(a_1, \ldots, a_k) \in \mathbb{R}^k : a_1 + \cdots + a_k = 0\}$ is an $\mathfrak{S}_k$-invariant subspace. Consider the following vector bundles

$$\begin{aligned}
\xi_{X,k}\colon &\quad \mathbb{R}^k \longrightarrow \mathrm{F}(X, k) \times_{\mathfrak{S}_k} \mathbb{R}^k \longrightarrow \mathrm{F}(X, k)/\mathfrak{S}_k,\\
\zeta_{X,k}\colon &\quad W_k \longrightarrow \mathrm{F}(X, k) \times_{\mathfrak{S}_k} W_k \longrightarrow \mathrm{F}(X, k)/\mathfrak{S}_k,\\
\tau_{X,k}\colon &\quad \mathbb{R} \longrightarrow \mathrm{F}(X, k)/\mathfrak{S}_k \times \mathbb{R} \longrightarrow \mathrm{F}(X, k)/\mathfrak{S}_k,
\end{aligned} \tag{5.2}$$

where $\tau_{X,k}$ is a trivial line bundle. An obvious decomposition holds:

$$\xi_{X,k} \cong \zeta_{X,k} \oplus \tau_{X,k}. \tag{5.3}$$

The bundle $\xi_{d,k}$ introduced in (4.12) is the pull-back vector bundle of the vector bundle $\xi_{\mathbb{R}^d,k}$ along the homotopy equivalence $C_d(k)/\mathfrak{S}_k \longrightarrow \mathrm{F}(\mathbb{R}^d, k)/\mathfrak{S}_k$ which is induced by the evaluation at centers of cubes map $\mathrm{ev}_{d,n}\colon C_d(n) \longrightarrow \mathrm{F}(\mathbb{R}^d, n)$ from Lemma 7.9.

Lemma 5.3 *An $\mathfrak{S}_k$-equivariant map $\mathrm{F}(X, k) \longrightarrow V_k(\mathbb{R}^N)$ exists if and only if the k-dimensional vector bundle $\xi_{X,k}$ admits an $(N-k)$-dimensional inverse.*

As a direct consequence of Lemmas 5.2 and 5.3 we get a criterion for the non-existence of a k-regular embedding in term of the dual Stiefel–Whitney class of the vector bundle $\xi_{X,k}$. We recall only a particular case of [15, Lem. 2.12 (2)] that we use here.

Lemma 5.4 *Let $d \geq 1$ and $k \geq 1$ be integers. If the dual Stiefel–Whitney class*

$$\overline{w}_{N-k+1}(\xi_{\mathbb{R}^d,k}) \neq 0$$

does not vanish, then there cannot be any k-regular embedding $\mathbb{R}^d \longrightarrow \mathbb{R}^N$.

The criterion for the non-existence of a k-regular embedding $X \longrightarrow \mathbb{R}^N$ given in the previous lemma motivates the study of Stiefel–Whitney classes of the vector bundle $\xi_{X,k}$. In [15, Thm. 2.13], based on the work of Hưng [64, (4.7)], the following theorem about the dual Stiefel–Whitney classes of $\xi_{\mathbb{R}^d,k}$ was proved.

Theorem 2.13. Let $k, d \geq 1$ be integers. Then the dual Stiefel–Whitney class

$$\overline{w}_{(d-1)(k-\alpha(k))}(\xi_{\mathbb{R}^d,k})$$

does not vanish.

The previous theorem and the criterion given in Lemma 5.4 implied [15, Thm. 2.1]:

Theorem 2.1. Let $k, d \geq 1$ be integers. There is no k-regular embedding $\mathbb{R}^d \longrightarrow \mathbb{R}^N$ for

$$N \leq d(k-\alpha(k)) + \alpha(k) - 1,$$

where $\alpha(k)$ denotes the number of ones in the dyadic presentation of k.

The proof of the key theorem [15, Thm. 2.13] was done in two steps. First, the special case when k is a power of 2 was established [15, Lem. 2.15], and then the general case for arbitrary integer $k \geq 2$ was derived [15, Lem. 2.17]. In the following we quote both lemmas as in they were written in the original.

Lemma 2.15. Let $d \geq 1$ be an integer, and $k = 2^m$ for some $m \geq 1$. Then the dual Stiefel–Whitney class

$$\overline{w}_{(d-1)(k-1)}(\xi_{\mathbb{R}^d,k})$$

does not vanish.

Lemma 2.17. Let $d, k \geq 1$ be integers. Then the dual Stiefel–Whitney class

$$\overline{w}_{(d-1)(k-\alpha(k))}(\xi_{\mathbb{R}^d,k})$$

does not vanish.

The proof of [15, Lem. 2.15] contains a gap at the place where [64, (4.7)] was applied and is not true for $d \geq 1$ not a power of 2. More precisely, in the proof stands:

The following decomposition of $H^*(\mathrm{F}(\mathbb{R}^d, k)/\mathfrak{S}_k)$ was proved by Hưng in [20, (4.7), page 279]:

$$(5) \quad H^*(\mathrm{F}(\mathbb{R}^d, k)/\mathfrak{S}_k) \cong \alpha^*_{d,k}\big(\ker(\mathrm{res}^{\mathfrak{S}_k}_{E_m}) \oplus \mathbb{F}_2[q_{m,m-1}, \dots, q_{m,1}]\big) \oplus \alpha^*_{d,k}(\langle q_{m,0}\rangle).$$

In particular, this implies that

$$\alpha^*_{d,k}\big(\ker(\mathrm{res}^{\mathfrak{S}_k}_{E_m}) \oplus \mathbb{F}_2[q_{m,m-1}, \dots, q_{m,1}]\big) \cap \alpha^*_{d,k}(\langle q_{m,0}\rangle) = \{0\}.$$

Consequently, [15, Lem. 2.15, Lem. 2.17, Thm. 2.13, Thm. 2.1] do not hold. Furthermore, a technical result stated in [15, Cor. 2.16] is incorrect. In the following we correct these gaps.

For the corrections we will combine different results we established so far. Many of them can be summarized in the commutative diagram (5.5). The cohomologies appearing in the diagram (5.5) are assumed to be with the field $\mathbb{F}_2$ coefficients.

First, we show that [15, Lem. 2.15, Lem. 2.17] and consequently [15, Thm. 2.13] hold, as stated, in the case when the dimension d is a power of 2. For that we use the following result from [14, Thm. 3.1]: For integers $d \geq 2$ and $k \geq 2$

$$\mathrm{height}(H^*(\mathrm{F}(\mathbb{R}^d, k)/\mathfrak{S}_k; \mathbb{F}_2)) \leq \min\{2^t : 2^t \geq d\}, \tag{5.4}$$

where $\mathrm{height}(\mathrm{F}(\mathbb{R}^d, k)/\mathfrak{S}_k; \mathbb{F}_2))$ is the height of the algebra $H^*(\mathrm{F}(\mathbb{R}^d, k)/\mathfrak{S}_k; \mathbb{F}_2)$. This result, obtained as an application of Kahn–Priddy [66, Thm. pp. 103], is the best possible as explained in [14, Sec. 3]. It is worth mentioning that a similar upper bound for the hight of the cohomology ring $H^*(\mathrm{F}(\mathbb{R}^d, k)/\mathfrak{S}_k; \mathbb{F}_p)$, where p is now an odd prime, was given in [14, Thm. 3.2].

Recall, for an algebra A over a field $\mathbb{F}$, the *height of an element* $a \in A\backslash A^*$ is a natural number or infinity

$$\mathrm{height}(a) := \min\{n \in \mathbb{N} : a^n = 0\}.$$

Here A^* denotes the group of all invertible elements of the algebra A. The *height of the algebra* A is defined to be $\mathrm{height}(A) := \max\{\mathrm{height}(a) : a \in A\backslash A^*\}$.

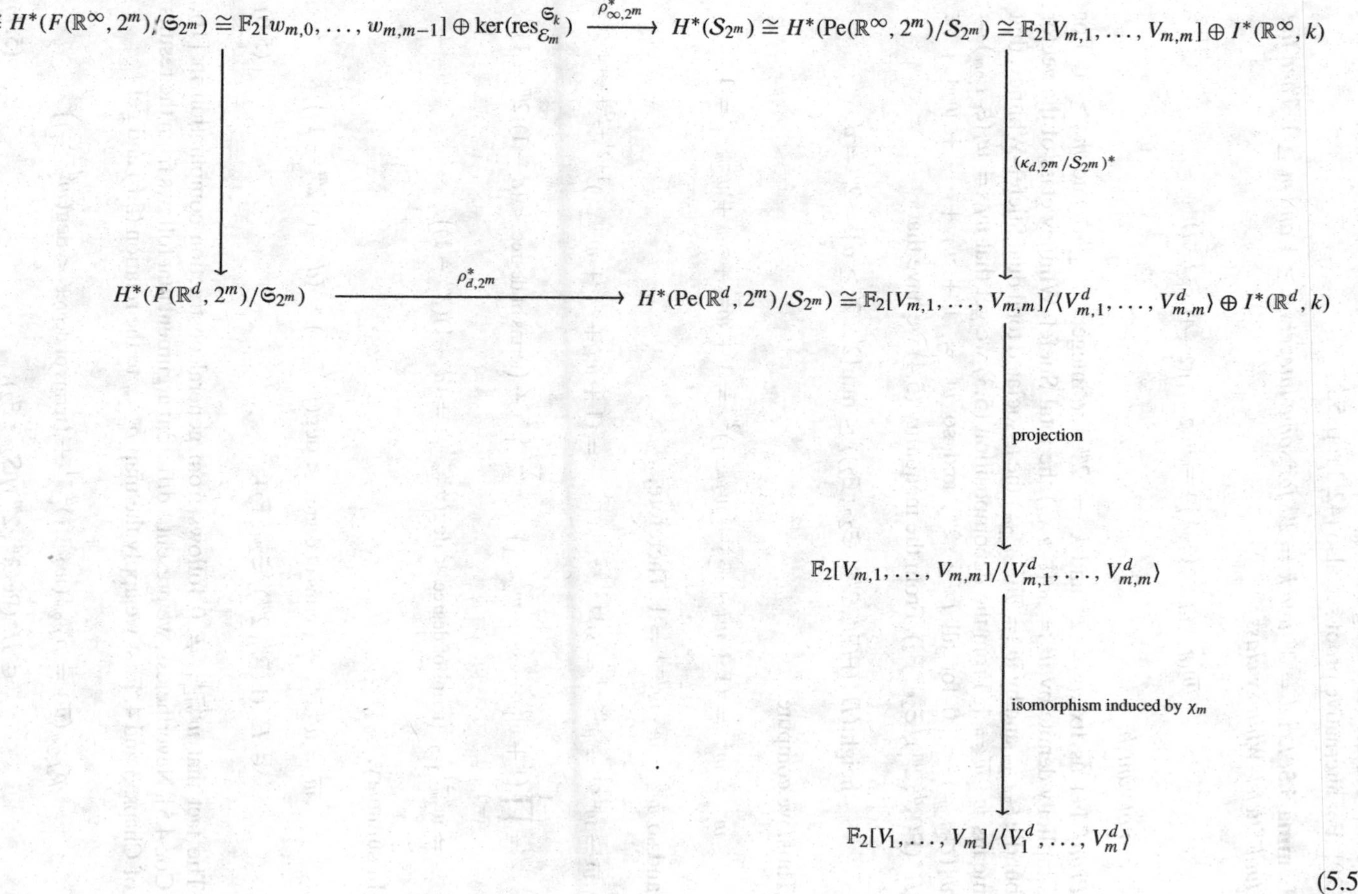
$m) \cong H^*(F(\mathbb{R}^\infty, 2^m)/\mathfrak{S}_{2^m}) \cong \mathbb{F}_2[w_{m,0}, \ldots, w_{m,m-1}] \oplus \ker(\mathrm{res}^{\mathfrak{S}_k}_{\mathcal{E}_m})$
$\rho^*_{\infty,2^m}$
$H^*(\mathcal{S}_{2^m}) \cong H^*(\mathrm{Pe}(\mathbb{R}^\infty, 2^m)/\mathcal{S}_{2^m}) \cong \mathbb{F}_2[V_{m,1}, \ldots, V_{m,m}] \oplus I^*(\mathbb{R}^\infty, k)$
$(\kappa_{d,2^m}/\mathcal{S}_{2^m})^*$
$H^*(F(\mathbb{R}^d, 2^m)/\mathfrak{S}_{2^m})$
$\rho^*_{d,2^m}$
$H^*(\mathrm{Pe}(\mathbb{R}^d, 2^m)/\mathcal{S}_{2^m}) \cong \mathbb{F}_2[V_{m,1}, \ldots, V_{m,m}]/\langle V^d_{m,1}, \ldots, V^d_{m,m}\rangle \oplus I^*(\mathbb{R}^d, k)$
projection
$\mathbb{F}_2[V_{m,1}, \ldots, V_{m,m}]/\langle V^d_{m,1}, \ldots, V^d_{m,m}\rangle$
isomorphism induced by χ_m
$\mathbb{F}_2[V_1, \ldots, V_m]/\langle V^d_1, \ldots, V^d_m\rangle$
(5.5)

We prove the following lemma which can be extracted from work of Chisholm [29]. For alternative proof see also [42, Prop. 5.2].

Lemma 5.5 *Let $d = 2^a$ and $k = 2^m$ for some integers $a \geq 1$ and $m \geq 1$. Then the dual Stiefel–Whitney class*

$$\overline{w}_{(d-1)(k-1)}(\xi_{\mathbb{R}^d,k}) = \overline{w}_{(2^a-1)(2^m-1)}(\xi_{\mathbb{R}^d,2^m})$$

does not vanish.

Proof Let us fix $d = 2^a$ and $k = 2^m$ for integers $a \geq 1$ and $m \geq 1$. For simplicity denote by $w := w(\xi_{\mathbb{R}^d,2^m})$ the total Stiefel–Whitney class of the vector bundle $\xi_{\mathbb{R}^d,2^m}$ and by $\overline{w} := \overline{w}(\xi_{\mathbb{R}^d,2^m})$ the associated total dual Stiefel–Whitney, that means $w \cdot \overline{w} = 1$. From the decomposition (5.3) we get that $w_i := w_i(\xi_{\mathbb{R}^d,2^m}) = w_i(\zeta_{\mathbb{R}^d,2^m}) = 0$ for all $i \geq 2^m$, and so $w = 1 + w_1 + \cdots + w_{2^m-1} \in H^*(\mathrm{F}(\mathbb{R}^d, 2^m)/\mathfrak{S}_{2^m}; \mathbb{F}_2)$. From the inequality (5.4) we have that

$$\mathrm{height}(H^*(\mathrm{F}(\mathbb{R}^d, 2^m)/\mathfrak{S}_{2^m}; \mathbb{F}_2)) \leq \min\{2^t : 2^t \geq d\} = 2^a = d.$$

Thus, we compute

$$w^d = w^{2^a} = (1 + w_1 + \cdots + w_{2^m-1})^{2^a} = 1 + w_1^{2^a} + \cdots + w_{2^m-1}^{2^a} = 1,$$

and so $w^d = w \cdot w^{d-1} = 1$. Therefore,

$$\begin{aligned}
\overline{w} = w^{d-1} &= w^{2^a-1} = w^{1+2^1+2^2+\cdots+2^{a-1}} = (1 + w_1 + \cdots + w_{2^m-1})^{1+2^1+2^2+\cdots+2^{a-1}} \\
&= \prod_{i=0}^{a-1} (1 + w_1^{2^i} + \cdots + w_{2^m-1}^{2^i}) = w_{2^m-1}^{\sum_{i=0}^{a-1} 2^i} + \big(\text{terms of degree} < (d-1)(2^m-1)\big) \\
&= w_{2^m-1}^{d-1} + \big(\text{terms of degree} < \deg(w_{2^m-1}^{d-1}) = (d-1)(2^m-1)\big).
\end{aligned}$$

In summary,

$$\begin{aligned}
\overline{w} &= w_{2^m-1}^{d-1} + \big(\text{terms of degree} < \deg(w_{2^m-1}^{d-1}) = (d-1)(2^m-1)\big) \\
&\in H^*(\mathrm{F}(\mathbb{R}^d, 2^m)/\mathfrak{S}_{2^m}; \mathbb{F}_2). \qquad (5.6)
\end{aligned}$$

The fact that $w_{2^m-1}^{d-1} \neq 0$ follows from general obstruction computation in [16, Cor. 4.5]. Nevertheless, we present a different argument that follows from the results of Chaps. 3 and 4. First we apply the map $\rho_{d,2^m}^*$ to the relation (5.6) and get

$$\begin{aligned}
\rho_{d,2^m}^*(\overline{w}) &= \rho_{d,2^m}^*(w_{2^m-1})^{d-1} + \big(\text{terms of degree} < \deg(w_{2^m-1}^{d-1})\big) \\
&\in H^*(\mathrm{Pe}(\mathbb{R}^d, 2^m)/\mathcal{S}_{2^m}; \mathbb{F}_2). \qquad (5.7)
\end{aligned}$$

Recall that according to Theorem 3.11 we have that

$$H^*(\mathrm{Pe}(\mathbb{R}^d, 2^m)/S_{2^m}; \mathbb{F}_2) \cong \mathbb{F}_2[V_{m,1}, \dots, V_{m,m}]/\langle V_{m,1}^d, \dots, V_{m,m}^d\rangle \oplus \mathrm{I}^*(\mathbb{R}^d, 2^m).$$

Furthermore, from the proof of Lemma 3.13, the commutative diagram (5.5) and the relation (3.32) we have that

$$\begin{aligned}\rho^*_{d,2^m}(w_{2^m-1}) = V_{m,1}\cdots V_{m,m} &\in \mathbb{F}_2[V_{m,1}, \dots, V_{m,m}]/\langle V_{m,1}^d, \dots, V_{m,m}^d\rangle \\ &\subseteq H^*(\mathrm{Pe}(\mathbb{R}^d, 2^m)/S_{2^m}; \mathbb{F}_2).\end{aligned}$$

Combining these facts we get

$$\begin{aligned}\rho^*_{d,2^m}(\overline{w}) &= (V_{m,1}\cdots V_{m,m})^{d-1} + \big(\text{terms of degree} < \deg(w_{2^m-1}^{d-1})\big) \\ &\in H^*(\mathrm{Pe}(\mathbb{R}^d, 2^m)/S_{2^m}; \mathbb{F}_2),\end{aligned}$$

and furthermore

$$\rho^*_{d,2^m}(\overline{w}_{(d-1)(2^m-1)}) = \rho^*_{d,2^m}(w_{2^m-1})^{d-1} = (V_{m,1}\cdots V_{m,m})^{d-1} \neq 0.$$

Consequently, $\overline{w}_{(d-1)(2^m-1)} \neq 0$, and the proof of the lemma is concluded. □

The extension of the previous lemma (for dimension $d = 2^a$) to the case of where $k \geq 1$ is an arbitrary integer (still for dimension $d = 2^a$) can be done as in the proof of [15, Lem. 2.17]. For the sake of completeness we give a detailed proof using the presentation from [15, Proof of Lem. 2.17]. Here $\alpha(k)$ denotes the number of 1s in the binary presentation of the integer $k \geq 1$.

Lemma 5.6 *Let $d = 2^a$ for some integer $a \geq 1$, and let $k \geq 1$ be an integer. Then the dual Stiefel–Whitney class*

$$\overline{w}_{(d-1)(k-\alpha(k))}(\xi_{\mathbb{R}^d,k}) = \overline{w}_{(2^a-1)(k-\alpha(k))}(\xi_{\mathbb{R}^d,k})$$

does not vanish.

Proof Let $r := \alpha(k)$ be the number of 1s in the binary presentation of the integer $k \geq 1$, and let $k = 2^{b_1} + \cdots + 2^{b_r}$ where $0 \leq b_1 < b_2 < \cdots < b_r$. Consider a morphism between vector bundles $\prod_{i=1}^r \xi_{\mathbb{R}^d,2^{b_i}}$ and $\xi_{\mathbb{R}^d,k}$ where the following commutative square is a pullback diagram:

$$\begin{array}{ccc}\prod_{i=1}^r \xi_{\mathbb{R}^d,2^{b_i}} & \xrightarrow{\ \Theta\ } & \xi_{\mathbb{R}^d,k} \\ \big\downarrow & & \big\downarrow \\ \prod_{i=1}^r \mathrm{F}(\mathbb{R}^d, 2^{b_i})/\mathfrak{S}_{2^{b_i}} & \xrightarrow{\ \theta\ } & \mathrm{F}(\mathbb{R}^d, k)/\mathfrak{S}_k.\end{array}$$

The map θ is induced, up to an equivariant homotopy, from a restriction of the little cubes operad structural map

$$(C_d(2^{b_1}) \times \cdots \times C_d(2^{b_r})) \times C_d(r) \longrightarrow C_d(2^{b_1} + \cdots + 2^{b_r}).$$

Alternatively, fix embeddings $e_i : \mathbb{R}^d \longrightarrow \mathbb{R}^d$ for $1 \leq i \leq r$ such that their images are pairwise disjoint open d-balls. They induces embeddings $\mathrm{F}(\mathbb{R}^d, \ell) \longrightarrow \mathrm{F}(\mathbb{R}^d, \ell)$ denoted by the same letter e_i for all natural numbers ℓ. Thus, the map θ is induced by the map $\prod_{i=1}^r \mathrm{F}(\mathbb{R}^d, 2^{b_i}) \longrightarrow \mathrm{F}(\mathbb{R}^d, k)$ defined by

$$\begin{aligned}&\big((x_{1,1}, \ldots, x_{1,2^{b_1}}), \ldots, (x_{r,1}, \ldots, x_{r,2^{b_r}})\big)\\ &\qquad\longmapsto e_1(x_{1,1}, \ldots, x_{1,2^{b_1}}) \times \cdots \times e_r(x_{r,1}, \ldots, x_{r,2^{b_r}}).\end{aligned}$$

The map Θ that covers θ is given by

$$\begin{aligned}&\big((x_{1,1}, \ldots, x_{1,2^{b_1}}; v_1), \ldots, (x_{r,1}, \ldots, x_{r,2^{b_r}}; v_r)\big)\\ &\qquad\longmapsto \big(e_1(x_{1,1}, \ldots, x_{1,2^{b_1}}) \times \cdots \times e_r(x_{r,1}, \ldots, x_{r,2^{b_r}}), v_1 \times \cdots \times v_r\big).\end{aligned}$$

Consequently, the pullback vector bundle is the product vector bundle $\theta^*\xi_{\mathbb{R}^d,k} \cong \prod_{i=1}^r \xi_{\mathbb{R}^d,2^{b_i}}$. Now the naturality property of the Stiefel–Whitney classes [80, Ax. 2, p. 37] implies that in cohomology we get

$$\theta^*(\overline{w}_{(d-1)(k-r)}(\xi_{\mathbb{R}^d,k})) = \overline{w}_{(d-1)(k-r)}\Big(\prod_{i=1}^r \xi_{\mathbb{R}^d,2^{b_i}}\Big).$$

The product formula for Stiefel–Whitney classes [80, Pr. 4-A, p. 54] implies that

$$\overline{w}\Big(\prod_{i=1}^r \xi_{\mathbb{R}^d,2^{b_i}}\Big) = \overline{w}(\xi_{\mathbb{R}^d,2^{b_1}}) \times \cdots \times \overline{w}(\xi_{\mathbb{R}^d,2^{b_r}}).$$

Now we compute

$$\begin{aligned}\theta^*(\overline{w}_{(d-1)(k-r)}(\xi_{\mathbb{R}^d,k})) &= \overline{w}_{(d-1)(k-r)}\Big(\prod_{i=1}^r \xi_{\mathbb{R}^d,2^{b_i}}\Big)\\ &= \sum_{s_1+\cdots+s_r=(d-1)(k-r)} \overline{w}_{s_1}(\xi_{\mathbb{R}^d,2^{b_1}}) \times \cdots \times \overline{w}_{s_r}(\xi_{\mathbb{R}^d,2^{b_r}}).\end{aligned}$$

From the Künneth formula [23, Thm. VI.3.2] we have that each term of the previous sum $\overline{w}_{s_1}(\xi_{\mathbb{R}^d,2^{b_1}}) \times \cdots \times \overline{w}_{s_r}(\xi_{\mathbb{R}^d,2^{b_r}})$ belongs to a different direct summand in the following direct decomposition of the $(d-1)(k-r)$th cohomology

$$H^{(d-1)(k-r)}\Big(\prod_{i=1}^{r} \mathrm{F}(\mathbb{R}^d, 2^{b_i})/\mathfrak{S}_{2^{b_i}}; \mathbb{F}_2\Big)$$
$$\cong \bigoplus_{s_1+\cdots+s_r=(d-1)(k-r)} H^{s_1}(\mathrm{F}(\mathbb{R}^d, 2^{b_1})/\mathfrak{S}_{2^{b_1}}; \mathbb{F}_2) \otimes \cdots \otimes H^{s_r}(\mathrm{F}(\mathbb{R}^d, 2^{b_r})/\mathfrak{S}_{2^{b_r}}; \mathbb{F}_2).$$

Thus, we have the equivalence

$$\overline{w}_{(d-1)(k-r)}\Big(\prod_{i=1}^{r} \xi_{\mathbb{R}^d,2^{b_i}}\Big) \neq 0 \Longleftrightarrow$$
$$\overline{w}_{s_1}(\xi_{\mathbb{R}^d,2^{b_1}}) \times \cdots \times \overline{w}_{s_r}(\xi_{\mathbb{R}^d,2^{b_r}}) \neq 0 \text{ for some } s_1 + \cdots + s_r = (d-1)(k-r).$$

Now, since d is a power of 2, Lemma 5.5 implies that $\overline{w}_{(d-1)(2^{b_i}-1)}(\xi_{\mathbb{R}^d,2^{b_i}}) \neq 0$ for all $1 \leq i \leq r$, and so

$$\overline{w}_{(d-1)(2^{b_1}-1)}(\xi_{\mathbb{R}^d,2^{b_1}}) \times \cdots \times \overline{w}_{(d-1)(2^{b_r}-1)}(\xi_{\mathbb{R}^d,2^{b_r}}) \neq 0.$$

Hence, $\theta^*(\overline{w}_{(d-1)(k-r)}(\xi_{\mathbb{R}^d,k})) \neq 0$, and furthermore $\overline{w}_{(d-1)(k-\alpha(k))}(\xi_{\mathbb{R}^d,k}) \neq 0$. □

Next we consider the case when d, the dimension of the Euclidean space $\mathbb{R}^d$, is not a power of 2. Like in the previous situation this is done in two separate steps depending whether k is power of 2 or not. Now we give two corrections of [15, Lem. 2.15]. In the first one, next lemma, we do not take into account the dyadic presentation of the dimensions d.

Lemma 5.7 *Let $d \geq 3$ be an integer that is not a power of 2, and let $k = 2^m$ for some integer $m \geq 1$. Then the dual Stiefel–Whitney class*

$$\overline{w}_{(d-1)k/2}(\xi_{\mathbb{R}^d,k}) = \overline{w}_{(d-1)2^{m-1}}(\xi_{\mathbb{R}^d,2^m})$$

does not vanish.

Proof Let us fix an integer $d \geq 3$ which is not a power of 2. Then there exists an integer $a \geq 2$ such that $2^{a-1} + 1 \leq d \leq 2^a - 1$. Furthermore, let $k = 2^m$ where $m \geq 1$ is an integer. Then $d-1$ can be presented as $d - 1 = 2^{a_1} + \cdots + 2^{a_q}$ where $0 \leq a_1 < \cdots < a_q = a - 1$. As before, for simplicity we again denote by $w = w(\xi_{\mathbb{R}^d,2^m})$ the total Stiefel–Whitney class of the vector bundle $\xi_{\mathbb{R}^d,2^m}$ and by $\overline{w} = \overline{w}(\xi_{\mathbb{R}^d,2^m})$ the associated total dual Stiefel–Whitney. Hence, $w \cdot \overline{w} = 1$. Now the inequality (5.4) implies that

$$\mathrm{height}(H^*(\mathrm{F}(\mathbb{R}^d, 2^m)/\mathfrak{S}_{2^m}; \mathbb{F}_2)) \leq \min\{2^t : 2^t \geq d\} = 2^a.$$

Consequently, $w^{2^a} = (1 + w_1 + \cdots + w_{2^m-1})^{2^a} = 1 + w_1^{2^a} + \cdots + w_{2^m-1}^{2^a} = 1$, and so

$$\begin{aligned}\overline{w} &= w^{2^a-1} \\ &= (1 + w_1 + \cdots + w_{2^m-1})^{1+2^1+2^2+\cdots+2^{a-1}} \\ &= \prod_{i=0}^{a-1} \big(1 + w_1^{2^i} + \cdots + w_{2^m-1}^{2^i}\big). \end{aligned} \tag{5.8}$$

Now we apply the monomorphism

$$\rho_{d,2^m}^* \colon H^*(\mathrm{F}(\mathbb{R}^d, 2^m)/\mathfrak{S}_{2^m}; \mathbb{F}_2) \longrightarrow H^*(\mathrm{Pe}(\mathbb{R}^d, 2^m)/S_{2^m}; \mathbb{F}_2)$$

from Theorem 4.1 to the equality (5.8) and use the decomposition of the cohomology

$$H^*(\mathrm{Pe}(\mathbb{R}^d, 2^m)/S_{2^m}; \mathbb{F}_2) \cong \mathbb{F}_2[V_{m,1}, \ldots, V_{m,m}]/\langle V_{m,1}^d, \ldots, V_{m,m}^d\rangle \oplus \mathrm{I}^*(\mathbb{R}^d, 2^m),$$

given in Theorem 3.11, to get that

$$\begin{aligned}\rho_{d,2^m}^*(\overline{w}) &= \prod_{i=0}^{a-1} \big(1 + D_{m,m-1}^{2^i} + \cdots + D_{m,0}^{2^i}\big) + R \\ &\in \mathbb{F}_2[V_{m,1}, \ldots, V_{m,m}]/\langle V_{m,1}^d, \ldots, V_{m,m}^d\rangle \oplus \mathrm{I}^*(\mathbb{R}^d, 2^m),\end{aligned}$$

where

- $D_{m,r} = (\kappa_{d,2^m}/S_{2^m})^*(D_{m,r}) = \rho_{d,2^m}^*(w_{2^m-2^r})$, for $0 \leq r \leq m-1$, with the obvious abuse of notation, see (3.28), and
- $R \in \mathrm{I}^*(\mathbb{R}^d, 2^m)$.

Let us denote by π the following composition of the maps from the diagram (5.5):

$$\begin{array}{c} \mathbb{F}_2[V_{m,1}, \ldots, V_{m,m}]/\langle V_{m,1}^d, \ldots, V_{m,m}^d\rangle \oplus I^*(\mathbb{R}^d, 2^m) \\ \Big\downarrow \text{projection} \\ \mathbb{F}_2[V_{m,1}, \ldots, V_{m,m}]/\langle V_{m,1}^d, \ldots, V_{m,m}^d\rangle \\ \Big\downarrow \chi_m \\ \mathbb{F}_2[V_1, \ldots, V_m]/\langle V_1^d, \ldots, V_m^d\rangle, \end{array}$$

where the first map is the projection on a direct summand and the second map is induced by the change of variables $\chi_m \in \mathrm{GL}_m(\mathbb{F}_2)$. Next we apply π on $\rho_{d,2^m}^*(\overline{w})$

and get:

$$\pi(\rho^*_{d,2^m}(\overline{w})) = \prod_{i=0}^{a-1}\left(1+\pi(D_{m,m-1})^{2^i}+\cdots+\pi(D_{m,0})^{2^i}\right).$$

Since the change of the variables χ_m transforms $\mathrm{L}_m(\mathbb{F}_2)$-invariants into $\mathrm{U}_m(\mathbb{F}_2)$-invariants and Dickson polynomials, $\mathrm{GL}_m(\mathbb{F}_2)$-invariants, can be presented in terms of $\mathrm{U}_m(\mathbb{F}_2)$-invariants, as explained in (8.3) and (8.4), we have

$$\begin{aligned}
\pi(\rho^*_{d,2^m}(\overline{w})) &= \prod_{i=0}^{a-1}\left(1+\pi(D_{m,m-1})^{2^i}+\cdots+\pi(D_{m,0})^{2^i}\right)\\
&= \prod_{i=0}^{a-1}\Big(1+(V_1^{2^{m-1}}+V_2^{2^{m-2}}+\cdots+V_m^{2^0})^{2^i}+\cdots+\\
&\quad \Big(\sum_{1\le j_1<\cdots<j_r\le m}(V_1\cdots V_{j_1-1})^{2^r}(V_{j_1+1}\cdots V_{j_2-1})^{2^{r-1}}\cdots(V_{j_r+1}\cdots V_m)^{2^0}\Big)^{2^i}\\
&\qquad\qquad +\cdots+\\
&\quad (V_1\cdots V_m)^{2^i}\Big)\\
&= \prod_{i=0}^{a-1}\Big(1+(V_1^{2^{i+m-1}}+V_2^{2^{i+m-2}}+\cdots+V_m^{2^i})+\cdots+\\
&\quad \Big(\sum_{1\le j_1<\cdots<j_r\le m}(V_1\cdots V_{j_1-1})^{2^{i+r}}(V_{j_1+1}\cdots V_{j_2-1})^{2^{i+r-1}}\cdots(V_{j_r+1}\cdots V_m)^{2^i}\Big)\\
&\qquad\qquad +\cdots+\\
&\quad (V_1\cdots V_m)^{2^i}\Big).
\end{aligned}$$

Since $d-1 = 2^{a_1}+\cdots+2^{a_q}$ and $0\le a_1<\cdots<a_q = a-1$, by choosing terms $V_m^{2^{a_\ell+0}}$ from factors in the product indexed by $i = a_1,\ldots,a_q$ we get that

$$\pi(\rho^*_{d,2^m}(\overline{w}_{(d-1)2^{m-1}})) = V_m^{d-1}+S \in \mathbb{F}_2[V_1,\ldots,V_m]/\langle V_1^d,\ldots,V_m^d\rangle,$$

where S denotes a sum of some monomials in $V_1,\ldots,V_m$ of degree $(d-1)2^{m-1} = (d-1)k/2$ which are all different from V_m^{d-1}. Consequently, $\pi(\rho^*_{d,k}(\overline{w}_{(d-1)k/2}))\neq 0$ and $\overline{w}_{(d-1)k/2}\neq 0$. □

Remark 5.8 In general, without analyzing the dyadic presentation of $d-1$ in more detail, we cannot give a better result. For example, in the case $k = 2^2 = 4$ and $d = 3$

we have

$$\begin{aligned}\pi(\rho_{3,4}^*(\overline{w})) &= (1+\pi(D_{2,1})+\pi(D_{2,0}))(1+\pi(D_{2,1})^2+\pi(D_{2,0})^2)\\ &= (1+V_1^2+V_2+V_1V_2)(1+V_1^4+V_2^2+V_1^2V_2^2)\\ &= (1+V_1^2+V_2+V_1V_2)(1+V_2^2+V_1^2V_2^2)\\ &= 1+(V_1^2+V_2)+V_1V_2+V_2^2 \in \mathbb{F}_2[V_1,V_2]/\langle V_1^3,V_2^3\rangle.\end{aligned}$$

Hence, $\pi(\rho_{3,4}^*(\overline{w}_4)) = V_2^2 \neq 0$ and $\pi(\rho_{3,4}^*(\overline{w}_i)) = 0$ for all $i > k(d-1)/2 = 4$.

Remark 5.9 On the other hand a better result can be obtained, as the following example will show. Let $k = 2^2 = 4$ and $d = 6$. We compute

$$\begin{aligned}\pi(\rho_{6,4}^*(\overline{w})) &= (1+\pi(D_{2,1})+\pi(D_{2,0}))(1+\pi(D_{2,1})^2+\pi(D_{2,0})^2)\\ &\quad (1+\pi(D_{2,1})^4+\pi(D_{2,0})^4)\\ &= (1+V_1^2+V_2+V_1V_2)(1+V_1^4+V_2^2+V_1^2V_2^2)\\ &\quad (1+V_1^8+V_2^4+V_1^4V_2^4)\\ &= 1+(V_1^2+V_2)+V_1V_2+(V_1^4+V_2^2)+(V_1^4V_2+V_2^3)+\\ &\quad (V_1V_2^3+V_1^5V_2)+(V_1^2V_2^3+V_1^4V_2^2+V_2^4)+V_1^3V_2^3+(V_1^2V_2^4+V_2^5)+\\ &\quad V_1V_2^5 \in \mathbb{F}_2[V_1,V_2]/\langle V_1^6,V_2^6\rangle.\end{aligned}$$

Thus, $\pi(\rho_{3,4}^*(\overline{w}_{10})) = V_1^2V_2^4+V_2^5 \neq 0$, as Lemma 5.7 predicts, but actually more is true $\pi(\rho_{3,4}^*(\overline{w}_{11})) = V_1V_2^5 \neq 0$. A natural question arises: Can Lemma 5.7 be improved?

Remark 5.10 The non-vanishing of the dual Stiefel–Whitney class $\overline{w}_{(d-1)k/2}(\xi_{\mathbb{R}^d,k})$ in the case when $d \geq 3$ is not a power of 2 and $k = 2^m$ which was established in Lemma 5.7 can be also obtain as follows. Consider the embedding

$$\underbrace{\mathrm{F}(\mathbb{R}^d,2)\times\cdots\times\mathrm{F}(\mathbb{R}^d,2)}_{2^{m-1}} \longrightarrow \mathrm{F}(\mathbb{R}^d,2^m)$$

induced by the fixed embeddings $e_i : \mathbb{R}^d \longrightarrow \mathbb{R}^d$ for $1 \leq i \leq 2^{m-1}$ such that their images are pairwise disjoint open d-balls. It induces a continuous map

$$\underbrace{\mathrm{F}(\mathbb{R}^d,2)/\mathbb{Z}_2\times\cdots\times\mathrm{F}(\mathbb{R}^d,2)/\mathbb{Z}_2}_{2^{m-1}} \longrightarrow \mathrm{F}(\mathbb{R}^d,2^m)/\mathfrak{S}_{2^m}.$$

Then the pull-back of the vector bundle $\xi_{\mathbb{R}^d,k}$ is isomorphic to the product vector bundle $\prod_{i=1}^{2^{m-1}} \xi_{\mathbb{R}^d,2}$. Now, the embedding $S^{d-1} \longrightarrow \mathrm{F}(\mathbb{R}^d,2)$ given by $x \longmapsto$

$(x, -x)$ induces a continuous map $\mathbb{RP}^{d-1} \longrightarrow \mathrm{F}(\mathbb{R}^d, 2)/\mathbb{Z}_2$ such that the pullback bundle of the vector bundle $\xi_{\mathbb{R}^d,2}$ is isomorphic to the Whitney sum of the tautological line vector bundle γ_1^{d-1} and a trivial line bundle. Thus, the pull-back of the vector bundle $\xi_{\mathbb{R}^d,k}$ along the composition

$$\underbrace{\mathbb{RP}^{d-1} \times \cdots \times \mathbb{RP}^{d-1}}_{2^{m-1}} \longrightarrow \underbrace{\mathrm{F}(\mathbb{R}^d, 2)/\mathbb{Z}_2 \times \cdots \times \mathrm{F}(\mathbb{R}^d, 2)/\mathbb{Z}_2}_{2^{m-1}} \longrightarrow \mathrm{F}(\mathbb{R}^d, 2^m)/\mathfrak{S}_{2^m}$$

is the product vector bundle $\gamma := \prod_{i=1}^{2^{m-1}} (\gamma_1^{d-1} \oplus \tau_1^{d-1})$. Here τ_1^{d-1} denotes the trivial line vector bundle over the real projective space $\mathbb{RP}^{d-1}$. Now we compute

$$\overline{w}(\gamma) = \overline{w}\Big(\prod_{i=1}^{2^{m-1}} (\gamma_1^{d-1} \oplus \tau_1^{d-1})\Big) = \bigtimes_{i=1}^{2^{m-1}} \overline{w}(\gamma_1^{d-1} \oplus \tau_1^{d-1}) = \bigtimes_{i=1}^{2^{m-1}} \overline{w}(\gamma_1^{d-1}).$$

Here "$\times$" denotes the cross product in cohomology. Since $w(\gamma_1^{d-1}) = 1 + u$, where $u \in H^*(\mathbb{RP}^{d-1}; \mathbb{F}_2) = \mathbb{F}_2[u]/\langle u^d \rangle$, we have that $\overline{w}(\gamma_1^{d-1}) = w(\gamma_1^{d-1})^{-1} = 1 + u + \cdots + u^{d-1}$. Therefore,

$$\overline{w}_{(d-1)2^{m-1}}(\gamma) = \overline{w}_{(d-1)k/2}(\gamma) = \underbrace{u \times \cdots \times u}_{2^{m-1}} \neq 0.$$

Having this simpler argument in mind it is natural to ask why we chose to present the more involved proof of the same fact. The reason lies in the method of the proof of Lemma 5.7 which is used once again for the proof of the next lemma.

Now, using a similar line of arguments as in the proof of Lemma 5.7 and taking into account the dyadic presentation of $d-1$ we get the following particular result.

Lemma 5.11 *Let $d \geq 6$ be an even integer such that for some $a \geq 3$ holds $2^{a-1} + 1 \leq d \leq 2^a - 1$, and let $k = 2^m$ for some integer $m \geq 1$. Then the dual Stiefel–Whitney class*

$$\overline{w}_{dk/2-1}(\xi_{\mathbb{R}^d,k}) = \overline{w}_{d2^{m-1}-1}(\xi_{\mathbb{R}^d,2^m})$$

does not vanish.

Proof Let us assume that $d-1 = 2^{a_1} + \cdots + 2^{a_q}$ where $0 = a_1 < \cdots < a_q = a-1$. Following the steps of the proof of Lemma 5.7 we reach again the equality:

$$\begin{aligned}\pi(\rho_{d,2^m}^*(\overline{w})) &= \prod_{i=0}^{a-1} \big(1 + \pi(D_{m,m-1})^{2^i} + \cdots + \pi(D_{m,0})^{2^i}\big) \qquad (5.9)\\ &= \prod_{i=0}^{a-1} \Big(1 + (V_1^{2^{m-1}} + V_2^{2^{m-2}} + \cdots + V_m^{2^0})^{2^i} + \cdots +\end{aligned}$$

$$\Big(\sum_{1\le j_1<\cdots<j_r\le m}(V_1\cdots V_{j_1-1})^{2^r}(V_{j_1+1}\cdots V_{j_2-1})^{2^{r-1}}\cdots(V_{j_r+1}\cdots V_m)^{2^0}\Big)^{2^i}$$

$$+\cdots+$$

$$(V_1\cdots V_m)^{2^i}\Big)$$

$$=\prod_{i=0}^{a-1}\Big(1+(V_1^{2^{i+m-1}}+V_2^{2^{i+m-2}}+\cdots+V_m^{2^i})+\cdots+$$

$$\Big(\sum_{1\le j_1<\cdots<j_r\le m}(V_1\cdots V_{j_1-1})^{2^{i+r}}(V_{j_1+1}\cdots V_{j_2-1})^{2^{i+r-1}}\cdots(V_{j_r+1}\cdots V_m)^{2^i}\Big)$$

$$+\cdots+$$

$$(V_1\cdots V_m)^{2^i}\Big).$$

Only now we show that

$$\begin{aligned}&\pi(\rho^*_{d,2^m}(\overline{w}_{d2^{m-1}-1}))\\ &=V_1\cdots V_{m-1}V_m^{d-1}+S\ \in\ \mathbb{F}_2[V_1,\ldots,V_m]/\langle V_1^d,\ldots,V_m^d\rangle,\end{aligned}$$

where S is a sum of monomials in $V_1,\ldots,V_m$ of degree $d2^{m-1}-1$ which are different from $V_1\cdots V_{m-1}V_m^{d-1}$. Hence, $\pi(\rho^*_{d,k}(\overline{w}_{d2^{m-1}-1}))\neq 0$, and consequently $\overline{w}_{d2^{m-1}-1}\neq 0$.

Indeed, observe that in every monomial of the ith factor of the product (5.9) which has the variable V_m with positive exponent this exponent is always the same and equal to 2^i, $0\le i\le a-1$. In particular, in the ith factor each monomial with the variable V_m is of the form $p_i(V_1,\ldots,V_{m-1})^{2^i}V_m^{2^i}$ where $p_i(V_1,\ldots,V_{m-1})$ is a monomial in variables $V_1,\ldots,V_{m-1}$. Now, when multiplying out the product (5.9) the monomial of the form $p(V_1,\ldots,V_{m-1})V_m^{d-1}$ can appear in the final result if and only if we take non-zero

- monomials $p_i(V_1,\ldots,V_{m-1})^{2^i}V_m^{2^i}$ from ith factors where $i\in\{a_1,\ldots,a_q\}$, and
- monomials $p'_j(V_1,\ldots,V_{m-1})$ from jth factors for $j\in\{0,\ldots,a-1\}\setminus\{a_1,\ldots,a_q\}$.

Thus, we have

$$p(V_1,\ldots,V_{m-1})V_m^{d-1}=\prod_{i\in\{a_1,\ldots,a_q\}}p_i(V_1,\ldots,V_{m-1})^{2^i}V_m^{2^i}\cdot\prod_{j\in\{0,\ldots,a-1\}\setminus\{a_1,\ldots,a_q\}}p'_j(V_1,\ldots,V_{m-1}).$$

Observe that, if $p_i(V_1, \ldots, V_{m-1}) \neq 1$ for some $i \in \{a_1, \ldots, a_q\}$, then there exists $1 \leq t \leq m-1$ such that $V_t \mid p_i(V_1, \ldots, V_{m-1})$. Hence, $V_t^{2^i} \mid p_i(V_1, \ldots, V_{m-1})^{2^i}$.

Now we want to understand in how many ways we can obtain the monomial $V_1 \ldots V_{m-1} V_m^{d-1}$ when we multiply out the product (5.9). This means that we need to find all possible p_is and p'_j's such that

$$V_1 \cdots V_{m-1} V_m^{d-1} = \prod_{i \in \{a_1, \ldots, a_q\}} p_i(V_1, \ldots, V_{m-1})^{2^i} V_m^{2^i} \cdot \prod_{j \in \{0, \ldots, a-1\} \setminus \{a_1, \ldots, a_q\}} p'_j(V_1, \ldots, V_{m-1}).$$

From the previous observation and the fact that $0 = a_1 < a_2 < \cdots < a_q = a-1$ we conclude that $p_{a_2}(V_1, \ldots, V_{m-1}) = \cdots = p_{a_q}(V_1, \ldots, V_{m-1}) = 1$. Thus, the previous equality becomes

$$V_1 \cdots V_{m-1} V_m^{d-1} = p_{a_1}(V_1, \ldots, V_{m-1})^{2^{a_1}} V_m^{d-1} \cdot \prod_{j \in \{0, \ldots, a-1\} \setminus \{a_1, \ldots, a_q\}} p'_j(V_1, \ldots, V_{m-1}).$$

Taking additionally into account that $a_1 = 0$ we have that

$$V_1 \cdots V_{m-1} V_m^{d-1} = p_0(V_1, \ldots, V_{m-1}) V_m^{d-1} \cdot \prod_{j \in \{1, \ldots, a-1\} \setminus \{a_2, \ldots, a_q\}} p'_j(V_1, \ldots, V_{m-1}).$$

Therefore, the monomial $V_1 \cdots V_{m-1} V_m^{d-1}$ can be obtained only in the case when we choose

$$\begin{aligned} p_0(V_1, \ldots, V_{m-1}) &= V_1 \cdots V_{m-1}, \\ p_i(V_1, \ldots, V_{m-1}) &= 1, \\ p'_j(V_1, \ldots, V_{m-1}) &= 1, \end{aligned}$$

for all $i \in \{a_2, \ldots, a_q\}$ and all $j \in \{1, \ldots, a-1\} \setminus \{a_2, \ldots, a_q\}$.

Indeed, if we assume that $p_{a_1}(V_1, \ldots, V_{m-1}) = p_0(V_1, \ldots, V_{m-1}) \neq 1$, then obviously we must have $p_0(V_1, \ldots, V_{m-1}) = V_1 \cdots V_{m-1}$. This is possible by taking monomial $V_1 \ldots V_{m-1} V_m$ in the a_0th factor of the product (5.9). On the other hand, if we assume that $p_{a_1}(V_1, \ldots, V_{m-1}) = p_0(V_1, \ldots, V_{m-1}) = 1$ then we should have that

$$\prod_{j \in \{1, \ldots, a-1\} \setminus \{a_2, \ldots, a_q\}} p'_j(V_1, \ldots, V_{m-1}) = V_1 \cdots V_{m-1}.$$

This is not possible since for every $p'_j(V_1, \ldots, V_{m-1}) \neq 1$ in the previous product there is $1 \leq t \leq m-1$ such that $V_t \mid p'_j(V_1, \ldots, V_{m-1})$. Looking closer at the typical monomials in the product (5.9) we see that more is true, actually $V_t^{2^j} \mid p'_j(V_1, \ldots, V_{m-1})$ and so

$$V_t^{2^j} \mid \prod_{j\in\{1,\ldots,a-1\}\backslash\{a_2,\ldots,a_q\}} p'_j(V_1, \ldots, V_{m-1}) = V_1 \cdots V_{m-1}.$$

Since $j \in \{1, \ldots, a-1\}\backslash\{a_2, \ldots, a_q\}$ we have that $2^j \geq 2$ and so $V_t^2 \mid V_1 \cdots V_{m-1}$; contradiction.

Hence, we showed that

$$\pi(\rho^*_{d,2^m}(\overline{w}_{d2^{m-1}-1})) = V_1 \cdots V_{m-1}V_m^{d-1} + S \in \mathbb{F}_2[V_1, \ldots, V_m]/\langle V_1^d, \ldots, V_m^d\rangle,$$

where S is a sum of monomials in $V_1, \ldots, V_m$ of degree $d2^{m-1} - 1$ which are different from $V_1 \cdots V_{m-1}V_m^{d-1}$. Thus, $\overline{w}_{d2^{m-1}-1} \neq 0$. □

Now we extend the previous two lemmas to the case of an arbitrary integer $k \geq 1$ and consequently correct [15, Lem. 2.17]. For an integer $k \geq 1$ let $\epsilon(k)$ be the reminder of k modulo 2, that is $\epsilon(k) = 1$ for k odd, and $\epsilon(k) = 0$ when k is even.

Lemma 5.12

(1) *Let $d \geq 3$ be an integer which is not a power of 2, and let $k \geq 1$ be an integer. Then the dual Stiefel–Whitney class*

$$\overline{w}_{(d-1)(k-\epsilon(k))/2}(\xi_{\mathbb{R}^d,k})$$

does not vanish.

(2) *Let $d \geq 6$ be an even integer which is not a power of 2, and let $k \geq 1$ be an integer. Then the dual Stiefel–Whitney class*

$$\overline{w}_{d(k-\epsilon(k))/2-\alpha(k)+\epsilon(k)}(\xi_{\mathbb{R}^d,k})$$

does not vanish.

Proof Let $r := \alpha(k)$ and $k = 2^{b_1} + \cdots + 2^{b_r}$ where $0 \leq b_1 < b_2 < \cdots < b_r$. As in the proof of Lemma 5.6 we consider a morphism between vector bundles $\prod_{i=1}^r \xi_{\mathbb{R}^d,2^{b_i}}$ and $\xi_{\mathbb{R}^d,k}$ where the following commutative square is a pullback diagram:

$$\begin{array}{ccc} \prod_{i=1}^r \xi_{\mathbb{R}^d,2^{b_i}} & \xrightarrow{\Theta} & \xi_{\mathbb{R}^d,k} \\ \downarrow & & \downarrow \\ \prod_{i=1}^r \mathrm{F}(\mathbb{R}^d, 2^{b_i})/\mathfrak{S}_{2^{b_i}} & \xrightarrow{\theta} & \mathrm{F}(\mathbb{R}^d, k)/\mathfrak{S}_k. \end{array}$$

The map θ is induced, up to an equivariant homotopy, from a restriction of the little cubes operad structural map

$$(C_d(2^{b_1}) \times \cdots \times C_d(2^{b_r})) \times C_d(r) \longrightarrow C_d(2^{b_1} + \cdots + 2^{b_r}),$$

as explained in the proof of Lemma 5.6. The naturality of the Stiefel–Whitney classes [80, Ax. 2, p. 37] implies that

$$\theta^*(\overline{w}_N(\xi_{\mathbb{R}^d,k})) = \overline{w}_N\Big(\prod_{i=1}^{r} \xi_{\mathbb{R}^d,2^{b_i}}\Big),$$

for any integer $N \geq 0$. Further on, the product formula [80, Pr. 4-A, p. 54] implies that

$$\overline{w}\Big(\prod_{i=1}^{r} \xi_{\mathbb{R}^d,2^{b_i}}\Big) = \overline{w}(\xi_{\mathbb{R}^d,2^{b_1}}) \times \cdots \times \overline{w}(\xi_{\mathbb{R}^d,2^{b_r}}).$$

Consequently,

$$\begin{aligned} \theta^*(\overline{w}_N(\xi_{\mathbb{R}^d,k})) &= \overline{w}_N\Big(\prod_{i=1}^{r} \xi_{\mathbb{R}^d,2^{b_i}}\Big) \\ &= \sum_{s_1+\cdots+s_r=N} \overline{w}_{s_1}(\xi_{\mathbb{R}^d,2^{b_1}}) \times \cdots \times \overline{w}_{s_r}(\xi_{\mathbb{R}^d,2^{b_r}}). \qquad (5.10) \end{aligned}$$

According to the Künneth formula [23, Thm. VI.3.2] each term, the cross product, $\overline{w}_{s_1}(\xi_{\mathbb{R}^d,2^{b_1}}) \times \cdots \times \overline{w}_{s_r}(\xi_{\mathbb{R}^d,2^{b_r}})$ in the previous sum belongs to a different direct summand of the cohomology

$$\begin{aligned} &H^N\Big(\prod_{i=1}^{r} \mathrm{F}(\mathbb{R}^d, 2^{b_i})/\mathfrak{S}_{2^{b_i}}; \mathbb{F}_2\Big) \\ &\cong \bigoplus_{s_1+\cdots+s_r=N} H^{s_1}(\mathrm{F}(\mathbb{R}^d, 2^{b_1})/\mathfrak{S}_{2^{b_1}}; \mathbb{F}_2) \otimes \cdots \otimes H^{s_r}(\mathrm{F}(\mathbb{R}^d, 2^{b_r})/\mathfrak{S}_{2^{b_r}}; \mathbb{F}_2). \end{aligned}$$

Hence the following equivalence holds

$$\begin{aligned} &\overline{w}_N\Big(\prod_{i=1}^{r} \xi_{\mathbb{R}^d,2^{b_i}}\Big) \neq 0 \\ &\Longleftrightarrow \overline{w}_{s_1}(\xi_{\mathbb{R}^d,2^{b_1}}) \times \cdots \times \overline{w}_{s_r}(\xi_{\mathbb{R}^d,2^{b_r}}) \neq 0 \text{ for some } s_1 + \cdots + s_r = N. \end{aligned}$$

To isolate a non-zero summand in (5.10) we use either Lemma 5.7 or Lemma 5.11.

Let $d \geq 3$ be an integer which is not a power of 2, and let $N = (d-1)(k-\epsilon(k))/2$. The Lemma 5.7 states that the dual Stiefel–Whitney class $\overline{w}_{(d-1)2^{b_i-1}}(\xi_{\mathbb{R}^d,2^{b_i}})$ does not vanish when $b_i \geq 1$. First consider the case when k is even. Hence, $\epsilon(k) = 0$, and since $k = 2^{b_1} + \cdots + 2^{b_r}$ we have that $1 \leq b_1 < b_2 < \cdots < b_r$. Now, since $N = (d-1)2^{b_1-1} + \cdots + (d-1)2^{b_r-1}$ the following summand in (5.10) does not vanish

$$\overline{w}_{(d-1)2^{b_1-1}}(\xi_{\mathbb{R}^d,2^{b_1}}) \times \cdots \times \overline{w}_{(d-1)2^{b_r-1}}(\xi_{\mathbb{R}^d,2^{b_r}}) \neq 0.$$

In the case when $\epsilon(k) = 1$, $k = 2^{b_1} + \cdots + 2^{b_r}$ and $0 = b_1 < b_2 < \cdots < b_r$, the summand in (5.10) that does not vanish is

$$\overline{w}_0(\xi_{\mathbb{R}^d,2^{b_1}}) \times \overline{w}_{(d-1)2^{b_2-1}}(\xi_{\mathbb{R}^d,2^{b_2}}) \times \cdots \times \overline{w}_{(d-1)2^{b_r-1}}(\xi_{\mathbb{R}^d,2^{b_r}}) \neq 0.$$

Consequently, $\theta^*(\overline{w}_{(d-1)(k-\epsilon(k))/2}(\xi_{\mathbb{R}^d,k})) \neq 0$, and so $\overline{w}_{(d-1)(k-\epsilon(k))/2}(\xi_{\mathbb{R}^d,k}) \neq 0$.

Let now $d \geq 6$ be an even integer which is not a power of 2, and let $N = d(k-\epsilon(k))/2 - \alpha(k) + \epsilon(k)$. The Lemma 5.11 states that the dual Stiefel–Whitney class $\overline{w}_{d2^{b_i-1}-1}(\xi_{\mathbb{R}^d,2^{b_i}})$ does not vanish when $b_i \geq 1$. Again we first consider the case when k is even, that is $\epsilon(k) = 0$. Since $k = 2^{b_1} + \cdots + 2^{b_r}$ we have that $1 \leq b_1 < b_2 < \cdots < b_r$. Now, since $N = (d2^{b_1-1} - 1) + \cdots + (d2^{b_r-1} - 1)$ the following summand in (5.10) does not vanish

$$\overline{w}_{d2^{b_1-1}-1}(\xi_{\mathbb{R}^d,2^{b_1}}) \times \cdots \times \overline{w}_{d2^{b_r-1}-1}(\xi_{\mathbb{R}^d,2^{b_r}}) \neq 0.$$

In the case of odd k we have that $\epsilon(k) = 1$, $k = 2^{b_1} + \cdots + 2^{b_r}$ and $0 = b_1 < b_2 < \cdots < b_r$. The summand in (5.10) that does not vanish is

$$\overline{w}_0(\xi_{\mathbb{R}^d,2^{b_1}}) \times \overline{w}_{d2^{b_2-1}-1}(\xi_{\mathbb{R}^d,2^{b_2}}) \times \cdots \times \overline{w}_{d2^{b_r-1}-1}(\xi_{\mathbb{R}^d,2^{b_r}}) \neq 0.$$

Consequently, $\theta^*(\overline{w}_{d(k-\epsilon(k))/2-\alpha(k)+\epsilon(k)}(\xi_{\mathbb{R}^d,k})) \neq 0$, and so the dual Stiefel–Whiney class $\overline{w}_{d(k-\epsilon(k))/2-\alpha(k)+\epsilon(k)}(\xi_{\mathbb{R}^d,k})$ does not vanish.

Summarizing the results obtained in Lemmas 5.5–5.7, 5.11 and 5.12 we get the following theorem which is a correction of [15, Thm. 2.13].

Theorem 5.13 *Let $k \geq 1$ and $d \geq 1$ be integers.*

(1) *If d is a power of 2, then the dual Stiefel–Whitney class*

$$\overline{w}_{(d-1)(k-\alpha(k))}(\xi_{\mathbb{R}^d,k})$$

does not vanish.

(2) *If d is not a power of 2, then the dual Stiefel–Whitney class*

$$\overline{w}_{(d-1)(k-\epsilon(k))/2}(\xi_{\mathbb{R}^d,k})$$

does not vanish.

(3) *If d is an even integer which is not a power of 2, then the dual Stiefel–Whitney class*

$$\overline{w}_{d(k-\epsilon(k))/2-\alpha(k)+\epsilon(k)}(\xi_{\mathbb{R}^d,k})$$

does not vanish.

Now, we use Theorem 5.13 and the criterion given in Lemma 5.4 to correct the result stated in [15, Thm. 2.1]. In this way we completed corrections of the invalid results in [15, Sec. 2].

Theorem 5.14 *Let $k \geq 1$ and $d \geq 1$ be integers. Denote by $\alpha(k)$ the number of 1s in the dyadic presentation of k, and by $\epsilon(k)$ the remainder of k modulo 2.*

(1) *If d is a power of 2, then there is no k-regular embedding $\mathbb{R}^d \longrightarrow \mathbb{R}^N$ for*

$$N \leq d(k - \alpha(k)) + \alpha(k) - 1.$$

(2) *If d is not a power of 2, then there is no k-regular embedding $\mathbb{R}^d \longrightarrow \mathbb{R}^N$ for*

$$N \leq \frac{1}{2}(d-1)(k - \epsilon(k)) + k - 1.$$

(3) *If d is an even integer which is not a power of 2, then is no k-regular embedding $\mathbb{R}^d \longrightarrow \mathbb{R}^N$ for*

$$N \leq \frac{1}{2}d(k - \epsilon(k)) + k - \alpha(k) + \epsilon(k) - 1.$$

5.2 ℓ-Skew Embeddings

In this section we revise [15, Sec. 3] and correct the related results [15, Thm. 3.1, Thm. 3.7]. In order for our presentation to be complete we recall basic definitions and necessary facts.

The affine subspaces $L_1, \ldots, L_\ell$ of the Euclidean space $\mathbb{R}^N$ are **affinely independent** if

$$\dim_{\text{aff}}(L_1 \cup \cdots \cup L_\ell) = (\dim_{\text{aff}} L_1 + 1) + \cdots + (\dim_{\text{aff}} L_\ell + 1) - 1.$$

In particular, any two lines in $\mathbb{R}^3$ are skew if and only if they are affinely independent.

Let M be a real smooth d-dimensional manifold. Then TM denotes he tangent bundle of M, and T_yM stands for the tangent space of M at the point $y \in M$. For a smooth map $f: M \longrightarrow \mathbb{R}^N$ we denote by $df: TM \longrightarrow T\mathbb{R}^N$ the differential map between associated tangent vector bundles induced by f. Further on, let

$\iota\colon T\mathbb{R}^N \longrightarrow \mathbb{R}^N$ denotes the map that sends a tangent vector $v \in T_x\mathbb{R}^N$ at the point $x \in \mathbb{R}^N$ to the point $x + v$. Here the standard identification $T_x\mathbb{R}^N = \mathbb{R}^N$ is assumed.

Definition 5.15 Let $\ell \geq 1$ be an integer, and let M be a real smooth d-dimensional manifold. A smooth embedding $f\colon M \longrightarrow \mathbb{R}^N$ is an **ℓ-skew embedding** if for every point $(y_1, \ldots, y_\ell) \in \mathrm{F}(M, \ell)$ the affine subspaces

$$(\iota \circ df_{y_1})(T_{y_1}M), \ldots, (\iota \circ df_{y_\ell})(T_{y_\ell}M)$$

of $\mathbb{R}^N$ are affinely independent.

Now, like in the case of k-regular embeddings, a criterion for non-existence of ℓ-skew embedding can be derived in terms of Stiefel–Whitney class of appropriate vector bundle over the configuration space. We recall [15, Lem. 3.6].

Lemma 5.16 *Let $d \geq 1$ and $\ell \geq 1$ be integers. If the dual Stiefel–Whitney class*

$$\overline{w}_{N-(d+1)\ell+2}(\xi_{\mathbb{R}^d,\ell}^{\oplus(d+1)})$$

does not vanish, then there is no ℓ-skew embedding $\mathbb{R}^d \longrightarrow \mathbb{R}^N$.

Motivated by the criterion in Lemma 5.16, based on the work of Hư'ng [64, (4.7)], a relevant study of Stiefel–Whitney classes of the vector bundle $\xi_{\mathbb{R}^d,\ell}^{\oplus(d+1)}$ was given in [15, Thm. 3.7].

> **Theorem 3.7.** Let $d, \ell \geq 1$ be integers. Then the dual Stiefel–Whitney class
>
> $$\overline{w}_{(2^{\gamma(d)}-d-1)(\ell-\alpha(\ell))}(\xi_{\mathbb{R}^d,\ell}^{\oplus(d+1)})$$
>
> does not vanish.

Here $\gamma(d) = \lfloor \log_2 d \rfloor + 1$ for $d \geq 1$. The result of the previous theorem in combination with [15, Lem. 3.6] directly implied the following result [15, Thm. 3.1].

> **Theorem 3.1.** Let $\ell, d \geq 2$ be integers. There is no ℓ-skew embedding $\mathbb{R}^d \longrightarrow \mathbb{R}^N$ for
>
> $$N \leq 2^{\gamma(d)}(\ell - \alpha(\ell)) + (d+1)\alpha(\ell) - 2,$$
>
> where $\alpha(\ell)$ denotes the number of ones in the dyadic presentation of l and $\gamma(d) = \lfloor \log_2 d \rfloor + 1$.

The proof of [15, Thm. 3.7] was based on some of the results presented in [15, Sec. 2]. In particular, in [15, Sec. 3.3.3] an incorrect result was used [15, Cor. 2.16]. Consequently, [15, Thm. 3.1] does not stand. In the following we give correct versions, first of [15, Thm. 3.7], and then of [15, Thm. 3.1].

Theorem 5.17 *Let $d \geq 1$ and $\ell \geq 2$ be integers.*

(1) *If $d = 2$, and if $\ell \geq 2$ is an integer, then*

$$\overline{w}_{\ell-\alpha(\ell)}\big(\xi_{\mathbb{R}^d,\ell}^{\oplus(d+1)}\big) = \overline{w}_{\ell-\alpha(\ell)}\big(\xi_{\mathbb{R}^2,\ell}^{\oplus 3}\big) \neq 0.$$

(2) *If $d \geq 1$ is an integer, and if $\ell = 2$, then*

$$\overline{w}_{2^{\gamma(d)}-d-1}\big(\xi_{\mathbb{R}^d,\ell}^{\oplus(d+1)}\big) = \overline{w}_{2^{\gamma(d)}-d-1}\big(\xi_{\mathbb{R}^d,2}^{\oplus(d+1)}\big) \neq 0.$$

(3) *If $d \geq 2$ is a power of 2, and if $\ell \geq 2$ is an integer, then*

$$\overline{w}_{(d-1)(\ell-\alpha(\ell))}\big(\xi_{\mathbb{R}^d,\ell}^{\oplus(d+1)}\big) \neq 0.$$

(4) *If $d + 1 \geq 2$ is a power of 2, and if $\ell \geq 2$ is an integer, then*

$$\overline{w}\big(\xi_{\mathbb{R}^d,\ell}^{\oplus(d+1)}\big) = w\big(\xi_{\mathbb{R}^d,\ell}^{\oplus(d+1)}\big) = 1.$$

(5) *If $d \geq 5$ is an integer which is not a power of 2, and in addition $d + 1$ is not a power of 2, and if $\ell \geq 3$ is an integer, then*

$$\overline{w}_{(2^{\gamma(d)}-d-1)(\ell-\epsilon(\ell))/2}\big(\xi_{\mathbb{R}^d,\ell}^{\oplus(d+1)}\big) \neq 0.$$

(6) *If $d \geq 5$ is an integer which is not a power of 2, $2^{\gamma(d)} - d - 1 = 2^{a_1}z$ where $a_1 \geq 0$ is an integer and $z \geq 1$ is an odd integer, and $d + 1$ is not a power of 2, and if $\ell \geq 3$ is an integer, then*

$$\overline{w}_{(2^{\gamma(d)}-d-1+2^{a_1})(\ell-\epsilon(\ell))/2-2^{a_1}\alpha(\ell)}\big(\xi_{\mathbb{R}^d,\ell}^{\oplus(d+1)}\big) \neq 0.$$

Proof We prove the theorem by discussing all cases separately.

(1) Let $d = 2$, and let $\ell \geq 2$ be an integer. Then from [38, Thm. 1] we get that the bundle $\xi_{\mathbb{R}^2,\ell}^{\oplus 2}$ is a trivial bundle. Consequently, $\overline{w}\big(\xi_{\mathbb{R}^2,\ell}^{\oplus 3}\big) = \overline{w}\big(\xi_{\mathbb{R}^2,\ell}\big)$. Since $d = 2$ is power of 2 we can use Lemma 5.6 to get that

$$\overline{w}_{\ell-\alpha(\ell)}\big(\xi_{\mathbb{R}^d,\ell}^{\oplus(d+1)}\big) = \overline{w}_{\ell-\alpha(\ell)}\big(\xi_{\mathbb{R}^2,\ell}^{\oplus 3}\big) = \overline{w}_{\ell-\alpha(\ell)}\big(\xi_{\mathbb{R}^2,\ell}\big) \neq 0.$$

(2) Let $d \geq 2$ be an integer and let $\ell = 2$. In this case the base space of the vector bundle $\xi_{\mathbb{R}^2,\ell}$ is the unordered configuration space $\mathrm{F}(\mathbb{R}^d, 2)/\mathfrak{S}_2$. The $\mathfrak{S}_2$-equivariant map $\mathrm{ecy}_{d,2}\colon S^{d-1} \longrightarrow \mathrm{F}(\mathbb{R}^d, 2)$ given by $x \longmapsto (x, -x)$ is an $\mathfrak{S}_2$-equivariant homotopy equivalence. Consequently, $\mathrm{ecy}_{d,2}$ induces homotopy equivalence $\rho_{d,2}\colon S^{d-1}/\mathfrak{S}_2 \longrightarrow \mathrm{F}(\mathbb{R}^d, 2)/\mathfrak{S}_2$. Recall that $S^{d-1}/\mathfrak{S}_2 \cong \mathbb{R}\mathrm{P}^{d-1}$. It can be directly checked that the vector bundle $\xi_{\mathbb{R}^d,2}$

over $F(\mathbb{R}^d, 2)/\mathfrak{S}_2$ pulls back to the vector bundle isomorphic to the Whitney sum of the tautological line bundle and trivial line bundle over the projective space $\mathbb{R}P^{d-1}$. Hence, if we denote the cohomology of the projective space by

$$H^*(\mathbb{R}P^{d-1}, \mathbb{F}_2) \cong H^*(F(\mathbb{R}^d, 2)/\mathfrak{S}_2, \mathbb{F}_2) = \mathbb{F}_2[w_1]/\langle w_1^d \rangle,$$

where $\deg(w_1) = 1$, we have that

$$w(\xi_{\mathbb{R}^d,2}) = 1 + w_1.$$

Now, the total Stiefel–Whitney class of the vector bundle $\xi_{\mathbb{R}^d,\ell}^{\oplus(d+1)}$ is

$$w\big(\xi_{\mathbb{R}^d,\ell}^{\oplus(d+1)}\big) = w\big(\xi_{\mathbb{R}^d,2}^{\oplus(d+1)}\big) = (1 + w_1)^{d+1}.$$

Now notice that $2^{\gamma(d)}$ is the minimal power of 2 that is greater that d. Therefore,

$$\begin{aligned} w\big(\xi_{\mathbb{R}^d,2}^{\oplus(d+1)}\big)(1 + w_1)^{2^{\gamma(d)}-d-1} &= (1 + w_1)^{d+1}(1 + w_1)^{2^{\gamma(d)}-d-1} \\ &= (1 + w_1)^{2^{\gamma(d)}} \\ &= 1. \end{aligned}$$

Thus, we have that

$$\overline{w}\big(\xi_{\mathbb{R}^d,\ell}^{\oplus(d+1)}\big) = \overline{w}\big(\xi_{\mathbb{R}^d,2}^{\oplus(d+1)}\big) = (1 + w_1)^{2^{\gamma(d)}-d-1} = 1 + \cdots + w_1^{2^{\gamma(d)}-d-1}.$$

Sine $2^{\gamma(d)} - d - 1 < d$ we have that $w_1^{2^{\gamma(d)}-d-1} \neq 0$, and so $\overline{w}_{2^{\gamma(d)}-d-1}\big(\xi_{\mathbb{R}^d,2}^{\oplus(d+1)}\big) \neq 0$.

(3) Let $d = 2^a$ for $a \geq 1$ an integer, and let $\ell \geq 2$ be an integer. From the decomposition of vector bundles (5.3) we have that

$$w(\xi_{\mathbb{R}^d,\ell}) = w(\zeta_{\mathbb{R}^d,\ell}) = 1 + w_1 + \cdots + w_{\ell-1},$$

where $w_i := w_i(\xi_{\mathbb{R}^d,\ell}) = w_i(\zeta_{\mathbb{R}^d,\ell})$. Consequently,

$$\begin{aligned} w\big(\xi_{\mathbb{R}^d,\ell}^{\oplus(d+1)}\big) &= (1 + w_1 + \cdots + w_{\ell-1})^{d+1} \\ &= (1 + w_1 + \cdots + w_{\ell-1})^{2^a}(1 + w_1 + \cdots + w_{\ell-1}) \\ &= (1 + w_1^{2^a} + \cdots + w_{\ell-1}^{2^a})(1 + w_1 + \cdots + w_{\ell-1}) \\ &= 1 + w_1 + \cdots + w_{\ell-1} \\ &= w(\xi_{\mathbb{R}^d,\ell}). \end{aligned}$$

Here we used that from (5.4) we know that

$$\text{height}((H^*(\mathrm{F}(\mathbb{R}^d,\ell)/\mathfrak{S}_\ell;\mathbb{F}_2)) \le \min\{2^t : 2^t \ge d\} = 2^a = d.$$

Thus, with the additional help of Lemma 5.6, we get that

$$\overline{w}_{(d-1)(\ell-\alpha(\ell))}\big(\xi_{\mathbb{R}^d,\ell}^{\oplus(d+1)}\big) = \overline{w}_{(d-1)(\ell-\alpha(\ell))}(\xi_{\mathbb{R}^d,\ell}) \neq 0.$$

(4) Let $d = 2^a - 1$ for $a \ge 1$ an integer, and let $\ell \ge 2$ be an integer. In the footsteps of the proof of the previous case we calculate:

$$\begin{aligned} w\big(\xi_{\mathbb{R}^d,\ell}^{\oplus(d+1)}\big) &= (1 + w_1 + \cdots + w_{\ell-1})^{d+1} \\ &= (1 + w_1 + \cdots + w_{\ell-1})^{2^a} \\ &= 1 + w_1^{2^a} + \cdots + w_{\ell-1}^{2^a} \\ &= 1. \end{aligned}$$

In this case we used the fact that

$$\text{height}(H^*(\mathrm{F}(\mathbb{R}^d,\ell)/\mathfrak{S}_\ell;\mathbb{F}_2)) \le \min\{2^t : 2^t \ge d\} = 2^a = d+1.$$

Consequently, $\overline{w}\big(\xi_{\mathbb{R}^d,\ell}^{\oplus(d+1)}\big) = 1$.

(5) This case is analyzed in two separate steps depending whether ℓ is a power of 2 or not.

(5A) Let $d \ge 5$ be an integer such that $2^{a-1} + 1 \le d \le 2^a - 2$ where $a \ge 3$ is an integer. We first consider the case when $\ell = 2^m$ for $m \ge 2$ an integer. The decomposition of vector bundles (5.3) implies that

$$w(\xi_{\mathbb{R}^d,2^m}) = w(\zeta_{\mathbb{R}^d,2^m}) = 1 + w_1 + \cdots + w_{2^m-1},$$

and consequently,

$$w\big(\xi_{\mathbb{R}^d,2^m}^{\oplus(d+1)}\big) = (1 + w_1 + \cdots + w_{2^m-1})^{d+1}.$$

From (5.4) we have that

$$\text{height}(H^*(\mathrm{F}(\mathbb{R}^d,2^m)/\mathfrak{S}_{2^m};\mathbb{F}_2)) \le \min\{2^t : 2^t \ge d\} = 2^a.$$

Therefore,

$$\begin{aligned} w\big(\xi_{\mathbb{R}^d,2^m}^{\oplus(d+1)}\big)(1+w_1+\cdots+w_{2^m-1})^{2^a-d-1} &= (1+w_1+\cdots+w_{2^m-1})^{2^a} \\ &= 1+w_1^{2^a}+\cdots+w_{2^m-1}^{2^a} \\ &= 1, \end{aligned}$$

and consequently,

$$\overline{w}\big(\xi_{\mathbb{R}^d,2^m}^{\oplus(d+1)}\big) = (1+w_1+\cdots+w_{2^m-1})^{2^a-d-1}.$$

Now, let $2^a-d-1 = 2^{a_1}+\cdots+2^{a_q}$ where $0 \le a_1 < \cdots < a_q \le a-2$. Then

$$\begin{aligned} \overline{w}\big(\xi_{\mathbb{R}^d,2^m}^{\oplus(d+1)}\big) &= (1+w_1+\cdots+w_{2^m-1})^{2^{a_1}+\cdots+2^{a_q}} \\ &= \prod_{i=1}^{q}\big(1+w_1^{2^{a_i}}+\cdots+w_{2^m-1}^{2^{a_i}}\big). \end{aligned} \tag{5.11}$$

Following the calculation in the proof of Lemma 5.7 we apply the monomorphism

$$\rho_{d,2^m}^*\colon\ H^*(\mathrm{F}(\mathbb{R}^d,2^m)/\mathfrak{S}_{2^m};\mathbb{F}_2) \longrightarrow H^*(\mathrm{Pe}(\mathbb{R}^d,2^m)/\mathcal{S}_{2^m};\mathbb{F}_2)$$

from Theorem 4.1 to the equality (5.11), and use the decomposition of the cohomology

$$H^*(\mathrm{Pe}(\mathbb{R}^d,2^m)/\mathcal{S}_{2^m};\mathbb{F}_2) \cong \mathbb{F}_2[V_{m,1},\ldots,V_{m,m}]/\langle V_{m,1}^d,\ldots,V_{m,m}^d\rangle \oplus \mathrm{I}^*(\mathbb{R}^d,2^m),$$

given in Theorem 3.11, to get that

$$\begin{aligned} \rho_{d,2^m}^*\big(\overline{w}\big(\xi_{\mathbb{R}^d,2^m}^{\oplus(d+1)}\big)\big) &= \prod_{i=1}^{q}\big(1+D_{m,m-1}^{2^{a_i}}+\cdots+D_{m,0}^{2^{a_i}}\big)+R \\ &\in\ H^*(\mathrm{Pe}(\mathbb{R}^d,2^m)/\mathcal{S}_{2^m};\mathbb{F}_2), \end{aligned} \tag{5.12}$$

where

- $D_{m,r} = (\kappa_{d,2^m}/\mathcal{S}_{2^m})^*(D_{m,r}) = \rho_{d,2^m}^*(w_{2^m-2^r})$, for $0 \le r \le m-1$, with the obvious abuse of notation, see (3.28), and
- $R \in \mathrm{I}^*(\mathbb{R}^d,2^m)$.

Let π be the following composition of the maps from the diagram (5.5):

$$\begin{array}{c}\mathbb{F}_2[V_{m,1},\ldots,V_{m,m}]/\langle V_{m,1}^d,\ldots,V_{m,m}^d\rangle\oplus I^*(\mathbb{R}^d,2^m)\\ \Big\downarrow \text{projection}\\ \mathbb{F}_2[V_{m,1},\ldots,V_{m,m}]/\langle V_{m,1}^d,\ldots,V_{m,m}^d\rangle\\ \Big\downarrow \chi_m\\ \mathbb{F}_2[V_1,\ldots,V_m]/\langle V_1^d,\ldots,V_m^d\rangle.\end{array}$$

The first map is the projection on a direct summand and the second map is induced by the change of variables $\chi_m\in\mathrm{GL}_m(\mathbb{F}_2)$. We apply π to (5.12) and have

$$\pi\big(\rho^*_{d,2^m}\big(\overline{w}\big(\xi_{\mathbb{R}^d,2^m}^{\oplus(d+1)}\big)\big)\big)=\prod_{i=1}^{q}\big(1+\pi(D_{m,m-1})^{2^i}+\cdots+\pi(D_{m,0})^{2^i}\big).$$

The change of the variables χ_m transforms $\mathrm{L}_m(\mathbb{F}_2)$-invariants into $\mathrm{U}_m(\mathbb{F}_2)$-invariants and Dickson polynomials, $\mathrm{GL}_m(\mathbb{F}_2)$-invariants, can be presented in terms of $\mathrm{U}_m(\mathbb{F}_2)$-invariants, as explained in (8.3) and (8.4). Hence, $p:=\pi\big(\rho^*_{d,2^m}\big(\overline{w}\big(\xi_{\mathbb{R}^d,2^m}^{\oplus(d+1)}\big)\big)\big)$ can be computed further as follows:

$$\begin{aligned}p&=\prod_{i=1}^{q}\big(1+\pi(D_{m,m-1})^{2^{a_i}}+\cdots+\pi(D_{m,0})^{2^{a_i}}\big)\\
&=\prod_{i=1}^{q}\Big(1+(V_1^{2^{m-1}}+V_2^{2^{m-2}}+\cdots+V_m^{2^0})^{2^{a_i}}+\cdots+\\
&\quad\Big(\sum_{1\le j_1<\cdots<j_r\le m}(V_1\cdots V_{j_1-1})^{2^r}(V_{j_1+1}\cdots V_{j_2-1})^{2^{r-1}}\cdots(V_{j_r+1}\cdots V_m)^{2^0}\Big)^{2^{a_i}}\\
&\qquad\qquad+\cdots+\\
&\quad(V_1\cdots V_m)^{2^{a_i}}\Big)\\
&=\prod_{i=1}^{q}\Big(1+(V_1^{2^{a_i+m-1}}+V_2^{2^{a_i+m-2}}+\cdots+V_m^{2^{a_i}})+\cdots+\\
&\quad\Big(\sum_{1\le j_1<\cdots<j_r\le m}(V_1\cdots V_{j_1-1})^{2^{a_i+r}}(V_{j_1+1}\cdots V_{j_2-1})^{2^{a_i+r-1}}\cdots(V_{j_r+1}\cdots V_m)^{2^{a_i}}\Big)\\
&\qquad\qquad+\cdots+\\
&\quad(V_1\cdots V_m)^{2^{a_i}}\Big).\end{aligned}$$

Since $2^a - d - 1 = 2^{a_1} + \cdots + 2^{a_q} \le d - 1$ by choosing terms $V_m^{2^{a_i}+0}$ from each factors in the product indexed by $i = 1, \ldots, q$ we get that

$$p = \pi\big(\rho^*_{d,2^m}(\overline{w}(\xi^{\oplus(d+1)}_{\mathbb{R}^d,2^m}))\big) = V_m^{2^a-d-1} + S \in \mathbb{F}_2[V_1, \ldots, V_m]/\langle V_1^d, \ldots, V_m^d\rangle,$$

where S is a sum of some monomials in $V_1, \ldots, V_m$ of degree $(2^a - d - 1)2^{m-1} = (2^a - d - 1)\ell/2$ which are all different from V_m^{d-1}. Hence, $p \neq 0$ and consequently $\overline{w}_{(2^a-d-1)\ell/2}\big(\xi^{\oplus(d+1)}_{\mathbb{R}^d,2^m}\big) = \overline{w}_{(2^a-d-1)(\ell-\epsilon(\ell))/2}\big(\xi^{\oplus(d+1)}_{\mathbb{R}^d,2^m}\big) \neq 0$.

(5B) Let $d \ge 5$ be an integer that is not a power of 2, and furthermore $d+1$ is not a power of 2. Now we consider the case when $\ell \ge 3$ is not a power of 2. Set $r := \alpha(\ell) \ge 2$ and $\ell = 2^{b_1} + \cdots + 2^{b_r}$ where $0 \le b_1 < b_2 < \cdots < b_r$. As in the proofs of Lemmas 5.6 and 5.12 we consider a morphism between vector bundles $\prod_{i=1}^r \xi_{\mathbb{R}^d,2^{b_i}}$ and $\xi_{\mathbb{R}^d,\ell}$ where the following commutative square is a pullback diagram:

$$\begin{array}{ccc} \prod_{i=1}^r \xi_{\mathbb{R}^d,2^{b_i}} & \xrightarrow{\ \Theta\ } & \xi_{\mathbb{R}^d,\ell} \\ \downarrow & & \downarrow \\ \prod_{i=1}^r \mathrm{F}(\mathbb{R}^d,2^{b_i})/\mathfrak{S}_{2^{b_i}} & \xrightarrow{\ \theta\ } & \mathrm{F}(\mathbb{R}^d,\ell)/\mathfrak{S}_\ell. \end{array}$$

That is $\theta^*\xi_{\mathbb{R}^d,\ell} \cong \prod_{i=1}^r \xi_{\mathbb{R}^d,2^{b_i}}$. The naturality of the Stiefel–Whitney classes [80, Ax. 2, p. 37] gives the equality

$$\theta^*\big(\overline{w}_{(2^{\gamma(d)}-d-1)(\ell-\epsilon(\ell))/2}(\xi^{\oplus(d+1)}_{\mathbb{R}^d,\ell})\big) = \overline{w}_{(2^{\gamma(d)}-d-1)(\ell-\epsilon(\ell))/2}\Big(\prod_{i=1}^r \xi^{\oplus(d+1)}_{\mathbb{R}^d,2^{b_i}}\Big).$$

The product formula [80, Pr. 4-A, p. 54] implies that

$$\overline{w}\Big(\prod_{i=1}^r \xi^{\oplus(d+1)}_{\mathbb{R}^d,2^{b_i}}\Big) = \overline{w}\big(\xi^{\oplus(d+1)}_{\mathbb{R}^d,2^{b_1}}\big) \times \cdots \times \overline{w}\big(\xi^{\oplus(d+1)}_{\mathbb{R}^d,2^{b_r}}\big).$$

Thus,

$$\begin{aligned} &\theta^*\big(\overline{w}_{(2^{\gamma(d)}-d-1)(\ell-\epsilon(\ell))/2}(\xi^{\oplus(d+1)}_{\mathbb{R}^d,\ell})\big) \\ &\quad= \overline{w}_{(2^{\gamma(d)}-d-1)(\ell-\epsilon(\ell))/2}\Big(\prod_{i=1}^r \xi^{\oplus(d+1)}_{\mathbb{R}^d,2^{b_i}}\Big) \\ &\quad= \sum_{s_1+\cdots+s_r=(2^{\gamma(d)}-d-1)(\ell-\epsilon(\ell))/2} \overline{w}_{s_1}(\xi^{\oplus(d+1)}_{\mathbb{R}^d,2^{b_1}}) \times \cdots \times \overline{w}_{s_r}(\xi^{\oplus(d+1)}_{\mathbb{R}^d,2^{b_r}}). \end{aligned} \tag{5.13}$$

The Künneth formula [23, Thm. VI.3.2] implies that each term

$$\overline{w}_{s_1}(\xi_{\mathbb{R}^d,2^{b_1}}^{\oplus(d+1)})\times\cdots\times\overline{w}_{s_r}(\xi_{\mathbb{R}^d,2^{b_r}}^{\oplus(d+1)})$$

in the previous sum belongs to a different direct summand of the cohomology

$$\begin{aligned}
&H^{(2^{\gamma(d)}-d-1)(\ell-\epsilon(\ell))/2}\Big(\prod_{i=1}^{r}\mathrm{F}(\mathbb{R}^d,2^{b_i})/\mathfrak{S}_{2^{b_i}};\mathbb{F}_2\Big)\\
\cong&\bigoplus_{s_1+\cdots+s_r=(2^{\gamma(d)}-d-1)(k-\epsilon(k))/2}H^{s_1}(\mathrm{F}(\mathbb{R}^d,2^{b_1})/\mathfrak{S}_{2^{b_1}};\mathbb{F}_2)\\
&\otimes\cdots\otimes H^{s_r}(\mathrm{F}(\mathbb{R}^d,2^{b_r})/\mathfrak{S}_{2^{b_r}};\mathbb{F}_2).
\end{aligned}$$

Therefore, the following equivalence holds

$$\begin{aligned}
&\overline{w}_{(2^{\gamma(d)}-d-1)(\ell-\epsilon(\ell))/2}\Big(\prod_{i=1}^{r}\xi_{\mathbb{R}^d,2^{b_i}}^{\oplus(d+1)}\Big)\neq 0\\
&\Longleftrightarrow\overline{w}_{s_1}(\xi_{\mathbb{R}^d,2^{b_1}}^{\oplus(d+1)})\times\cdots\times\overline{w}_{s_r}(\xi_{\mathbb{R}^d,2^{b_r}}^{\oplus(d+1)})\neq 0\\
&\qquad\text{for some } s_1+\cdots+s_r=(2^{\gamma(d)}-d-1)(\ell-\epsilon(\ell))/2.
\end{aligned}$$

To isolate a non-zero summand in (5.13) we use the previous case of this theorem which states that $\overline{w}_{(2^{\gamma(d)}-d-1)2^{b_i-1}}(\xi_{\mathbb{R}^d,2^{b_i}}^{\oplus(d+1)})\neq 0$ when $b_i\geq 1$. We discuss two separate cases. Let us assume that ℓ be even, or in other words $\epsilon(\ell)=0$. Since $\ell=2^{b_1}+\cdots+2^{b_r}$ it follows that $1\leq b_1<b_2<\cdots<b_r$. Hence, the following summand in (5.13) does not vanish

$$\overline{w}_{(2^{\gamma(d)}-d-1)2^{b_1-1}}(\xi_{\mathbb{R}^d,2^{b_1}}^{\oplus(d+1)})\times\cdots\times\overline{w}_{(2^{\gamma(d)}-d-1)2^{b_r-1}}(\xi_{\mathbb{R}^d,2^{b_r}}^{\oplus(d+1)})\neq 0.$$

When $\epsilon(\ell)=1$ we have that $0=b_1<b_2<\cdots<b_r$, and the summand in (5.13) which does not vanish is

$$\overline{w}_0(\xi_{\mathbb{R}^d,2^{b_1}}^{\oplus(d+1)})\times\overline{w}_{(2^{\gamma(d)}-d-1)2^{b_2-1}}(\xi_{\mathbb{R}^d,2^{b_2}}^{\oplus(d+1)})\times\cdots\times\overline{w}_{(2^{\gamma(d)}-d-1)2^{b_r-1}}(\xi_{\mathbb{R}^d,2^{b_r}}^{\oplus(d+1)})\neq 0.$$

In summary, $\theta^*(\overline{w}_{(2^{\gamma(d)}-d-1)(\ell-\epsilon(\ell))/2}(\xi_{\mathbb{R}^d,\ell}^{\oplus(d+1)}))\neq 0$ and therefore

$$\overline{w}_{(2^{\gamma(d)}-d-1)(\ell-\epsilon(\ell))/2}(\xi_{\mathbb{R}^d,\ell}^{\oplus(d+1)})\neq 0.$$

(6) In this case we follow the footsteps of the proof of the previous claim. For completeness reasons we discuss all steps of the proof. Again we distinguish case when ℓ is a power of 2 from the case when ℓ is not a power of 2

(6A) Let $d \geq 5$ be an integer such that $2^{a-1}+1 \leq d \leq 2^a - 2$ where $\gamma(d) = a \geq 3$ is an integer. Take $\ell = 2^m$ for $m \geq 2$ an integer. Using the decomposition of vector bundles (5.3) we get that

$$w\big(\xi_{\mathbb{R}^d,2^m}^{\oplus(d+1)}\big) = w\big(\zeta_{\mathbb{R}^d,2^m}^{\oplus(d+1)}\big) = (1+w_1+\cdots+w_{2^m-1})^{d+1}.$$

From (5.4) the height of the algebra $H^*(\mathrm{F}(\mathbb{R}^d,2^m)/\mathfrak{S}_{2^m};\mathbb{F}_2)$ is known:

$$\mathrm{height}(H^*(\mathrm{F}(\mathbb{R}^d,2^m)/\mathfrak{S}_{2^m};\mathbb{F}_2)) \leq \min\{2^t : 2^t \geq d\} = 2^a.$$

Therefore,

$$\overline{w}\big(\xi_{\mathbb{R}^d,2^m}^{\oplus(d+1)}\big) = (1+w_1+\cdots+w_{2^m-1})^{2^a-d-1}.$$

Let $2^a - d - 1 = 2^{a_1}+\cdots+2^{a_q}$ for $0 \leq a_1 < \cdots < a_q \leq a-2$. Then

$$\begin{aligned}\overline{w}\big(\xi_{\mathbb{R}^d,2^m}^{\oplus(d+1)}\big) &= (1+w_1+\cdots+w_{2^m-1})^{2^{a_1}+\cdots+2^{a_q}}\\ &= \prod_{i=1}^{q}\big(1+w_1^{2^{a_i}}+\cdots+w_{2^m-1}^{2^{a_i}}\big).\end{aligned} \tag{5.14}$$

We apply the monomorphism

$$\rho_{d,2^m}^*\colon H^*(\mathrm{F}(\mathbb{R}^d,2^m)/\mathfrak{S}_{2^m};\mathbb{F}_2) \longrightarrow H^*(\mathrm{Pe}(\mathbb{R}^d,2^m)/S_{2^m};\mathbb{F}_2)$$

from Theorem 4.1 to the equality (5.14). Using the decomposition of the cohomology

$$H^*(\mathrm{Pe}(\mathbb{R}^d,2^m)/S_{2^m};\mathbb{F}_2) \cong \mathbb{F}_2[V_{m,1},\ldots,V_{m,m}]/\langle V_{m,1}^d,\ldots,V_{m,m}^d\rangle \oplus \mathrm{I}^*(\mathbb{R}^d,2^m),$$

given in Theorem 3.11, we have

$$\rho_{d,2^m}^*(\overline{w}\big(\xi_{\mathbb{R}^d,2^m}^{\oplus(d+1)}\big)) = \prod_{i=1}^{q}\big(1+D_{m,m-1}^{2^{a_i}}+\cdots+D_{m,0}^{2^{a_i}}\big)+R \tag{5.15}$$

where

- $D_{m,r} = (\kappa_{d,2^m}/S_{2^m})^*(D_{m,r}) = \rho_{d,2^m}^*(w_{2^m-2^r})$, for $0 \leq r \leq m-1$, with the obvious abuse of notation, see (3.28), and
- $R \in \mathrm{I}^*(\mathbb{R}^d,2^m)$.

Furthermore, let π denote the following composition of the maps from the diagram (5.5):

$$\begin{array}{c}
\mathbb{F}_2[V_{m,1},\ldots,V_{m,m}]/\langle V_{m,1}^d,\ldots,V_{m,m}^d\rangle \oplus I^*(\mathbb{R}^d,2^m)\\
\big\downarrow \text{projection}\\
\mathbb{F}_2[V_{m,1},\ldots,V_{m,m}]/\langle V_{m,1}^d,\ldots,V_{m,m}^d\rangle\\
\big\downarrow \chi_m\\
\mathbb{F}_2[V_1,\ldots,V_m]/\langle V_1^d,\ldots,V_m^d\rangle,
\end{array}$$

The first map is the projection on a direct summand and the second map is induced by the change of variables $\chi_m \in \mathrm{GL}_m(\mathbb{F}_2)$. Applying π to (5.15) we get

$$\pi\big(\rho_{d,2^m}^*\big(\overline{w}\big(\xi_{\mathbb{R}^d,2^m}^{\oplus(d+1)}\big)\big)\big)=\prod_{i=1}^{q}\big(1+\pi(D_{m,m-1})^{2^i}+\cdots+\pi(D_{m,0})^{2^i}\big).$$

Recall that the change of the variables χ_m transforms $\mathrm{L}_m(\mathbb{F}_2)$-invariants into $\mathrm{U}_m(\mathbb{F}_2)$-invariants and Dickson polynomials, $\mathrm{GL}_m(\mathbb{F}_2)$-invariants, can be presented in terms of $\mathrm{U}_m(\mathbb{F}_2)$-invariants, as explained in (8.3) and (8.4). Now, $p := \pi\big(\rho_{d,2^m}^*\big(\overline{w}\big(\xi_{\mathbb{R}^d,2^m}^{\oplus(d+1)}\big)\big)\big)$ can be expressed as follows:

$$\begin{aligned}
p &= \prod_{i=1}^{q}\big(1+\pi(D_{m,m-1})^{2^{a_i}}+\cdots+\pi(D_{m,0})^{2^{a_i}}\big)\\
&= \prod_{i=1}^{q}\Big(1+(V_1^{2^{m-1}}+V_2^{2^{m-2}}+\cdots+V_m^{2^0})^{2^{a_i}}+\cdots+\\
&\quad \Big(\sum_{1\le j_1<\cdots<j_r\le m}(V_1\cdots V_{j_1-1})^{2^r}(V_{j_1+1}\cdots V_{j_2-1})^{2^{r-1}}\cdots(V_{j_r+1}\cdots V_m)^{2^0}\Big)^{2^{a_i}}\\
&\qquad\qquad +\cdots+\\
&\quad (V_1\cdots V_m)^{2^{a_i}}\Big)\\
&= \prod_{i=1}^{q}\Big(1+(V_1^{2^{a_i+m-1}}+V_2^{2^{a_i+m-2}}+\cdots+V_m^{2^{a_i}})+\cdots+ \qquad (5.16)\\
&\quad \Big(\sum_{1\le j_1<\cdots<j_r\le m}(V_1\cdots V_{j_1-1})^{2^{a_i+r}}(V_{j_1+1}\cdots V_{j_2-1})^{2^{a_i+r-1}}\cdots(V_{j_r+1}\cdots V_m)^{2^{a_i}}\Big)\\
&\qquad\qquad +\cdots+\\
&\quad (V_1\cdots V_m)^{2^{a_i}}\Big).
\end{aligned}$$

Now we want to show that

$$\pi\left(\rho^*_{d,2^m}\left(\overline{w}_{(2^{\gamma(d)}-d-1+2^{a_1})2^{m-1}-2^{a_1}}\left(\xi^{\oplus(d+1)}_{\mathbb{R}^d,2^m}\right)\right)\right)$$
$$= (V_1\cdots V_{m-1})^{2^{a_1}} V_m^{2^{\gamma(d)}-d-1} + S \in \mathbb{F}_2[V_1,\ldots,V_m]/\langle V_1^d,\ldots,V_m^d\rangle.$$

Here S is a sum of monomials in $V_1,\ldots,V_m$ of degree $(2^{\gamma(d)} - d - 1 + 2^{a_1})2^{m-1} - 2^{a_1}$ which are different from the monomial $(V_1\cdots V_{m-1})^{2^{a_1}} V_m^{2^{\gamma(d)}-d-1} + S$. Hence,

$$\pi\left(\rho^*_{d,2^m}\left(\overline{w}_{(2^{\gamma(d)}-d-1+2^{a_1})2^{m-1}-2^{a_1}}\left(\xi^{\oplus(d+1)}_{\mathbb{R}^d,2^m}\right)\right)\right) \neq 0,$$

and consequently

$$\overline{w}_{(2^{\gamma(d)}-d-1+2^{a_1})2^{m-1}-2^{a_1}}\left(\xi^{\oplus(d+1)}_{\mathbb{R}^d,2^m}\right) \neq 0.$$

Indeed, observe that in every monomial of the ith factor of the product (5.16) which has the variable V_m, with a positive exponent, the variable V_m has always the same exponent equal to 2^{a_i}, $1 \le i \le q$. In particular, in the ith factor each monomial with the variable V_m is of the form $p_i(V_1,\ldots,V_{m-1})^{2^{a_i}} V_m^{2^{a_i}}$ where $p_i(V_1,\ldots,V_{m-1})$ is a monomial in variables $V_1,\ldots,V_{m-1}$. Now, when multiplying out the product (5.16), since $2^{\gamma(d)} - d - 1 = 2^{a_1} + \cdots + 2^{a_q}$, the monomial of the form $p(V_1,\ldots,V_{m-1})V_m^{2^{\gamma(d)}-d-1}$ can appear in the final result if and only if we take from each factor a non-zero monomial of the form $p_i(V_1,\ldots,V_{m-1})^{2^{a_i}} V_m^{2^{a_i}}$. Thus, we have

$$p(V_1,\ldots,V_{m-1})V_m^{2^{\gamma(d)}-d-1} = \prod_{i=1}^{q} p_i(V_1,\ldots,V_{m-1})^{2^{a_i}} V_m^{2^{a_i}}$$

Observe that, if $p_i(V_1,\ldots,V_{m-1}) \neq 1$ for some $1 \le i \le q$, then there exists $1 \le t \le q$ such that $V_t \mid p_i(V_1,\ldots,V_{m-1})$. Hence, $V_t^{2^{a_i}} \mid p_i(V_1,\ldots,V_{m-1})^{2^{a_i}}$.

Now we want to count in how many ways we can obtain the monomial

$$(V_1\cdots V_{m-1})^{2^{a_1}} V_m^{2^{\gamma(d)}-d-1}$$

when we multiply out the product (5.16). This means that we need to find all possible p_is and p'_j's such that

$$(V_1\cdots V_{m-1})^{2^{a_1}} V_m^{2^{\gamma(d)}-d-1} = \prod_{i=1}^{q} p_i(V_1,\ldots,V_{m-1})^{2^{a_i}} V_m^{2^{a_i}}$$

From the previous observation and the fact that $0 \leq a_1 < a_2 < \cdots < a_q \leq a-2$ we conclude that $p_{a_2}(V_1, \ldots, V_{m-1}) = \cdots = p_{a_q}(V_1, \ldots, V_{m-1}) = 1$. Thus, the previous equality becomes

$$(V_1 \cdots V_{m-1})^{2^{a_1}} V_m^{2^{\gamma(d)}-d-1} = p_1(V_1, \ldots, V_{m-1})^{2^{a_i}} V_m^{2^{\gamma(d)}-d-1}.$$

Therefore, the monomial $(V_1 \cdots V_{m-1})^{2^{a_1}} V_m^{2^{\gamma(d)}-d-1}$ can be obtained for

$$p_{a_1}(V_1, \ldots, V_{m-1}) = V_1 \cdots V_{m-1},$$

and

$$p_{a_2}(V_1, \ldots, V_{m-1}) = \cdots = p_{a_q}(V_1, \ldots, V_{m-1}) = 1.$$

Hence, we completed the proof of the non-vanishing of the dual class:

$$\overline{w}_{(2^{\gamma(d)}-d-1+2^{a_1})2^{m-1}-2^{a_1}}\big(\xi_{\mathbb{R}^d,2^m}^{\oplus(d+1)}\big) \neq 0.$$

(6B) Let $d \geq 5$ be an integer that is not a power of 2, and furthermore $d+1$ is not a power of 2. Consider the case when $\ell \geq 3$ is not a power of 2. Set $r := \alpha(\ell) \geq 2$ and $\ell = 2^{b_1} + \cdots + 2^{b_r}$ where $0 \leq b_1 < b_2 < \cdots < b_r$. As many times before we consider a morphism between vector bundles $\prod_{i=1}^{r} \xi_{\mathbb{R}^d,2^{b_i}}$ and $\xi_{\mathbb{R}^d,\ell}$ where the following commutative square is a pullback diagram:

$$\begin{array}{ccc} \prod_{i=1}^{r} \xi_{\mathbb{R}^d,2^{b_i}} & \xrightarrow{\ \ \Theta\ \ } & \xi_{\mathbb{R}^d,\ell} \\ \big\downarrow & & \big\downarrow \\ \prod_{i=1}^{r} \mathrm{F}(\mathbb{R}^d, 2^{b_i})/\mathfrak{S}_{2^{b_i}} & \xrightarrow{\ \ \theta\ \ } & \mathrm{F}(\mathbb{R}^d, \ell)/\mathfrak{S}_{\ell}. \end{array}$$

In particular, $\theta^*\xi_{\mathbb{R}^d,\ell} \cong \prod_{i=1}^{r} \xi_{\mathbb{R}^d,2^{b_i}}$. The naturality of the Stiefel–Whitney classes [80, Ax. 2, p. 37] gives the equality

$$\theta^*(\overline{w}(\xi_{\mathbb{R}^d,\ell}^{\oplus(d+1)})) = \overline{w}\Big(\prod_{l=1}^{r} \xi_{\mathbb{R}^d,2^{b_i}}^{\oplus(d+1)}\Big),$$

while the product formula [80, Pr. 4-A, p. 54] implies that

$$\overline{w}\Big(\prod_{i=1}^{r} \xi_{\mathbb{R}^d,2^{b_i}}^{\oplus(d+1)}\Big) = \overline{w}\big(\xi_{\mathbb{R}^d,2^{b_1}}^{\oplus(d+1)}\big) \times \cdots \times \overline{w}\big(\xi_{\mathbb{R}^d,2^{b_r}}^{\oplus(d+1)}\big).$$

Thus, for every integer $N \geq 0$

$$\theta^*(\overline{w}_N(\xi_{\mathbb{R}^d,\ell}^{\oplus(d+1)})) = \overline{w}_N\Big(\prod_{i=1}^{r}\xi_{\mathbb{R}^d,2^{b_i}}^{\oplus(d+1)}\Big)$$
$$= \sum_{s_1+\cdots+s_r=N} \overline{w}_{s_1}(\xi_{\mathbb{R}^d,2^{b_1}}^{\oplus(d+1)}) \times \cdots \times \overline{w}_{s_r}(\xi_{\mathbb{R}^d,2^{b_r}}^{\oplus(d+1)}). \tag{5.17}$$

The Künneth formula [23, Thm. VI.3.2] implies that each term

$$\overline{w}_{s_1}(\xi_{\mathbb{R}^d,2^{b_1}}^{\oplus(d+1)}) \times \cdots \times \overline{w}_{s_r}(\xi_{\mathbb{R}^d,2^{b_r}}^{\oplus(d+1)})$$

in the previous sum belongs to a different direct summand of the cohomology

$$H^N\big(\prod_{i=1}^{r}\mathrm{F}(\mathbb{R}^d, 2^{b_i})/\mathfrak{S}_{2^{b_i}}; \mathbb{F}_2\big)$$
$$\cong \bigoplus_{s_1+\cdots+s_r=N} H^{s_1}(\mathrm{F}(\mathbb{R}^d, 2^{b_1})/\mathfrak{S}_{2^{b_1}}; \mathbb{F}_2) \otimes \cdots \otimes H^{s_r}(\mathrm{F}(\mathbb{R}^d, 2^{b_r})/\mathfrak{S}_{2^{b_r}}; \mathbb{F}_2).$$

Therefore, the following equivalence holds

$$\overline{w}_N\Big(\prod_{i=1}^{r}\xi_{\mathbb{R}^d,2^{b_i}}^{\oplus(d+1)}\Big) \neq 0 \Longleftrightarrow \overline{w}_{s_1}(\xi_{\mathbb{R}^d,2^{b_1}}^{\oplus(d+1)}) \times \cdots \times \overline{w}_{s_r}(\xi_{\mathbb{R}^d,2^{b_r}}^{\oplus(d+1)}) \neq 0$$
$$\text{for some } s_1 + \cdots + s_r = N.$$

To isolate a non-zero summand in (5.17) for

$$N = (2^{\gamma(d)} - d - 1 + 2^{a_1})(\ell - \epsilon(\ell))/2 - 2^{a_1}\alpha(\ell)$$

we use the previous case of this theorem which states that for $b_i \geq 1$:

$$\overline{w}_{(2^{\gamma(d)}-d-1+2^{a_1})2^{b_i-1}-2^{a_1}}(\xi_{\mathbb{R}^d,2^{b_i}}^{\oplus(d+1)}) \neq 0.$$

We discuss two separate cases. Let ℓ be even, or $\epsilon(\ell) = 0$. Since $\ell = 2^{b_1} + \cdots + 2^{b_r}$ it follows that $1 \leq b_1 < b_2 < \cdots < b_r$. Thus, the following summand in (5.17) does not vanish

$$\overline{w}_{(2^{\gamma(d)}-d-1+2^{a_1})2^{b_1-1}-2^{a_1}}(\xi_{\mathbb{R}^d,2^{b_1}}^{\oplus(d+1)}) \times \cdots \times \overline{w}_{(2^{\gamma(d)}-d-1+2^{a_1})2^{b_r-1}-2^{a_1}}(\xi_{\mathbb{R}^d,2^{b_r}}^{\oplus(d+1)}) \neq 0.$$

When $\epsilon(\ell) = 1$ we have that $0 = b_1 < b_2 < \cdots < b_r$, and the summand in (5.17) which does not vanish is

$$\overline{w}_0(\xi_{\mathbb{R}^d,2^{b_1}}^{\oplus(d+1)}) \times \overline{w}_{(2^{\gamma(d)}-d-1+2^{a_1})2^{b_2-1}-2^{a_1}}(\xi_{\mathbb{R}^d,2^{b_1}}^{\oplus(d+1)}) \times \cdots \times \overline{w}_{(2^{\gamma(d)}-d-1+2^{a_1})2^{b_r-1}-2^{a_1}}(\xi_{\mathbb{R}^d,2^{b_r}}^{\oplus(d+1)}) \neq 0.$$

In summary,

$$\overline{w}_{(2^{\gamma(d)}-d-1+2^{a_1})(\ell-\epsilon(\ell))/2-2^{a_1}\alpha(\ell)}\big(\xi_{\mathbb{R}^d,\ell}^{\oplus(d+1)}\big) \neq 0.$$

□

Now, we use Theorem 5.17 and the criterion from Lemma 5.16 to correct the result stated in [15, Thm. 3.1]. In this way we completed corrections of the invalid claims in [15, Sec. 3].

Theorem 5.18 *Let $\ell \geq 1$ and $d \geq 2$ be integers.*

(1) *If $d = 2$, and if $\ell \geq 2$ is an integer, then there is no ℓ-skew embedding $\mathbb{R}^2 \longrightarrow \mathbb{R}^N$ for*

$$N \leq 4\ell - \alpha(\ell) - 2.$$

(2) *If $d \geq 1$ is an integer, and if $\ell = 2$, then there is no 2-skew embedding $\mathbb{R}^d \longrightarrow \mathbb{R}^N$ for*

$$N < 2^{\gamma(d)} + d - 1.$$

(3) *If $d \geq 2$ is a power of 2, and if $\ell \geq 2$ is an integer, then there is no ℓ-skew embedding $\mathbb{R}^d \longrightarrow \mathbb{R}^N$ for*

$$N \leq 2d\ell - (d-1)\alpha(\ell) - 2.$$

(4) *If $d + 1 \geq 2$ is a power of 2, and if $\ell \geq 2$ is an integer, then this method does not produce any non-trivial result about the existence of ℓ-skew embeddings $\mathbb{R}^d \longrightarrow \mathbb{R}^N$.*

(5) *If $d \geq 5$ is an integer which is not a power of 2, and in addition $d + 1$ is not a power 2, and if $\ell \geq 3$ is an integer, then there is no ℓ-skew embedding $\mathbb{R}^d \longrightarrow \mathbb{R}^N$ for*

$$N \leq \frac{1}{2}\big(2^{\gamma(d)} - d - 1\big)(\ell - \epsilon(\ell)) + (d+1)\ell - 2.$$

(6) *If $d \geq 5$ is an integer which is not a power of 2, $2^{\gamma(d)} - d - 1 = 2^{a_1}z$ where $a_1 \geq 0$ is an integer and $z \geq 1$ is an odd integer, and $d+1$ is not a power of 2, and if $\ell \geq 3$ is an integer, then there is no ℓ-skew embedding $\mathbb{R}^d \longrightarrow \mathbb{R}^N$ for*

$$N \leq \frac{1}{2}(2^{\gamma(d)} - d - 1 + 2^{a_1})(\ell - \epsilon(\ell)) - 2^{a_1}\alpha(\ell) + (d+1)\ell - 2.$$

5.3 k-Regular-ℓ-Skew Embeddings

In this section we revise [15, Sec. 4] and correct the related results [15, Thm. 4.1, Thm. 4.8]. First we recall some basic notions on k-regular-ℓ-skew embeddings.

Definition 5.19 Let $k \geq 1$ and $\ell \geq 1$ be an integer, and let M be a real smooth d-dimensional manifold. A smooth embedding $f\colon M \longrightarrow \mathbb{R}^N$ is **k-regular-ℓ-skew embedding** if for every $(x_1, \ldots, x_k, y_1, \ldots, y_\ell)$ in $\mathrm{F}(M, k+\ell)$ the affine subspaces

$$\{f(x_1)\}, \ldots, \{f(x_k)\}, (\iota \circ df_{y_1})(T_{y_1}M), \ldots, (\iota \circ df_{y_\ell})(T_{y_\ell}M)$$

of $\mathbb{R}^N$ are affinely independent.

Now, like in the case of k-regular embeddings and ℓ-skew embeddings a criterion for non-existence of k-regular-ℓ-skew embedding can be derived in terms of Stiefel–Whitney class of appropriate vector bundle over the relevant configuration space. We state a consequence of [15, Lem. 4.6, Lem. 4.7] used for the proof of [15, Thm. 4.1].

Lemma 5.20 *Let $d \geq 1$, $k \geq 1$ and $\ell \geq 1$ be integers. If the dual Stiefel–Whitney class*

$$\overline{w}_{N-(d+1)\ell-k+2}(\xi_{\mathbb{R}^d,k} \times \xi_{\mathbb{R}^d,\ell}^{\oplus(d+1)})$$

does not vanish, then there is no k-regular-ℓ-skew embedding $\mathbb{R}^d \longrightarrow \mathbb{R}^N$.

Like in the case of k-regular embeddings and ℓ-skew embeddings, now based on invalid results in [15, Thm. 2.13, Thm. 3.7] the following theorem was proved [15, Thm. 4.8]:

> **Theorem 4.8.** Let $\ell, k, d \geq 2$ be integers. The dual Stiefel–Whitney class
>
> $$\overline{w}_{(d-1)(k-\alpha(k))+(2^{\gamma(d)}-d-1)(\ell-\alpha(\ell))}(\xi_{\mathbb{R}^d,k} \times \xi_{\mathbb{R}^d,\ell}^{\oplus(d+1)})$$
>
> does not vanish.

The result of the previous theorem in combination with [15, Lem. 4.6, Lem. 4.4] directly implied the following result [15, Thm. 4.1].

Theorem 4.1. et $\ell, d \geq 2$ be integers. There is no k-regular-ℓ-skew embedding $\mathbb{R}^d \longrightarrow \mathbb{R}^N$ for

$$N \leq (d-1)(k-\alpha(k)) + (2^{\gamma(d)} - d - 1)(\ell - \alpha(\ell)) + (d+1)\ell + k - 2,$$

where $\alpha(c)$ denotes the number of ones in the dyadic presentation of c, and $\gamma(d) := \lfloor \log_2 d \rfloor + 1$.

The proof of [15, Thm. 4.8] was based on incorrect results [15, Cor. 2.13, Cor. 3.7]. Since we corrected these results we can now give correct versions, first of [15, Thm. 4.8], and then of [15, Thm. 4.1].

Theorem 5.21 *Let $d \geq 1$, $k \geq 1$, and $\ell \geq 1$ be integers.*

(1) *If $d = 2$, and if $k \geq 1$ and $\ell \geq 1$ are integers, then*

$$\overline{w}_{k-\alpha(k)+\ell-\alpha(\ell)}\big(\xi_{\mathbb{R}^d,k} \times \xi_{\mathbb{R}^d,\ell}^{\oplus(d+1)}\big) \neq 0.$$

(2) *If $d \geq 2$ is a power of 2, $\ell = 2$, and if $k \geq 1$ is an integer, then*

$$\overline{w}_{(d-1)(k-\alpha(k)+1)}\big(\xi_{\mathbb{R}^d,k} \times \xi_{\mathbb{R}^d,\ell}^{\oplus(d+1)}\big) \neq 0.$$

(3) *If $d \geq 3$ is not a power of 2, $\ell = 2$, and if $k \geq 1$ is an integer, then*

$$\overline{w}_{(d-1)(k-\epsilon(k))/2+2^{\gamma(d)}-d-1}\big(\xi_{\mathbb{R}^d,k} \times \xi_{\mathbb{R}^d,\ell}^{\oplus(d+1)}\big) \neq 0.$$

(4) *If $d \geq 2$ is a power of 2, and if $k \geq 1$ and $\ell \geq 1$ are integers, then*

$$\overline{w}_{(d-1)(k-\alpha(k)+\ell-\alpha(\ell))}\big(\xi_{\mathbb{R}^d,k} \times \xi_{\mathbb{R}^d,\ell}^{\oplus(d+1)}\big) \neq 0.$$

(5) *If $d+1 \geq 2$ is a power of 2, and if $k \geq 1$ and $\ell \geq 1$ are integers, then*

$$\overline{w}_{(d-1)(k-\epsilon(k))/2}\big(\xi_{\mathbb{R}^d,k} \times \xi_{\mathbb{R}^d,\ell}^{\oplus(d+1)}\big) \neq 0.$$

(6) *If $d \geq 5$ is an integer which is not a power of 2, and in addition $d+1$ is a not power of 2, and if $k \geq 1$ and $\ell \geq 1$ are integers, then*

$$\overline{w}_{(d-1)(k-\epsilon(k))/2+(2^{\gamma(d)}-d-1)(\ell-\epsilon(\ell))/2}\big(\xi_{\mathbb{R}^d,k} \times \xi_{\mathbb{R}^d,\ell}^{\oplus(d+1)}\big) \neq 0.$$

(7) *If $d \geq 6$ be an even integer which is not a power of 2, $2^{\gamma(d)} - d - 1 = 2^{a_1} z$ where $a_1 \geq 0$ is an integer and $z \geq 1$ is an odd integer, and if $k \geq 1$ and $\ell \geq 3$ are integers, then*

$$\overline{w}_{d(k-\epsilon(k))/2-\alpha(k)+\epsilon(k)+(2^{\gamma(d)}-d-1+2^{a_1})(\ell-\epsilon(\ell))/2-2^{a_1}\alpha(\ell)}\big(\xi_{\mathbb{R}^d,k} \times \xi_{\mathbb{R}^d,\ell}^{\oplus(d+1)}\big) \neq 0.$$

Proof In order to compute the dual Stiefel–Whitney class of the product vector bundle $\xi_{\mathbb{R}^d,k} \times \xi_{\mathbb{R}^d,\ell}^{\oplus(d+1)}$ we use the product formula [80, Problem 4-A, page 54]:

$$\overline{w}\big(\xi_{\mathbb{R}^d,k} \times \xi_{\mathbb{R}^d,\ell}^{\oplus(d+1)}\big) = \overline{w}(\xi_{\mathbb{R}^d,k}) \times \overline{w}\big(\xi_{\mathbb{R}^d,\ell}^{\oplus(d+1)}\big),$$

where "×" on the right hand side denotes the cross product in cohomology. In particular, for a fixed integer $r \geq 0$ we have that

$$\overline{w}_r\big(\xi_{\mathbb{R}^d,k} \times \xi_{\mathbb{R}^d,\ell}^{\oplus(d+1)}\big) = \sum_{i+j=r} \overline{w}_i(\xi_{\mathbb{R}^d,k}) \times \overline{w}_j\big(\xi_{\mathbb{R}^d,\ell}^{\oplus(d+1)}\big)$$
$$\in H^r(\mathrm{F}(\mathbb{R}^d,k)/\mathfrak{S}_k \times \mathrm{F}(\mathbb{R}^d,\ell)/\mathfrak{S}_\ell;\mathbb{F}_2).$$

The Künneth formula [23, Thm. VI.3.2] implies that each of the terms $\overline{w}_i(\xi_{\mathbb{R}^d,k}) \times \overline{w}_j\big(\xi_{\mathbb{R}^d,\ell}^{\oplus(d+1)}\big)$ in the previous sum belongs to a different direct summand of the cohomology

$$\begin{aligned} &H^r(\mathrm{F}(\mathbb{R}^d,k)/\mathfrak{S}_k \times \mathrm{F}(\mathbb{R}^d,\ell)/\mathfrak{S}_\ell;\mathbb{F}_2)\\ &\cong \bigoplus_{i+j=r} H^i(\mathrm{F}(\mathbb{R}^d,k)/\mathfrak{S}_k;\mathbb{F}_2)\otimes H^j(\mathrm{F}(\mathbb{R}^d,\ell)/\mathfrak{S}_\ell;\mathbb{F}_2). \end{aligned}$$

Therefore, the following equivalence holds

$$\overline{w}_r\big(\xi_{\mathbb{R}^d,k} \times \xi_{\mathbb{R}^d,\ell}^{\oplus(d+1)}\big) \neq 0 \Longleftrightarrow \overline{w}_i(\xi_{\mathbb{R}^d,k}) \times \overline{w}_j\big(\xi_{\mathbb{R}^d,\ell}^{\oplus(d+1)}\big) \neq 0 \text{ for some } i+j=r.$$

Now, using Theorems 5.13 and 5.17 to prove all cases of the theorem.

(1) Let $d = 2$, and let $k \geq 1$ and $\ell \geq 1$ be integers. Then

$$\overline{w}_{k-\alpha(k)}(\xi_{\mathbb{R}^d,k}) \neq 0 \qquad \text{and} \qquad \overline{w}_{\ell-\alpha(\ell)}\big(\xi_{\mathbb{R}^d,\ell}^{\oplus(d+1)}\big) \neq 0,$$

and consequently $\overline{w}_{k-\alpha(k)+\ell-\alpha(\ell)}\big(\xi_{\mathbb{R}^d,k} \times \xi_{\mathbb{R}^d,\ell}^{\oplus(d+1)}\big) \neq 0$.

(2) Let $d \geq 2$ be a power of 2, $\ell = 2$, and let $k \geq 1$ be an integer. Then

$$\overline{w}_{(d-1)(k-\alpha(k))}(\xi_{\mathbb{R}^d,k}) \neq 0 \qquad \text{and} \qquad \overline{w}_{d-1}\big(\xi_{\mathbb{R}^d,\ell}^{\oplus(d+1)}\big) \neq 0,$$

and consequently $\overline{w}_{(d-1)(k-\alpha(k)+1)}\big(\xi_{\mathbb{R}^d,k} \times \xi_{\mathbb{R}^d,\ell}^{\oplus(d+1)}\big) \neq 0$.

(3) Let $d \geq 2$ be not a power of 2, $\ell = 2$, and let $k \geq 1$ be an integer. Then

$$\overline{w}_{(d-1)(k-\epsilon(k))/2}(\xi_{\mathbb{R}^d,k}) \neq 0 \qquad \text{and} \qquad \overline{w}_{2^{\gamma(d)}-d-1}\big(\xi_{\mathbb{R}^d,\ell}^{\oplus(d+1)}\big) \neq 0,$$

and consequently $\overline{w}_{(d-1)(k-\epsilon(k))/2+2^{\gamma(d)}-d-1}\big(\xi_{\mathbb{R}^d,k} \times \xi_{\mathbb{R}^d,\ell}^{\oplus(d+1)}\big) \neq 0$.

(4) Let $d \geq 2$ be not a power of 2, and let $k \geq 1$ and $\ell \geq 1$ be integers. Then

$$\overline{w}_{(d-1)(k-\alpha(k))}(\xi_{\mathbb{R}^d,k}) \neq 0 \qquad \text{and} \qquad \overline{w}_{(d-1)(\ell-\alpha(\ell))}\big(\xi_{\mathbb{R}^d,\ell}^{\oplus(d+1)}\big) \neq 0,$$

and consequently $\overline{w}_{(d-1)(k-\alpha(k)+\ell-\alpha(\ell))}\big(\xi_{\mathbb{R}^d,k} \times \xi_{\mathbb{R}^d,\ell}^{\oplus(d+1)}\big) \neq 0$.

(5) Let $d+1 \geq 2$ be a power of 2, and let $k \geq 1$ and $\ell \geq 1$ be integers. Then

$$\overline{w}_{(d-1)(k-\epsilon(k))/2}(\xi_{\mathbb{R}^d,k}) \neq 0 \qquad \text{and} \qquad \overline{w}_0\big(\xi_{\mathbb{R}^d,\ell}^{\oplus(d+1)}\big) \neq 0,$$

and consequently $\overline{w}_{(d-1)(k-\epsilon(k))/2}\big(\xi_{\mathbb{R}^d,k} \times \xi_{\mathbb{R}^d,\ell}^{\oplus(d+1)}\big) \neq 0$

(6) Let $d \geq 5$ be an integer which is not a power of 2, and in addition $d+1$ is a not power of 2, and let $k \geq 1$ and $\ell \geq 1$ be integers. Then

$$\overline{w}_{(d-1)(k-\epsilon(k))/2}(\xi_{\mathbb{R}^d,k}) \neq 0 \qquad \text{and} \qquad \overline{w}_{(2^{\gamma(d)}-d-1)(\ell-\epsilon(\ell))/2}\big(\xi_{\mathbb{R}^d,\ell}^{\oplus(d+1)}\big) \neq 0,$$

and consequently $\overline{w}_{(d-1)(k-\epsilon(k))/2+(2^{\gamma(d)}-d-1)(\ell-\epsilon(\ell))/2}\big(\xi_{\mathbb{R}^d,k} \times \xi_{\mathbb{R}^d,\ell}^{\oplus(d+1)}\big) \neq 0$.

(7) Let $d \geq 6$ be an even integer which is not a power of 2, $2^{\gamma(d)} - d - 1 = 2^{a_1}z$ where $a_1 \geq 0$ is an integer and $z \geq 1$ is an odd integer, and let $k \geq 1$ and $\ell \geq 3$ be integers. Then

$$\overline{w}_{d(k-\epsilon(k))/2-\alpha(k)+\epsilon(k)}(\xi_{\mathbb{R}^d,k}) \neq 0 \quad \text{and} \quad \overline{w}_{d(k-\epsilon(k))/2-\alpha(k)+\epsilon(k)}(\xi_{\mathbb{R}^d,k}) \neq 0,$$

and consequently

$$\overline{w}_{d(k-\epsilon(k))/2-\alpha(k)+\epsilon(k)+(2^{\gamma(d)}-d-1+2^{a_1})(\ell-\epsilon(\ell))/2-2^{a_1}\alpha(\ell)}\big(\xi_{\mathbb{R}^d,k} \times \xi_{\mathbb{R}^d,\ell}^{\oplus(d+1)}\big) \neq 0.$$

Like in the previous situation, we use Theorem 5.21 and the criterion from Lemma 5.20 to correct the result stated in [15, Thm. 4.1]. In this way we completed corrections of the invalid results in [15, Sec. 4].

Theorem 5.22 *Let $d \geq 1$, $k \geq 1$, and $\ell \geq 1$ be integers.*

(1) *If $d = 2$, and if $k \geq 1$ and $\ell \geq 1$ are integers, then there is no k-regular-ℓ-skew embedding $\mathbb{R}^2 \longrightarrow \mathbb{R}^N$ for*

$$N \leq (d+1)\ell + 2k - \alpha(k) + \ell - \alpha(l) - 2.$$

(2) *If $d \geq 2$ is a power of 2, $\ell = 2$, and if $k \geq 1$ is an integer, then there is no k-regular-2-skew embedding $\mathbb{R}^d \longrightarrow \mathbb{R}^N$ for*

$$N \leq (d+1)\ell + k - 2 + (d-1)(k - \alpha(k) + 1).$$

(3) *If $d \geq 2$ is not a power of 2, $\ell = 2$, and if $k \geq 1$ is an integer, then there is no k-regular-2-skew embedding $\mathbb{R}^d \longrightarrow \mathbb{R}^N$ for*

$$N \leq (d+1)\ell + k - 2 + \frac{1}{2}(d-1)(k - \epsilon(k)) + 2^{\gamma(d)} - d - 1.$$

(4) *If $d \geq 2$ is not a power of 2, and if $k \geq 1$ and $\ell \geq 1$ are integers, then there is no k-regular-ℓ-skew embedding $\mathbb{R}^2 \longrightarrow \mathbb{R}^N$ for*

$$N \leq (d+1)\ell + k - 2 + (d-1)(k - \alpha(k) + \ell - \alpha(\ell)).$$

(5) *If $d+1 \geq 2$ is a power of 2, and if $k \geq 1$ and $\ell \geq 1$ are integers, then there is no k-regular-ℓ-skew embedding $\mathbb{R}^d \longrightarrow \mathbb{R}^N$ for*

$$N \leq (d+1)\ell + k - 2 + \frac{1}{2}(d-1)(k - \epsilon(k)).$$

(6) *If $d \geq 5$ is an integer which is not a power of 2, and in addition $d+1$ is a not power of 2, and if $k \geq 1$ and $\ell \geq 1$ are integers, then there is no k-regular-ℓ-skew embedding $\mathbb{R}^d \longrightarrow \mathbb{R}^N$ for*

$$N \leq (d+1)\ell + k - 2 + \frac{1}{2}(d-1)(k - \epsilon(k)) + \frac{1}{2}\big(2^{\gamma(d)} - d - 1\big)(\ell - \alpha(\ell)).$$

(7) *If $d \geq 6$ be an even integer which is not a power of 2, $2^{\gamma(d)} - d - 1 = 2^{a_1}z$ where $a_1 \geq 0$ is an integer and $z \geq 1$ is an odd integer, and if $k \geq 1$ and $\ell \geq 3$ are integers, then there is no k-regular-ℓ-skew embedding $\mathbb{R}^d \longrightarrow \mathbb{R}^N$ for*

$$N \leq (d+1)\ell + k - 2 + \frac{1}{2}d(k - \epsilon(k)) - \alpha(k) + \epsilon(k)$$
$$+\frac{1}{2}\big(2^{\gamma(d)} - d - 1 + 2^{a_1}\big)(\ell - \epsilon(\ell)) - 2^{a_1}\alpha(\ell).$$

This concludes all corrections of the paper [15].

5.4 Complex Highly Regular Embeddings

A gap in the decomposition [64, (4.7)] created incorrectness that we have already discussed in the study of real highly regular embeddings [15], which in turn implied two results on complex highly regular embeddings [14, Thm. 5.1, Thm. 6.1] that now also need to be corrected. For completeness we recall some basic notions about complex highly regular embeddings.

First we introduce a notion of complex k-regular embedding in a similar way as in the case of real k-regular embedding, see Sect. 5.1.

Definition 5.23 Let $k \geq 1$ be an integer, and let X be a topological space. A continuous map $f : X \longrightarrow \mathbb{C}^N$ is a **complex k-regular embedding** if for every $(x_1, \dots, x_k) \in \mathrm{F}(X, k)$ the vectors $f(x_1), \dots, f(x_n)$ of the complex vector space $\mathbb{C}^N$ are linearly independent.

Next we introduce a notion of complex ℓ-regular embedding. For a real analogue consult Sect. 5.2. A collection of complex affine subspaces $\{L_1, \dots, L_\ell\}$ of the complex vector space $\mathbb{C}^N$ is **affinely independent** if the following equality holds

$$\dim_{\mathrm{aff}}^{\mathbb{C}} \operatorname{span}(L_1 \cup \dots \cup L_\ell) = (\dim_{\mathrm{aff}}^{\mathbb{C}} L_1 + 1) + \dots + (\dim_{\mathrm{aff}}^{\mathbb{C}} L_\ell + 1) - 1.$$

Let M be a complex d-dimensional manifold. The associated complex tangent bundle of M is denote by TM. For a point $y \in M$ we denote by T_yM the corresponding tangent space to M. If $f : M \longrightarrow \mathbb{C}^N$ is a smooth complex map, then $df : TM \longrightarrow T\mathbb{C}^N$ denotes the complex differential map between tangent complex vector bundles induced by f. Furthermore, let $\iota : T\mathbb{C}^N \longrightarrow \mathbb{C}^N$ be the map which sends a tangent vector $v \in T_x\mathbb{C}^N$ at a point $x \in \mathbb{C}^N$ to the sum $x + v$, where the standard identification $T_x\mathbb{C}^N = \mathbb{C}^N$ is assumed.

Definition 5.24 Let $\ell \geq 1$ be an integer, and let M be a smooth complex d-dimensional manifold. A smooth complex embedding $f : M \longrightarrow \mathbb{C}^N$ is an **complex ℓ-skew embedding** if for every $(y_1, \dots, y_\ell) \in \mathrm{F}(M, \ell)$ the collection of complex affine subspaces $\{(\iota \circ df_{y_1})(T_{y_1}M), \dots, (\iota \circ df_{y_\ell})(T_{y_\ell}M)\}$ of $\mathbb{C}^N$ is affinely independent.

In the following we will work with complex vector bundles that are analogues of real vector bundles introduced in (5.2), see also [14, Sec. 4]. Hence, consider the complex vector bundles:

$$\begin{aligned}
\xi_{X,k}^{\mathbb{C}} &: \quad \mathbb{C}^k \longrightarrow \mathrm{F}(X, k) \times_{\mathfrak{S}_k} \mathbb{C}^k \longrightarrow \mathrm{F}(X, k)/\mathfrak{S}_k, \\
\zeta_{X,k}^{\mathbb{C}} &: \quad W_k^{\mathbb{C}} \longrightarrow \mathrm{F}(X, k) \times_{\mathfrak{S}_k} W_k^{\mathbb{C}} \longrightarrow \mathrm{F}(X, k)/\mathfrak{S}_k, \\
\tau_{X,k}^{\mathbb{C}} &: \quad \mathbb{C} \longrightarrow \mathrm{F}(X, k)/\mathfrak{S}_k \times \mathbb{C} \longrightarrow \mathrm{F}(X, k)/\mathfrak{S}_k,
\end{aligned} \tag{5.18}$$

where $W_k^{\mathbb{C}} = \{(b_1, \dots, b_k) \in \mathbb{C}^k : b_1 + \dots + b_k = 0\}$ is an $\mathfrak{S}_k$-invariant subspace of $\mathbb{C}^k$. It is obvious that on the level of underlying real vector bundles the following bundle isomorphisms hold:

$$\xi_{X,k}^{\mathbb{C}} \cong \xi_{X,k}^{\oplus 2}, \qquad \zeta_{X,k}^{\mathbb{C}} \cong \zeta_{X,k}^{\oplus 2}, \qquad \tau_{X,k}^{\mathbb{C}} \cong \tau_{X,k}^{\oplus 2}.$$

Like in the real case a criterion for an existence of complex k-regular embeddings and of complex ℓ-skew embeddings can be phrased in terms of the vector bundle $\xi_{X,k}^{\mathbb{C}}$. Let us recall the relevant special cases of [14, Lem. 5.7, Lem. 6.6].

Lemma 5.25 *Let $k \geq 1$, $\ell \geq 1$, $d \geq 1$ and $N \geq 1$ be integers.*

(1) *If there exists a complex k-regular embedding* $\mathbb{R}^d \longrightarrow \mathbb{C}^N$, *then the complex vector bundle* $\xi^{\mathbb{C}}_{\mathbb{R}^d,k}$ *admits an* $(N-k)$*-dimensional complex inverse.*
(2) *If there exists a complex* ℓ*-skew embedding* $\mathbb{C}^d \longrightarrow \mathbb{C}^N$, *then the complex vector bundle* $(\xi^{\mathbb{C}}_{\mathbb{C}^d,\ell})^{\oplus(d+1)}$ *admits an* $(N-(d+1)\ell+1)$*-dimensional complex inverse.*

The result we prove next corrects [14, Thm. 5.1]. In particular, we follow the outline of the proof of [14, Thm. 5.1] and alternate at a single place.

Theorem 5.26 *Let* $d \geq 1$ *be an integer. There is no complex k-regular embedding* $\mathbb{R}^d \longrightarrow \mathbb{C}^N$ *for* $N < \frac{1}{2}(M+k)$, *where*

$$M := \begin{cases} (d-1)(k-\alpha(k)), & d \text{ is a power of } 2, \\ (d-1)(k-\epsilon(k)), & d \text{ is not a power of } 2. \end{cases}$$

Proof Consider a complex k-regular embedding $\mathbb{R}^d \longrightarrow \mathbb{C}^N$. From Lemma 5.25 the complex vector bundle $\xi^{C}_{\mathbb{R}^d,k}$ admits an $(N-k)$-dimensional complex inverse. Hence, the real vector bundle $\xi^{\oplus 2}_{\mathbb{R}^d,k}$ admits a $2(N-k)$-dimensional real inverse. In particular, the real vector bundle $\xi_{\mathbb{R}^d,k}$ admits a $(2N-k)$-dimensional real inverse. Since $\overline{w}_M(\xi_{\mathbb{R}^d,k}) \neq 0$, Theorem 5.13, we have that

$$2N-k \geq M \quad \Longleftrightarrow \quad N \geq \tfrac{1}{2}(M+k).$$

Remark 5.27 Notice that, compared to [14, Thm. 5.1], a correction was needed for the case when d is not a power of 2.

Next we correct [14, Thm. 6.1]. Again we proceed in the footsteps of the proof of [14, Thm. 6.1].

Theorem 5.28 *Let* $d \geq 1$ *and* $\ell \geq 1$ *be integers. There is no complex* ℓ*-skew embedding* $\mathbb{C}^d \longrightarrow \mathbb{C}^N$ *for*

$$N \leq d + \frac{1}{2}(M-\ell-2),$$

where

$$M := \begin{cases} 2^{\gamma(d)+1} - 2d - 1, & \ell = 2, \\ (2d-1)(\ell-\alpha(\ell)), & d \text{ is a power of } 2, \\ (2^{\gamma(d)+1} - 2d - 1)(\ell-\epsilon(\ell))/2, & d \geq 3 \text{ is not power of } 2. \end{cases}$$

Proof Consider a complex ℓ-skew embedding $\mathbb{C}^d \longrightarrow \mathbb{C}^N$. According to Lemma 5.25 the complex vector bundle $(\xi^{\mathbb{C}}_{\mathbb{C}^d,\ell})^{\oplus(d+1)}$ admits an $(N-(d+1)\ell+1)$-dimensional complex inverse. Hence, the real vector bundle $\xi^{\oplus 2(d+1)}_{\mathbb{R}^{2d},\ell}$ admits a

$2(N-(d+1)\ell+1)$-dimensional real inverse. Consequently, the real vector bundle $\xi_{\mathbb{R}^{2d},\ell}^{\oplus(2d+1)}$ admits a $(2N-2(d+1)\ell+2+\ell)$-dimensional real inverse. From Theorem 5.17 follows that $\overline{w}_M(\xi_{\mathbb{R}^{2d},\ell}^{\oplus(2d+1)}) \neq 0$ and therefore:

$$2N-2(d+1)\ell+2+\ell \geq M \quad \Longleftrightarrow \quad N \geq \frac{1}{2}\big(2(d+1)-2-\ell+M\big).$$

This completes the corrections of the paper [14].

Chapter 6
More Bounds for Highly Regular Embeddings

In this chapter we present computations that yield additional bounds for the existence of highly regular embeddings. For this we present an alternative approach in the study of Stiefel–Whitney classes of the vector bundle $\xi_{\mathbb{R}^d,2^m}$. In particular, we utilize a specific decomposition of the pull-back vector bundle $\rho^*_{d,2^m}\xi_{\mathbb{R}^d,2^m}$ of the vector bundle $\xi_{\mathbb{R}^d,2^m}$ via the map $\rho_{d,2^m}\colon\ \mathrm{Pe}(\mathbb{R}^d,2^m)/S_{2^m} \longrightarrow \mathrm{F}(\mathbb{R}^d,2^m)/\mathfrak{S}_{2^m}$.

First, we introduce several real S_{2^m}-representations and consider properties of the associated vector bundles.

6.1 Examples of S_{2^m}-Representations and Associated Vector Bundles

In this text the real vector space $\mathbb{R}^k$ is often considered as the real k-dimensional $\mathfrak{S}_k$-representation with the action defined by

$$\pi\cdot(a_1,\dots,a_k) := (a_{\pi^{-1}(1)},\dots,a_{\pi^{-1}(k)}),$$

where $\pi\in\mathfrak{S}_k$ and $(a_1,\dots,a_k)\in\mathbb{R}^k$. Furthermore, its $(n-1)$-dimensional vector subspace

$$W_k := \{(a_1,\dots,a_k)\in\mathbb{R}^k : a_1+\cdots+a_k=0\}$$

is an $\mathfrak{S}_k$-subrepresentation. For every subgroup G of $\mathfrak{S}_k$ both vector spaces $\mathbb{R}^k$ and W_k become real G-representations via the corresponding inclusion homomorphism $G\hookrightarrow\mathfrak{S}_k$.

P. V. M. Blagojević et al., *Equivariant Cohomology of Configuration Spaces Mod 2*, Lecture Notes in Mathematics 2282, https://doi.org/10.1007/978-3-030-84138-6_6

6.1.1 Examples of $\mathcal{S}_{2^m}$-Representations

Let $k = 2^m$ for some integer $m \geq 1$. Recall that the group $\mathcal{S}_{2^m}$, as a subgroup of $\mathfrak{S}_{2^m}$, was introduced in Definition 2.1 and the corresponding inclusion homomorphism was denotes by $\iota_m : \mathcal{S}_{2^m} \longrightarrow \mathfrak{S}_{2^m}$. Inductively, we can describe the group $\mathcal{S}_{2^m}$ via the exact sequence of groups

$$1 \longrightarrow \mathcal{S}_{2^{m-1}} \times \mathcal{S}_{2^{m-1}} \xrightarrow{\vartheta_m} \mathcal{S}_{2^m} \xrightarrow{\varsigma_m} \mathbb{Z}_2 \longrightarrow 1, \tag{6.1}$$

where the inclusion map ϑ_m is defined in the expected way.

Now we define inductively a sequence of real $\mathcal{S}_{2^m}$-representations. Let $M_m[m]$ be the $\mathcal{S}_{2^m}$-representation obtained as the pull-back of the real 1-dimensional $\mathbb{Z}_2$-representation W_2 via the surjection ς_m from the exact sequence (6.1). This means that $M_m[m] = W_2$ as a vector space, and that $\pi \cdot v := \varsigma_m(\pi) \cdot v$ for every $v \in M_m[m]$ and every $\pi \in \mathcal{S}_{2^m}$.

Assume that for every integer $1 \leq i \leq m-1$ we have defined the sequence of $\mathcal{S}_{2^i}$-representations $M_i[1], \ldots, M_i[i]$ with $\dim(M_i[j]) = 2^{i-j}, 1 \leq j \leq i$. Next, for $i = m$ and $1 \leq j \leq m-1$, we define the $\mathcal{S}_{2^m}$-representation $M_m[j]$ to be, on the level of vector spaces, the direct sum $M_m[j] := M_{m-1}[j] \oplus M_{m-1}[j]$. The action of $\mathcal{S}_{2^m}$ on $M_m[j]$ is given by:

$$(h_1, h_2) \cdot (v_1, v_2) := (h_1 \cdot v_1, h_2 \cdot v_2),$$
$$(h_1, h_2, \omega) \cdot (v_1, v_2) := (h_2 \cdot v_2, h_1 \cdot v_1),$$

where $(h_1, h_2) \in \mathcal{S}_{2^{m-1}} \times \mathcal{S}_{2^{m-1}} \subseteq (\mathcal{S}_{2^{m-1}} \times \mathcal{S}_{2^{m-1}}) \rtimes \mathbb{Z}_2$, ω is the generator of the subgroup $\mathbb{Z}_2 \subseteq (\mathcal{S}_{2^{m-1}} \times \mathcal{S}_{2^{m-1}}) \rtimes \mathbb{Z}_2$, and $(v_1, v_2) \in M_{m-1}[j] \oplus M_{m-1}[j]$. It follows directly that $\dim(M_m[j]) = 2^{m-j}$ for all $1 \leq j \leq m$.

The vector space W_{2^m}, now considered as an $\mathcal{S}_{2^m}$-representation, we denote by L_m. The following decomposition of $\mathcal{S}_{2^m}$-representations holds.

Lemma 6.1 *For every integer $m \geq 1$ there is an isomorphism of real $\mathcal{S}_{2^m}$-representations*

$$L_m \cong M_m[1] \oplus \cdots \oplus M_m[m].$$

6.1.2 Associated Vector Bundles

Let $m \geq 1$ be an integer, and let $d \geq 1$ be an integer, or $d = \infty$. To every $\mathcal{S}_{2^m}$-representation introduced in the previous section we associate real vector bundles in the following way:

$$\lambda_{d,m} : \quad L_m \longrightarrow \mathrm{Pe}(\mathbb{R}^d, 2^m) \times_{\mathcal{S}_{2^m}} L_m \longrightarrow \mathrm{Pe}(\mathbb{R}^d, 2^m)/\mathcal{S}_{2^m}, \tag{6.2}$$

$$\mu_{d,m}[j]: \quad M_m[j] \longrightarrow \mathrm{Pe}(\mathbb{R}^d, 2^m) \times_{\mathcal{S}_{2^m}} M_m[j] \longrightarrow \mathrm{Pe}(\mathbb{R}^d, 2^m)/\mathcal{S}_{2^m}, \tag{6.3}$$

where $1 \le j \le m$. Note that

- $\lambda_{1,d} = \mu_{1,d}[1]$ is the Hopf line bundle over $\mathbb{RP}^{d-1} \cong \mathrm{Pe}(\mathbb{R}^d, 2)/\mathcal{S}_2$,
- $\mu_{d,m}[m]$ is the pull-back vector bundle of the line bundle $\lambda_{d,1}$ via the map

$$\big(\mathrm{Pe}(\mathbb{R}^d, 2^{m-1})/\mathcal{S}_{2^{m-1}} \times \mathrm{Pe}(\mathbb{R}^d, 2^{m-1})/\mathcal{S}_{2^{m-1}}\big) \times_{\mathbb{Z}_2} \mathrm{Pe}(\mathbb{R}^d, 2) \longrightarrow \mathrm{Pe}(\mathbb{R}^d, 2)/\mathcal{S}_2,$$

- $\lambda_{d,m}$ and $\mu_{d,m}[j]$ are pull-backs of $\lambda_{\infty,m}$ and $\mu_{\infty,m}[j]$ along the map

$$\kappa_{d,2^m}/\mathcal{S}_{2^m}: \mathrm{Pe}(\mathbb{R}^d, 2^m)/\mathcal{S}_{2^m} \longrightarrow \mathrm{Pe}(\mathbb{R}^\infty, 2^m)/\mathcal{S}_{2^m},$$

- $\lambda_{d,m}$ is the pull-back of $\zeta_{\mathbb{R}^d,2^m}$ via the map

$$\rho_{d,2^m}: \mathrm{Pe}(\mathbb{R}^d, 2^m)/\mathcal{S}_{2^m} \longrightarrow \mathrm{F}(\mathbb{R}^d, 2^m)/\mathfrak{S}_{2^m}.$$

The vector bundles $\mu_{d,m}[j]$ for different values of m, are connected via the wreath square of bundles operations as follows:

$$S^2\mu_{\infty,m-1}[j] \cong \mu_{\infty,m}[j] \qquad \text{and} \qquad S^{2,d}\mu_{d,m-1}[j] \cong \mu_{d,m}[j], \tag{6.4}$$

for all integers $d \ge 2$, $m \ge 1$, and all $1 \le j \le m-1$. For details about wreath square operations see Chap. 9.

The decomposition from Lemma 6.1 implies the following Whitney sum decomposition on the vector bundle $\lambda_{d,m}$.

Lemma 6.2 *For every integer $m \ge 1$, and every integer $d \ge 1$ or $d = \infty$, there is an isomorphism of real $\mathcal{S}_{2^m}$-representations*

$$\lambda_{d,m} \cong \mu_{d,m}[1] \oplus \cdots \oplus \mu_{d,m}[m].$$

Using the relation (3.32) and the decomposition from the previous lemma we get the following equalities.

Lemma 6.3 *For every integer $m \ge 1$ and every integer $1 \le j \le m$:*

$$w_{2^m-1}(\lambda_{m,\infty}) = D_{m,0} \qquad \textit{and} \qquad w_{2^{m-j}}(\mu_{m,\infty}[j]) = V_{m,m-j}.$$

From the fact that $w(\zeta_{\mathbb{R}^d,2^m}) = w(\xi_{\mathbb{R}^d,2^m})$, Lemmas 5.5, 5.7, and Theorem 4.1, the injectivity of the homomorphism

$$\rho^*_{d,2^m}: H^*(\mathrm{F}(\mathbb{R}^d, 2^m)/\mathfrak{S}_{2^m}; \mathbb{F}_2) \longrightarrow H^*(\mathrm{Pe}(\mathbb{R}^d, 2^m)/\mathcal{S}_{2^m}; \mathbb{F}_2),$$

we deduce the following facts.

Lemma 6.4 *Let $m \geq 1$ be an integer.*

(1) *If $d = 2^a$ for an integer $a \geq 1$, then*

$$\begin{aligned}\overline{w}_{(d-1)(2^m-1)}(\lambda_{d,m}) &= \overline{w}_{(d-1)(2^m-1)}(\rho^*_{d,2^m}\zeta_{\mathbb{R}^d,2^m})\\ &= \overline{w}_{(d-1)(2^m-1)}(\rho^*_{d,2^m}\xi_{\mathbb{R}^d,2^m})\\ &= \rho^*_{d,2^m}\big(\overline{w}_{(d-1)(2^m-1)}(\xi_{\mathbb{R}^d,2^m})\big) \neq 0.\end{aligned}$$

(2) *If $d \geq 2$ is an integer, then*

$$\begin{aligned}\overline{w}_{(d-1)2^{m-1}}(\lambda_{d,m}) &= \overline{w}_{(d-1)2^{m-1}}(\rho^*_{d,2^m}\zeta_{\mathbb{R}^d,2^m})\\ &= \overline{w}_{(d-1)2^{m-1}}(\rho^*_{d,2^m}\xi_{\mathbb{R}^d,2^m})\\ &= \rho^*_{d,2^m}\big(\overline{w}_{(d-1)2^{m-1}}(\xi_{\mathbb{R}^d,2^m})\big) \neq 0.\end{aligned}$$

Furthermore, from the fact that $\mu_{d,m}[m]$ is the pull-back of the line bundle $\lambda_{d,1}$, and the description of the cohomology of $\mathrm{Pe}(\mathbb{R}^d, 2^m)/S_{2^m}$ given in Chap. 3 we compute $\mu_{d,m}[m]$.

Lemma 6.5 *Let $d \geq 1$ and $m \geq 1$ be integers. Then*

$$w(\mu_{d,m}[m]) = 1 + f,$$

where the class $f \in H^1(\mathrm{Pe}(\mathbb{R}^d, 2^m)/S_{2^m}; \mathbb{F}_2)$ was denoted by $V_{m,1}$ in the proof of Lemma 3.10. In particular $f^d = 0$.

6.2 The Key Lemma and its Consequences

Let $m \geq 1$ and $d \geq 1$ be integers. For arbitrary non-negative integers $r_1, \ldots, r_m$ we define the vector bundle $\psi_{d,m}[r_1, \ldots, r_m]$ by:

$$\psi_{d,m}[r_1, \ldots, r_m] := \mu_{d,m}[1]^{\oplus r_1} \oplus \cdots \oplus \mu_{d,m}[m]^{\oplus r_m}.$$

Now we will prove the key technical lemma of this chapter.

Lemma 6.6 (The Key Lemma) *Let $m \geq 1$, $d \geq 1$ and ℓ be integers, and let $r_1, \ldots, r_m$ be non-negative integers. If the binomial coefficient*

$$\binom{(r_1 2^{m-2} + \cdots + r_{m-1}2^0) + r_m - (d-1+\ell)(2^{m-1}-1) - \ell}{d-1} \tag{6.5}$$

is odd, then

$$w_{(d-1)(2^m-1)}(\lambda_{d,m}^{-\ell} \oplus \psi_{d,m}[r_1, \dots, r_m]) \neq 0$$

as an element of the cohomology group $H^{(d-1)(2^m-1)}(\mathrm{Pe}(\mathbb{R}^d, 2^m)/\mathcal{S}_{2^m}; \mathbb{F}_2)$. *Here* $\lambda_{d,m}^{-\ell}$ *denotes an inverse of the vector bundle* $\lambda_{d,m}^{\oplus\ell}$.

Proof We prove the claim of the lemma by induction on $m \geq 1$. In the case $m = 1$ the condition (6.5) reads $\binom{r_1-\ell}{d-1} \neq 0$ in $\mathbb{F}_2$. Furthermore, $\psi_{d,1}[r_1] = \mu_{d,1}[1]^{\oplus r_1}$ and $\lambda_{d,1} \cong \mu_{d,1}[1]$. Consequently, $\lambda_{d,1}^{-\ell} \oplus \psi_{d,1}[r_1] \cong \mu_{d,1}[1]^{\oplus(r_1-\ell)}$, and so

$$w(\lambda_{d,1}^{-\ell} \oplus \psi_{d,1}[r_1]) = w(\mu_{d,1}[1]^{\oplus(r_1-\ell)}) = w(\mu_{d,1}[1])^{r_1-\ell} = (1+f)^{r_1-\ell}$$

in $H^*(\mathrm{Pe}(\mathbb{R}^d, 2)/\mathcal{S}_2; \mathbb{F}_2) \cong H^*(\mathbb{R}\mathrm{P}^{d-1}; \mathbb{F}_2) = \mathbb{F}_2[f]/\langle f\rangle$. Here we use the fact that the vector bundle $\mu_{d,1}[1]$ is the Hopf line bundle, and so $w(\mu_{d,1}[1]) = 1 + f$. Hence, we have that

$$w_{(d-1)(2^1-1)}(\lambda_{d,1}^{-\ell} \oplus \psi_{d,1}[r_1]) = w_{d-1}(\lambda_{d,1}^{-\ell} \oplus \psi_{d,1}[r_1]) = \binom{r_1-\ell}{d-1} w_1^{d-1} \neq 0.$$

Let $m \geq 2$, and assume, as an induction hypothesis, that

$$w_{(d-1)(2^{m-1}-1)}(\lambda_{d,m-1}^{-\ell} \oplus \psi_{d,m-1}[r_1, \dots, r_{m-1}]) \neq 0$$

as an element of $H^{(d-1)(2^{m-1}-1)}(\mathrm{Pe}(\mathbb{R}^d, 2^{m-1})/\mathcal{S}_{2^{m-1}}; \mathbb{F}_2)$. Then there exists a vector bundle ϖ of dimension $(d-1)(2^{m-1}-1)$ over $\mathrm{Pe}(\mathbb{R}^d, 2^{m-1})/\mathcal{S}_{2^{m-1}}$ with the property that for some integers $N_0 \geq \ell$ and $N_1 \geq 1$ there is an isomorphism of vector bundles

$$\begin{aligned}\varpi \oplus \lambda_{d,m-1}^{\oplus\ell} \oplus \tau_{d,m-1}^{\oplus N_0} &\cong \psi_{d,m-1}[r_1, \dots, r_{m-1}] \oplus \tau_{d,m-1}^{\oplus N_1} \\ &\cong \mu_{d,m-1}[1]^{\oplus r_1} \oplus \dots \oplus \mu_{d,m-1}[m-1]^{\oplus r_{m-1}} \oplus \tau_{d,m-1}^{\oplus N_1}.\end{aligned} \tag{6.6}$$

Here $\tau_{d,m-1}$ denotes the trivial line bundle over $\mathrm{Pe}(\mathbb{R}^d, 2^{m-1})/\mathcal{S}_{2^{m-1}}$. Consequently,

$$w_{(d-1)(2^{m-1}-1)}(\varpi) = w_{(d-1)(2^{m-1}-1)}(\lambda_{d,m-1}^{-\ell} \oplus \psi_{d,m-1}[r_1, \dots, r_{m-1}]) \neq 0 \tag{6.7}$$

does not vanish—by induction hypothesis.

Now we apply $(d-1)$-partial wreath square $S^{2,d}$ to the isomorphism of vector bundles (6.6) and get the following isomorphism of vector bundles

$$\begin{aligned} &S^{2,d}\varpi \oplus \lambda_{d,m}^{\oplus\ell} \oplus \tau_{d,m}^{\oplus N_0} \oplus \mu_{d,m}[m]^{\oplus N_0-\ell} \\ &\quad\cong \mu_{d,m}[1]^{\oplus r_1} \oplus \cdots \oplus \mu_{d,m}[m-1]^{\oplus r_{m-1}} \oplus \tau_{d,m}^{\oplus N_1} \oplus \mu_{d,m}[m]^{\oplus N_1}, \end{aligned} \tag{6.8}$$

over the base space $S^{2,d}\,\mathrm{Pe}(\mathbb{R}^d, 2^{m-1})/\mathcal{S}_{2^{m-1}} \cong \mathrm{Pe}(\mathbb{R}^d, 2^m)/\mathcal{S}_{2^m}$. Indeed, by direct inspection and using (6.4) and (9.1), we get that

$$S^{2,d}\tau_{d-1,m} \cong \tau_{d,m} \oplus \mu_{d,m}[m] \qquad \text{and} \qquad S^{2,d}\mu_{d,m-1}[j] \cong \mu_{d,m}[j]$$

for all $1 \le j \le m-1$. Consequently, from Lemma 6.2 we get that

$$S^2\lambda_{d,m-1} \cong S^2(\mu_{d,m-1}[1]\oplus\cdots\oplus\mu_{d,m-1}[m-1]) \cong \mu_{d,m}[1]\oplus\cdots\oplus\mu_{d,m}[m-1].$$

Collecting all these facts together we compute

$$\begin{aligned} &\varpi \oplus \lambda_{d,m-1}^{\oplus\ell} \oplus \tau_{d,m-1}^{\oplus N_0} = \\ &\qquad \varpi \oplus (\mu_{d,m-1}[1] \oplus \cdots \oplus \mu_{d,m-1}[m-1])^{\oplus\ell} \oplus \tau_{d,m-1}^{\oplus N_0} \\ &\qquad\qquad \Big\downarrow S^{2,d} \\ &S^{2,d}\varpi \oplus (\mu_{d,m}[1] \oplus \cdots \oplus \mu_{d,m}[m-1])^{\oplus\ell} \oplus \tau_{d,m}^{N_0} \oplus \mu_{d,m}[m]^{\oplus N_0} \\ &\qquad \cong S^{2,d}\varpi \oplus \lambda_{d,m}^{\oplus\ell} \oplus \tau_{d,m}^{N_0} \oplus \mu_m[m]^{\oplus N_0-\ell}. \end{aligned}$$

On the other hand directly follows that

$$\begin{aligned} &\mu_{d,m-1}[1]^{\oplus r_1} \oplus \cdots \oplus \mu_{d,m-1}[m-1]^{\oplus r_{m-1}} \oplus \tau_{d,m-1}^{\oplus N_1} \\ &\qquad\qquad \Big\downarrow S^{2,d} \\ &\qquad \mu_{d,m}[1]^{\oplus r_1} \oplus \cdots \oplus \mu_{d,m}[m-1]^{\oplus r_{m-1}} \oplus \tau_{d,m}^{\oplus N_1} \oplus \mu_{d,m}[m]^{\oplus N_1}. \end{aligned}$$

Thus, the isomorphism (6.8) holds.

For an arbitrary vector bundle η over $\mathrm{Pe}(\mathbb{R}^d, 2^m)/\mathcal{S}_{2^m}$ the isomorphism of vector bundles (6.8) yields the isomorphism

$$\begin{aligned}&\eta \oplus S^{2,d}\varpi \oplus \lambda_{d,m}^{\oplus \ell} \oplus \tau_{d,m}^{\oplus N_0} \oplus \mu_{d,m}[m]^{\oplus N_0-\ell}\\ &\cong \eta \oplus \mu_{d,m}[1]^{\oplus r_1} \oplus \cdots \oplus \mu_{d,m}[m-1]^{\oplus r_{m-1}} \oplus \tau_{d,m}^{\oplus N_1} \oplus \mu_{d,m}[m]^{\oplus N_1}.\end{aligned} \tag{6.9}$$

Take η to be stable equivalent to the vector bundle $\mu_{d,m}[m]^{\oplus r_m+N_0-\ell-N_1}$. Then the vector bundles

$$\eta \oplus S^{2,d}\varpi \qquad \text{and} \qquad \lambda_{d,m}^{-\ell} \oplus \psi_{d,m}[r_1, \dots, r_m]$$

are stable equivalent. Indeed, if η is stable equivalent to $\mu_{d,m}[m]^{\oplus r_m+N_0-\ell-N_1}$, then there exist integers $M_1, M_2 \geq 0$, and an isomorphism

$$\eta \oplus \tau_{d,m}^{\oplus M_1} \cong \mu_{d,m}[m]^{\oplus r_m+N_0-\ell-N_1} \oplus \tau_{d,m}^{\oplus M_2}.$$

Now the isomorphism (6.9) implies that

$$\begin{aligned}&\eta \oplus S^{2,d}\varpi \oplus \lambda_{d,m}^{\oplus \ell} \oplus \tau_{d,m}^{\oplus N_0+M_1} \oplus \mu_{d,m}[m]^{\oplus N_0-\ell}\\ &\cong (\mu_{d,m}[1]^{\oplus r_1} \oplus \cdots \oplus \mu_{d,m}[m-1]^{\oplus r_{m-1}}) \oplus \mu_{d,m}[m]^{\oplus r_m} \oplus\\ &\mu_{d,m}[m]^{\oplus N_0-\ell-N_1} \oplus \tau_{d,m}^{\oplus N_1+M_2} \oplus \mu_{d,m}[m]^{\oplus N_1}.\end{aligned}$$

Hence,

$$\begin{aligned}&(\eta \oplus S^{2,d}\varpi) \oplus \lambda_{d,m}^{\oplus \ell} \oplus \tau_{d,m}^{\oplus N_0+M_1} \oplus \mu_{d,m}[m]^{\oplus N_0-\ell}\\ &\cong \psi_{d,m}[r_1, \dots, r_m] \oplus \tau_{d,m}^{\oplus N_1+M_2} \oplus \mu_{d,m}[m]^{\oplus N_0-\ell},\end{aligned}$$

and consequently the vector bundles $\eta \oplus S^{2,d}\varpi$ and $\lambda_{d,m}^{-\ell} \oplus \psi_{d,m}[r_1, \dots, r_m]$ are stable equivalent. In particular,

$$w(\lambda_{d,m}^{-\ell}\oplus\psi_{d,m}[r_1, \dots, r_m]) = w(\eta\oplus S^{2,d}\varpi) = w(\mu_{d,m}[m]^{\oplus r_m+N_0-\ell-N_1}\oplus S^{2,d}\varpi).$$

Now we continue computation of the total Stiefel–Whitney class of the vector bundle $\lambda_{d,m}^{-\ell} \oplus \psi_{d,m}[r_1, \dots, r_m]$ as follows:

$$\begin{aligned}w(\lambda_{d,m}^{-\ell} \oplus \psi_{d,m}[r_1, \dots, r_m]) &= w(\mu_{d,m}[m]^{\oplus r_m+N_0-\ell-N_1} \oplus S^{2,d}\varpi)\\ &= w(\mu_{d,m}[m])^{\oplus r_m+N_0-\ell-N_1} \cdot w(S^{2,d}\varpi)\\ &\overset{\text{Lemma 6.5}}{=} (1+f)^{r_m+N_0-\ell-N_1} \cdot w(S^{2,d}\varpi)\end{aligned}$$

$$
\begin{aligned}
&= (1+f)^{r_m+N_0-\ell-N_1}\cdot s^{2,d}(\varpi)\\
&= \Big(\sum_{i=0}^{r_m+N_0-\ell-N_1}\binom{r_m+N_0-\ell-N_1}{i}f^i\Big)\cdot s^{2,d}(\varpi)\\
&\overset{f^d=0}{=} \Big(\sum_{i=0}^{d-1}\binom{r_m+N_0-\ell-N_1}{i}f^i\Big)\cdot s^{2,d}(\varpi).
\end{aligned}
$$

From Corollary 9.2 applied to the $(d-1)(2^{m-1}-1)$-dimensional vector bundle ϖ we have that

$$
\begin{aligned}
s^{2,d}(\varpi) = &\sum_{0\le r<s\le N} T(w_r(\varpi)\otimes w_s(\varpi))\\
&+\sum_{0\le r\le N}\ \sum_{0\le j\le \min\{N-r,d-1\}}\binom{N-r}{j}P(w_r(\varpi))f^j,
\end{aligned}
$$

where $N=(d-1)(2^{m-1}-1)$.

Combining the last two equations we get

$$
\begin{aligned}
w(\lambda_{d,m}^{-\ell}\oplus\psi_{d,m}[r_1,\dots,r_m]) = &\Big(\sum_{i=0}^{d-1}\binom{r_m+N_0-\ell-N_1}{i}f^i\Big)\\
&\cdot\Big(\sum_{0\le r<s\le N} T(w_r(\varpi)\otimes w_s(\varpi))\\
&+\sum_{0\le r\le N}\ \sum_{0\le j\le \min\{N-r,d-1\}}\binom{N-r}{j}P(w_r(\varpi))f^j\Big)\\
&\overset{(9.5)}{=}\sum_{0\le r<s\le N} T(w_r(\varpi)\otimes w_s(\varpi))\\
&+\sum_{i=0}^{d-1}\sum_{0\le r\le N}\ \sum_{0\le j\le \min\{N-r,d-1\}}\\
&\binom{r_m+N_0-\ell-N_1}{i}\binom{N-r}{j}P(w_r(\varpi))\,f^{i+j}.
\end{aligned}
$$

Note that

- $\deg(T(w_r(\varpi)\otimes w_s(\varpi)))\le 2N-1=(d-1)(2^m-2)-1$, and
- $\deg(P(w_r(\varpi))f^{i+j})\le 2N+d-1=(d-1)(2^m-1)$.

Consequently, we have that

$$w_{(d-1)(2^m-1)}(\lambda_{d,m}^{-\ell} \oplus \psi_{d,m}[r_1, \ldots, r_m]) = \binom{r_m + N_0 - \ell - N_1}{d-1} P(w_N(\varpi))\, f^{d-1}.$$

Before making final arguments that the Stiefel–Whitney class

$$w_{(d-1)(2^m-1)}(\lambda_{d,m}^{-\ell} \oplus \psi_{d,m}[r_1, \ldots, r_m])$$

does not vanish let us review all the assumptions we have:

- $\binom{(r_1 2^{m-2}+\cdots+r_{m-1}2^0)+r_m-(d-1+\ell)(2^{m-1}-1)-\ell}{d-1} \neq 0 \in \mathbb{F}_2$—the assumption (6.5) from the statement of the lemma, and
- $w_{(d-1)(2^{m-1}-1)}(\varpi) \neq 0$—the induction hypothesis (6.7).

Hence, by the induction hypothesis we know that the class $P(w_N(\varpi))\, f^{d-1}$ does not vanish. For the binomial coefficient $\binom{r_m+N_0-\ell-N_1}{d-1}$ note that the isomorphism of vector bundles (6.6), by evaluating dimensions, implies that

$$N_0 - N_1 = (r_1 2^{m-2} + \cdots + r_{m-1}2^0) - (d-1+\ell)(2^{m-1}-1),$$

or in other words

$$r_m + N_0 - \ell - N_1 = (r_1 2^{m-2} + \cdots + r_{m-1}2^0) + r_m - (d-1+\ell)(2^{m-1}-1) - \ell.$$

Therefore, by the assumption (6.5)

$$\binom{r_m + N_0 - \ell - N_1}{d-1} \neq 0 \in \mathbb{F}_2.$$

This completes the proof that

$$w_{(d-1)(2^m-1)}(\lambda_{d,m}^{-\ell} \oplus \psi_{d,m}[r_1, \ldots, r_m]) \neq 0.$$

Remark 6.7 Observe that the proof of the previous lemma actually yields the following equivalence:

$$\binom{(r_1 2^{m-2} + \cdots + r_{m-1}2^0) + r_m - (d-1+\ell)(2^{m-1}-1) - \ell}{d-1} \neq 0 \in \mathbb{F}_2$$

if and only if

$$w_{(d-1)(2^m-1)}(\lambda_{d,m}^{-\ell} \oplus \psi_{d,m}[r_1, \ldots, r_m]) \neq 0.$$

Remark 6.8 The proof of Lemma 6.6 can be simplified as follows. In the step when made a choice of the vector bundle η, to be stable equivalent to the vector bundle

$\mu_m[m]^{\oplus r_m+N_0-\ell-N_1}$, we could have in addition asked that η is in addition $(d-1)$-dimensional. (Note that $\mu_m[m]$ is a pull-back vector bundle of the vector bundle $\lambda_{d,1}$ over $\mathbb{R}\mathrm{P}^{d-1}$.) Then the Stiefel–Whitney class

$$w_{(d-1)(2^m-1)}(\lambda_{d,m}^{-\ell} \oplus \psi_{d,m}[r_1, \ldots, r_m]) = w_{(d-1)(2^m-1)}(\eta \oplus S^{2,d}\varpi)$$

is actually the mod 2 Euler class $\mathfrak{e}(\eta \oplus S^{2,d}\varpi)$ of the vector bundle $\eta \oplus S^{2,d}\varpi$. Indeed, $\dim(\eta \oplus S^{2,d}\varpi) = (d-1)(2^m-1)$. Thus, using the product formula for Euler classes [80, Prop. 9.6] we have that

$$w_{(d-1)(2^m-1)}(\eta \oplus S^{2,d}\varpi) = \mathfrak{e}(\eta \oplus S^{2,d}\varpi) = \mathfrak{e}(\eta) \cdot \mathfrak{e}(S^{2,d}\varpi).$$

Here $\mathfrak{e}(\cdot)$ denotes the mod 2 Euler class of the respected vector bundle. Now from Corollary 9.3 and the description of the cohomology $H^*(\mathrm{Pe}(\mathbb{R}^d, 2^{m-1})/S_{2^{m-1}}; \mathbb{F}_2)$ of the base space we have that

$$\begin{aligned} w_{(d-1)(2^m-1)}(\eta \oplus S^{2,d}\varpi) &= \mathfrak{e}(\eta) \cdot \mathfrak{e}(S^{2,d}\varpi) \\ &= \binom{r_m + N_0 - \ell - N_1}{d-1} P(w_N(\varpi))\, f^{d-1} \neq 0. \end{aligned}$$

Thus, there was no need to evaluate the total Stiefel–Whitney class of the vector bundle $\eta \oplus S^{2,d}\varpi$ in full.

After proving Lemma 6.6 we want to discuss how to utilize it. In other words, for which integer parameters $d \geq 2$, $m \geq 1$, $r_1, \ldots, r_m \geq 0$ the assumption (6.5) is satisfied.

Lemma 6.9 *Let $d \geq 2$, $m \geq 1$ and ℓ be integers, and let $d = 2^t + e$ for some integers $t \geq 1$ and $0 \leq e \leq 2^t - 1$. If*

(1) $\ell = 1$, $r_1 = 0$, *and* $r_2 = \cdots = r_m = 2e$, *or*
(2) $\ell = d + 1$, *and* $r_1 = \cdots = r_m = 2e$, *or*
(3) $\ell = -(d - 1 + k2^{t+1})$ *for some integer* k, *and* $r_1 = \cdots = r_m = 0$,

then

$$\binom{(r_1 2^{m-2} + \cdots + r_{m-1} 2^0) + r_m - (d-1+\ell)(2^{m-1}-1) - \ell}{d-1} = 1 \in \mathbb{F}_2.$$

Proof

(1) Let $\ell = 1$, $r_1 = 0$, and $r_2 = \cdots = r_m = 2e$. Then

$$\begin{aligned} &(r_1 2^{m-2} + \cdots + r_{m-1} 2^0) + r_m - (d-1+\ell)(2^{m-1}-1) - \ell \\ &= 2e(2^{m-3} + \cdots + 2^0) + 2e - d(2^{m-1}-1) - 1 \end{aligned}$$

$$= 2^{m-1}(e-d) + 2e + d - 1$$

$$= 2^{m-1}(e-d) + d - 1 = \begin{cases} d - 1 - 2^{m-1+t}, & m \geq 2, \\ -1, & m = 1. \end{cases}$$

Now, $\binom{-1}{d-1} =_{\mathbb{F}_2} 1 \in \mathbb{F}_2$ and $\binom{d-1-2^{m-1+t}}{d-1} =_{\mathbb{F}_2} \binom{d-1}{d-1} =_{\mathbb{F}_2} 1 \in \mathbb{F}_2$, because for $m \geq 2$ we have that $m - 1 + t \geq t + 1$ and $d - 1 < 2^{m+1}$.

(2) Let $\ell = d + 1$, and $r_1 = \cdots = r_m = 2e$. Then

$$(r_1 2^{m-2} + \cdots + r_{m-1} 2^0) + r_m - (d - 1 + \ell)(2^{m-1} - 1) - \ell$$

$$= 2e(2^{m-2} + \cdots + 2^0) + 2e - 2d(2^{m-1} - 1) - d - 1$$
$$= 2^m(e-d) + d - 1$$
$$= d - 1 - 2^{m+t}.$$

Now $\binom{d-1-2^{m+t}}{d-1} =_{\mathbb{F}_2} 1 \in \mathbb{F}_2$, because $d - 1 < 2^{t+1} \leq 2^{t+m}$, and in the ring of formal power series $\mathbb{F}_2[[T]]$ the following equality holds

$$\begin{aligned}(1+T)^{d-1-2^{m+t}} &= (1+T)^{d-1}(1+T)^{-2^{m+t}} \\ &= (1+T)^{d-1}(1+T^{2^{m+t}})^{-1} \\ &= (1+T)^{d-1} \sum_{j \geq 0} T^{j2^{m+t}}.\end{aligned}$$

(3) Let $\ell = -(d - 1 + j2^{t+1})$ for some integer $k \geq 1$, and $r_1 = \cdots = r_m = 0$. Then

$$(r_1 2^{m-2} + \cdots + r_{m-1} 2^0) + r_m - (d - 1 + \ell)(2^{m-1} - 1) - \ell$$
$$= -(d - 1 + \ell)(2^{m-1} - 1) - \ell = d - 1 + k2^{m+t}$$

Hence, $\binom{d-1+k2^{m+t}}{d-1} =_{\mathbb{F}_2} 1 \in \mathbb{F}_2$, because $d - 1 < 2^{m+t}$, and in the ring $\mathbb{F}_2[[T]]$ the following equality holds

$$\begin{aligned}(1+T)^{d-1+k2^{m+t}} &= (1+T)^{d-1}(1+T)^{k2^{m+t}} \\ &= (1+T)^{d-1}(1+T^{2^{m+t}})^k \\ &= (1+T)^{d-1} \sum_{j=0}^{k} \binom{k}{j} T^{j2^{m+t}}.\end{aligned}$$

Now from Lemmas 6.6 and 6.9 we get the following particular results.

Corollary 6.10 *Let $d \geq 2$, $m \geq 1$ and k be integers, and let $d = 2^t + e$ for some integers $t \geq 1$ and $0 \leq e \leq 2^t - 1$. Then*

(1) $w_{(d-1)(2^m-1)}(\lambda_{d,m}^{-1} \oplus (\mu_{d,m}[2] \oplus \cdots \oplus \mu_{d,m}[m])^{\oplus 2e}) \neq 0$,

(2) $w_{(d-1)(2^m-1)}(\lambda_{d,m}^{-(d+1)} \oplus (\mu_{d,m}[1] \oplus \cdots \oplus \mu_{d,m}[m])^{\oplus 2e}) \neq 0$, *and*

(3) $w_{(d-1)(2^m-1)}(\lambda_{d,m}^{d-1+k2^{t+1}}) \neq 0$.

Remark 6.11 In the case (1) of the previous corollary for $e = 0$ we get an alternative proof of Lemma 5.5. On the other hand, in the case (2) for $e = 0$ we have a particular case of Theorem 5.17 (3).

6.3 Additional Bounds for the Existence of Highly Regular Embeddings

In this section we use consequences of the Key Lemma to derive further bounds for the existence of highly regular embeddings.

First, we use Corollary 6.10 to get specific results which are relevant for the study of highly regular embeddings.

Corollary 6.12 *Let $d \geq 2$ and $m \geq 1$ be integers, and let $d = 2^t + e$ for some integers $t \geq 1$ and $0 \leq e \leq 2^t - 1$. Then exist integers a and b with the property that*

$$(d-1)(2^m-1) - 2(2^{m-1}-1)e \leq a \leq (d-1)(2^m-1),$$

$$(d-1)(2^m-1) - 2(2^m-1)e \leq b \leq (d-1)(2^m-1),$$

and in addition, the Stiefel–Whitney classes

$$w_a(\lambda_{d,m}^{-1}) \neq 0 \qquad \textit{and} \qquad w_b(\lambda_{d,m}^{-(d+1)}) \neq 0 \tag{6.10}$$

do not vanish.

Proof From Corollary 6.10 (i) we have that

$$w_{(d-1)(2^m-1)}(\lambda_{d,m}^{-1} \oplus (\mu_{d,m}[2] \oplus \cdots \oplus \mu_{d,m}[m])^{\oplus 2e}) \neq 0. \tag{6.11}$$

On the other hand

$$\begin{aligned} &w_{(d-1)(2^m-1)}(\lambda_{d,m}^{-1} \oplus (\mu_{d,m}[2] \oplus \cdots \oplus \mu_{d,m}[m])^{\oplus 2e}) \\ &= \sum_{i=0}^{(d-1)(2^m-1)} w_i(\lambda_{d,m}^{-1}) \cdot w_{(d-1)(2^m-1)-i}((\mu_{d,m}[2] \oplus \cdots \oplus \mu_{d,m}[m])^{\oplus 2e}), \end{aligned}$$

where

$$\dim(\mu_{d,m}[2] \oplus \cdots \oplus \mu_{d,m}[m])^{\oplus 2e} = 2e \cdot \dim(\mu_{d,m}[2] \oplus \cdots \oplus \mu_{d,m}[m])$$
$$= 2e(2^{m-2} + \cdots + 2^0) = 2e(2^{m-1} - 1). \tag{6.12}$$

Now (6.11) and (6.12) imply that at least one Stiefel–Whitney class $w_i(\lambda_{d,m}^{-1}) \neq 0$ does not vanish for

$$(d-1)(2^m - 1) - i \leq 2e(2^{m-1} - 1) = 2e \cdot \dim(\mu_{d,m}[2] \oplus \cdots \oplus \mu_{d,m}[m])$$

and $i \leq (d-1)(2^m - 1)$. Hence we proved the first claim of (6.10).

For the second part of (6.10) recall Corollary 6.10 (ii):

$$w_{(d-1)(2^m-1)}(\lambda_{d,m}^{-(d+1)} \oplus (\mu_{d,m}[1] \oplus \cdots \oplus \mu_{d,m}[m])^{\oplus 2e}) \neq 0. \tag{6.13}$$

Similarly,

$$w_{(d-1)(2^m-1)}(\lambda_{d,m}^{-(d+1)} \oplus (\mu_{d,m}[1] \oplus \cdots \oplus \mu_{d,m}[m])^{\oplus 2e})$$
$$= \sum_{i=0}^{(d-1)(2^m-1)} w_i(\lambda_{d,m}^{-(d+1)}) \cdot w_{(d-1)(2^m-1)-i}((\mu_{d,m}[1] \oplus \cdots \oplus \mu_{d,m}[m])^{\oplus 2e}),$$

where

$$\dim(\mu_{d,m}[1] \oplus \cdots \oplus \mu_{d,m}[m])^{\oplus 2e} = 2e \cdot \dim(\mu_{d,m}[1] \oplus \cdots \oplus \mu_{d,m}[m])$$
$$= 2e(2^{m-1} + \cdots + 2^0) = 2e(2^{m-1} - 1). \tag{6.14}$$

In the same way, now (6.13) and (6.14), imply that there exists at least one Stiefel–Whitney class $w_i(\lambda_{d,m}^{-(d+1)}) \neq 0$ which does not vanish where

$$(d-1)(2^m - 1) - i \leq 2e(2^m - 1) = 2e \cdot \dim(\mu_{d,m}[1] \oplus \cdots \oplus \mu_{d,m}[m])$$

and $i \leq (d-1)(2^m - 1)$. Thus, we proved the second part of (6.10). □

Remark 6.13 Since the vector bundle $\lambda_{d,m}$ is the pull-back of the vector bundle $\zeta_{\mathbb{R}^d,2^m}$, and $w(\zeta_{\mathbb{R}^d,2^m}) = w(\xi_{\mathbb{R}^d,2^m})$, the previous corollary implies, under identical assumptions on parameters, that

$$\overline{w}_a(\xi_{\mathbb{R}^d,2^m}) \neq 0 \qquad \text{and} \qquad \overline{w}_b(\xi_{\mathbb{R}^d,2^m}^{\oplus d+1}) \neq 0.$$

Next, we use Corollary 6.12 to get further insights on the dual Stiefel–Whitney classes $\overline{w}(\xi_{\mathbb{R}^d,k})$ and $\overline{w}(\xi_{\mathbb{R}^d,k}^{\oplus d+1})$, but this time without restricting to the case when k is a power of 2.

Corollary 6.14 *Let $d \geq 2$ and $k \geq 1$ be integers, and let $d = 2^t + e$ for some integers $t \geq 1$ and $0 \leq e \leq 2^t - 1$. Then exist integers A and B with the property that*

$$(d - e - 1)(k - \alpha(k)) + e(\alpha(k) - \epsilon(k)) \leq A \leq (d - 1)(k - 1),$$

$$(d - 2e - 1)(k - \alpha(k)) \leq B \leq (d - 1)(k - 1),$$

and in addition, the dual Stiefel–Whitney classes

$$\overline{w}_A(\xi_{\mathbb{R}^d,k}) \neq 0 \qquad \text{and} \qquad \overline{w}_B(\xi_{\mathbb{R}^d,k}^{\oplus d+1}) \neq 0 \tag{6.15}$$

do not vanish. Recall that $\epsilon(k) = 1$ for k odd, and $\epsilon(k) = 0$ for k even.

Proof Let $r := \alpha(k)$ be the number of 1s in the binary presentation of the integer $k \geq 1$, and let $k = 2^{k_1} + \cdots + 2^{k_r}$ where $0 \leq k_1 < k_2 < \cdots < k_r$. Like in the proof of Lemma 5.6, consider a morphism between vector bundles $\prod_{i=1}^r \xi_{\mathbb{R}^d,2^{k_i}}$ and $\xi_{\mathbb{R}^d,k}$ where the following commutative square is a pullback diagram:

$$\begin{array}{ccc} \prod_{i=1}^r \xi_{\mathbb{R}^d,2^{k_i}} & \xrightarrow{\ \Theta\ } & \xi_{\mathbb{R}^d,k} \\ \downarrow & & \downarrow \\ \prod_{i=1}^r \mathrm{F}(\mathbb{R}^d, 2^{k_i})/\mathfrak{S}_{2^{k_i}} & \xrightarrow{\ \theta\ } & \mathrm{F}(\mathbb{R}^d, k)/\mathfrak{S}_k. \end{array}$$

Recall, that the map θ is induced, up to an equivariant homotopy, from a restriction of the little cubes operad structural map.

Thus, we have that $\theta^*\xi_{\mathbb{R}^d,k} \cong \prod_{i=1}^r \xi_{\mathbb{R}^d,2^{k_i}}$, and consequently

$$\theta^*(\overline{w}(\xi_{\mathbb{R}^d,k})) = \overline{w}\Big(\prod_{i=1}^r \xi_{\mathbb{R}^d,2^{k_i}}\Big) \quad \text{and} \quad \theta^*(\overline{w}(\xi_{\mathbb{R}^d,k}^{\oplus d+1})) = \overline{w}\Big(\prod_{i=1}^r \xi_{\mathbb{R}^d,2^{k_i}}^{\oplus d+1}\Big). \tag{6.16}$$

The product formula for Stiefel–Whitney classes [80, Pr. 4-A, p. 54] implies that

$$\overline{w}\Big(\prod_{i=1}^r \xi_{\mathbb{R}^d,2^{k_i}}\Big) = \overline{w}(\xi_{\mathbb{R}^d,2^{k_1}}) \times \cdots \times \overline{w}(\xi_{\mathbb{R}^d,2^{k_r}}),$$

and similarly

$$\overline{w}\Big(\prod_{i=1}^r \xi_{\mathbb{R}^d,2^{k_i}}^{\oplus d+1}\Big) = \overline{w}(\xi_{\mathbb{R}^d,2^{k_1}}^{\oplus d+1}) \times \cdots \times \overline{w}(\xi_{\mathbb{R}^d,2^{k_r}}^{\oplus d+1}).$$

From Corollary 6.12 and for every $0 \leq k_1 < k_2 < \cdots < k_r$ there exist integers $a_1, \ldots, a_r$, and integers $b_1, \ldots, b_r$, with the property that

$$\begin{cases} a_1 = 0, & k_1 = 0, \\ a_1 \geq (d-1)(2^{k_1}-1) - 2(2^{k_1-1}-1)e, & k_1 \geq 1, \\ a_i \geq (d-1)(2^{k_i}-1) - 2(2^{k_i-1}-1)e, & 2 \leq i \leq r, \end{cases}$$

and

$$\begin{cases} b_1 = (d-1)(2^{k_1}-1) - 2(2^{k_1}-1)e = 0, & k_1 = 0, \\ b_1 \geq (d-1)(2^{k_1}-1) - 2(2^{k_1}-1)e, & k_1 \geq 1, \\ b_i \geq (d-1)(2^{k_i}-1) - 2(2^{k_i}-1)e, & 2 \leq i \leq r, \end{cases}$$

and in addition the dual Stiefel– Whitney classes

$$\overline{w}_{a_i}(\xi_{\mathbb{R}^d,2^{k_i}}) \neq 0 \qquad \text{and} \qquad \overline{w}_{b_i}(\xi_{\mathbb{R}^d,2^{k_i}}^{\oplus d+1}) \neq 0.$$

do not vanish.

Let us denote the sums of integers $a_1, \ldots, a_r$ and $b_1, \ldots, b_r$ by

$$A := \sum_{i=0}^{r} a_i \geq (d-e-1)(k-\alpha(k)) + e(\alpha(k) - \epsilon(k)),$$

and

$$B := \sum_{i=0}^{r} b_i \geq (d-2e-1)(k-\alpha(k)).$$

Now, from the product formula for Stiefel–Whitney classes and the Künneth formula we conclude that

$$\overline{w}_A\Big(\prod_{i=1}^{r} \xi_{\mathbb{R}^d,2^{k_i}}\Big) \neq 0 \qquad \text{and} \qquad \overline{w}_B\Big(\prod_{i=1}^{r} \xi_{\mathbb{R}^d,2^{k_i}}^{\oplus d+1}\Big) \neq 0.$$

(For details of the last argument consult for example the proof of Lemma 5.6.) Finally, from the pull-backs (6.16) we conclude that

$$\overline{w}_A(\xi_{\mathbb{R}^d,k}) \neq 0 \qquad \text{and} \qquad \overline{w}_B(\xi_{\mathbb{R}^d,k}^{\oplus d+1}) \neq 0.$$

Combining the results of Corollary 6.14 and the approach used in the proof of Theorem 5.21 we get the following result.

Corollary 6.15 *Let $d \geq 2$, $k \geq 1$ and $\ell \geq 1$ be integers, and let $d = 2^t + e$ for some integers $t \geq 1$ and $0 \leq e \leq 2^t - 1$. Then exists an integer C with the property that*

$$\begin{aligned}(d-e-1)(k-\alpha(k)) + e(\alpha(k)-\epsilon(k)) + (d-2e-1)(\ell-\alpha(\ell))\\ \leq C \leq (d-1)(k+\ell-2),\end{aligned}$$

and in addition, the dual Stiefel–Whitney class

$$\overline{w}_C(\xi_{\mathbb{R}^d,k} \times \xi_{\mathbb{R}^d,\ell}^{\oplus d+1}) \neq 0 \tag{6.17}$$

does not vanish.

Proof We start as in the proof of Theorem 5.21. To compute the dual Stiefel–Whitney class of the product vector bundle $\xi_{\mathbb{R}^d,k} \times \xi_{\mathbb{R}^d,\ell}^{\oplus(d+1)}$ we apply the product formula [80, Problem 4-A, page 54]:

$$\overline{w}\big(\xi_{\mathbb{R}^d,k} \times \xi_{\mathbb{R}^d,\ell}^{\oplus(d+1)}\big) = \overline{w}(\xi_{\mathbb{R}^d,k}) \times \overline{w}\big(\xi_{\mathbb{R}^d,\ell}^{\oplus(d+1)}\big).$$

Hence, for fixed $r \geq 0$ we have that

$$\begin{aligned}\overline{w}_r\big(\xi_{\mathbb{R}^d,k} \times \xi_{\mathbb{R}^d,\ell}^{\oplus(d+1)}\big) &= \sum_{i+j=r} \overline{w}_i(\xi_{\mathbb{R}^d,k}) \times \overline{w}_j\big(\xi_{\mathbb{R}^d,\ell}^{\oplus(d+1)}\big)\\ &\in H^r(\mathrm{F}(\mathbb{R}^d,k)/\mathfrak{S}_k \times \mathrm{F}(\mathbb{R}^d,\ell)/\mathfrak{S}_\ell;\mathbb{F}_2).\end{aligned}$$

From the Künneth formula [23, Thm. VI.3.2] we get that each of the terms $\overline{w}_i(\xi_{\mathbb{R}^d,k}) \times \overline{w}_j\big(\xi_{\mathbb{R}^d,\ell}^{\oplus(d+1)}\big)$ in the previous sum belongs to a different direct summand of the cohomology

$$\begin{aligned}&H^r(\mathrm{F}(\mathbb{R}^d,k)/\mathfrak{S}_k \times \mathrm{F}(\mathbb{R}^d,\ell)/\mathfrak{S}_\ell;\mathbb{F}_2)\\ &\cong \bigoplus_{i+j=r} H^i(\mathrm{F}(\mathbb{R}^d,k)/\mathfrak{S}_k;\mathbb{F}_2) \otimes H^j(\mathrm{F}(\mathbb{R}^d,\ell)/\mathfrak{S}_\ell;\mathbb{F}_2).\end{aligned}$$

Therefore, the following equivalence holds:

$$\overline{w}_r\big(\xi_{\mathbb{R}^d,k} \times \xi_{\mathbb{R}^d,\ell}^{\oplus(d+1)}\big) \neq 0 \Longleftrightarrow \overline{w}_i(\xi_{\mathbb{R}^d,k}) \times \overline{w}_j\big(\xi_{\mathbb{R}^d,\ell}^{\oplus(d+1)}\big) \neq 0 \text{ for some } i+j=r.$$

Now, from Corollary 6.14 we know that there are integers A and B such that

$$(d-e-1)(k-\alpha(k)) + e(\alpha(k)-\epsilon(k)) \leq A \leq (d-1)(k-1),$$

$$(d-2e-1)(\ell-\alpha(\ell)) \leq B \leq (d-1)(\ell-1),$$

and

$$\overline{w}_A(\xi_{\mathbb{R}^d,k}) \neq 0 \quad \text{and} \quad \overline{w}_B(\xi_{\mathbb{R}^d,\ell}^{\oplus d+1}) \neq 0.$$

Consequently, for $C = A + B$ we have that $\overline{w}_A(\xi_{\mathbb{R}^d,k}) \times \overline{w}_B\big(\xi_{\mathbb{R}^d,\ell}^{\oplus(d+1)}\big) \neq 0$ and so $\overline{w}_C(\xi_{\mathbb{R}^d,k} \times \xi_{\mathbb{R}^d,\ell}^{\oplus d+1}) \neq 0$. Obviously, we have that

$$\begin{aligned}(d-e-1)(k-\alpha(k)) + e(\alpha(k)-\epsilon(k)) + (d-2e-1)(\ell-\alpha(\ell))\\ \leq C \leq (d-1)(k+\ell-2),\end{aligned}$$

and we have completed the proof of the corollary. □

Finally, using the criteria in Lemmas 5.4, 5.16 and 5.20 in combination with Corollaries 6.14 and 6.15 we get the strongest lower bounds for the existence of highly regular embeddings.

Theorem 6.16 *Let $d \geq 2$, $k \geq 1$ and $\ell \geq 1$ be integers, and let $d = 2^t + e$ for some integers $t \geq 1$ and $0 \leq e \leq 2^t - 1$. Then*

(1) *there is no k-regular embedding $\mathbb{R}^d \longrightarrow \mathbb{R}^N$ if*

$$N \leq (d-e-1)(k-\alpha(k)) + e(\alpha(k)-\epsilon(k)) + k - 1,$$

(2) *there is no ℓ-skew embedding $\mathbb{R}^d \longrightarrow \mathbb{R}^N$ if*

$$N \leq (d-2e-1)(\ell-\alpha(\ell)) + (d+1)\ell - 2,$$

(3) *there is no k-regular-ℓ-skew embedding $\mathbb{R}^d \longrightarrow \mathbb{R}^N$ if*

$$\begin{aligned}N \leq (d-e-1)(k-\alpha(k)) + e(\alpha(k)-\epsilon(k))\\ + (d-2e-1)(\ell-\alpha(\ell)) + (d+1)\ell + k - 2.\end{aligned}$$

In order to explain the strength of the previous theorem we demonstrate that Theorem 5.14, Theorem 5.18 and Theorem 5.22 are consequences of Theorem 6.16.

Corollary 6.17 *Let $d \geq 2$ and $k \geq 1$ be integers, and let $d = 2^t + e$ for some integers $t \geq 1$ and $0 \leq e \leq 2^t - 1$. Then*

$$\text{Theorem 6.16(1)} \quad \Longrightarrow \quad \text{Theorem 5.14.}$$

Proof We check that Theorem 6.16(1) implies all three cases of Theorem 5.14 independently.

(1) In Theorem 5.14(1) it is stated that for d being a power of 2 there is no k-regular embedding $\mathbb{R}^d \longrightarrow \mathbb{R}^N$ when $N \leq d(k-\alpha(k))+\alpha(k)-1$. The same results follows from Theorem 6.16(1) by taking $e=0$ and observing that in this case

$$\begin{aligned}
&(d-e-1)(k-\alpha(k))+e(\alpha(k)-\epsilon(k))+k-1\\
&\quad=(d-1)(k-\alpha(k))+k-1\\
&\quad=(d-1)(k-\alpha(k))+k-\alpha(k)+\alpha(k)-1\\
&\quad=d(k-\alpha(k))+\alpha(k)-1.
\end{aligned}$$

(2) Next, in Theorem 5.14(2) it is claimed that for d being not a power of 2 there is no k-regular embedding $\mathbb{R}^d \longrightarrow \mathbb{R}^N$ when $N \leq \frac{1}{2}(d-1)(k-\epsilon(k))+k-1$. Since d is not a power of 2 we have that $e>0$. Set that $k=2k'+\epsilon(k)$. Hence, $\alpha(k)=\alpha(k')+\epsilon(k)$. Now, the claim of Theorem 5.14(2) follows from Theorem 6.16(1) since the following difference is non-negative:

$$\begin{aligned}
&\big((d-e-1)(k-\alpha(k))+e(\alpha(k)-\epsilon(k))+k-1\big)\\
&\quad-\big(\frac{1}{2}(d-1)(k-\epsilon(k))+k-1\big)\\
&=(2^t-1-\frac{1}{2}(2^t+e-1))k-(2^t-1-e)\alpha(k)\\
&+\big(\frac{1}{2}(2^t+e-1)-e\big)\epsilon(k)\\
&=\frac{1}{2}(2^t-e-1)k-(2^t-1-e)\alpha(k)+\frac{1}{2}\big(2^t-e-1\big)\epsilon(k)\\
&=\frac{1}{2}\big(2^t-e-1\big)(k-2\alpha(k)+\epsilon(k))=\big(2^t-e-1\big)(k'-\alpha(k'))\geq 0.
\end{aligned}$$

(3) Finally, Theorem 5.14(3) says that for d being even and not a power of 2 there is no k-regular embedding $\mathbb{R}^d \longrightarrow \mathbb{R}^N$ when $N \leq \frac{1}{2}d(k-\epsilon(k))+k-\alpha(k)+\epsilon(k)-1$. We see that this result is also a consequence of Theorem 6.16(1) by showing that following difference is non-negative:

$$\begin{aligned}
&\big((d-e-1)(k-\alpha(k))+e(\alpha(k)-\epsilon(k))+k-1\big)\\
&\quad-\big(\frac{1}{2}d(k-\epsilon(k))+k-\alpha(k)+\epsilon(k)-1\big)\\
&=(2^t-1)k-(2^t-1)\alpha(k)+e\alpha(k)-e\epsilon(k)-\frac{1}{2}\big(2^t+e\big)k\\
&+\frac{1}{2}\big(2^t+e\big)\epsilon(k)+\alpha(k)-\epsilon(k)\\
&=\frac{1}{2}\big(2^t-e-2\big)k-(2^t-2-e)\alpha(k)+\frac{1}{2}\big(2^t-e-2\big)\epsilon(k)\\
&=\frac{1}{2}\big(2^t-e-2\big)(k+\epsilon(k)-2\alpha(k))=\big(2^t-e-2\big)(k'-\alpha(k'))\geq 0.
\end{aligned}$$

Here we have that $e \leq 2^t - 2$ and $\epsilon(k) = 0$, because e is even. As in the previous computation we set $k = 2k' + \epsilon(k) = 2k'$, and so $\alpha(k) = \alpha(k') + \epsilon(k) = \alpha(k')$. □

Corollary 6.18 *Let $d \geq 2$ and $\ell \geq 1$ be integers, and let $d = 2^t + e$ for some integers $t \geq 1$ and $0 \leq e \leq 2^t - 1$. Then*

$$\text{Theorem 6.16(2)} \quad \Longrightarrow \quad \text{Theorem 5.18.}$$

Proof We discuss each part of Theorem 5.18 separately.

(1) In Theorem 5.18(1) it is stated that for $d = 2$ and $\ell \geq 2$ there is no ℓ-skew embedding $\mathbb{R}^2 \longrightarrow \mathbb{R}^N$ when $N \leq 4\ell - \alpha(\ell) - 2$. The same conclusion follows from Theorem 6.16(2) because in this case $d = 2^t + e = 2 \Leftrightarrow t = 1, e = 0$, and the following difference vanishes:

$$\begin{aligned}&\big((d - 2e - 1)(\ell - \alpha(\ell)) + (d + 1)\ell - 2\big) - \big(4\ell - \alpha(\ell) - 2\big)\\ &= (\ell - \alpha(\ell) + 3\ell - 2) - (4\ell - \alpha(\ell) - 2) = 0.\end{aligned}$$

(2) In Theorem 5.18(2) we have claimed that for $d \geq 1$ and $\ell = 2$ there is no 2-skew embedding $\mathbb{R}^d \longrightarrow \mathbb{R}^N$ when $N \leq 2^{\gamma(d)} + d - 1$. Since we have assumed that $d = 2^t + e$ and $0 \leq e \leq 2^t - 1$ note that $2^{\gamma(d)} = 2^{t+1}$. Again, the case result can be deduced from Theorem 6.16(2) because the following difference vanishes:

$$\begin{aligned}&\big((d - 2e - 1)(\ell - \alpha(\ell)) + (d + 1)\ell - 2\big) - \big(2^{\gamma(d)} + d - 1\big) =\\ &d - 2e - 1 + 2d + 2 - 2 - 2^{\gamma(d)} - d - 1 = 2d - 2^{\gamma(d)} - 2e =\\ &2^{t+1} + 2e - 2^{t+1} - 2e = 0.\end{aligned}$$

(3) In Theorem 5.18(3) we considered the case when $d = 2^t + 0 \geq 2$ is a power of 2 and $\ell \geq 2$. We proved that there is no ℓ-skew embedding $\mathbb{R}^d \longrightarrow \mathbb{R}^N$ when $N \leq 2d\ell - (d - 1)\alpha(\ell) - 2$. To see that Theorem 6.16(2) implies this result we show that the following difference vanishes:

$$\begin{aligned}&\big((d - 2e - 1)(\ell - \alpha(\ell)) + (d + 1)\ell - 2\big) - \big(2d\ell - (d - 1)\alpha(\ell) - 2\big)\\ &= (d - 1)\ell - (d - 1)\alpha(\ell) + (d + 1)\ell - 2d\ell + (d - 1)\alpha(\ell) = 0.\end{aligned}$$

(4) In Theorem 5.18(4) we analysed the case when $d + 1 \geq 2$ is a power of 2 and $\ell \geq 2$. We showed that the method based of the computation of dual Stiefel–Whitney class does not produce any non-trivial result about the existence of ℓ-skew embeddings $\mathbb{R}^d \longrightarrow \mathbb{R}^N$. In this situation also Theorem 6.16(2) does not give any relevant result.

(5) In Theorem 5.18(5) we studied the case where both $d \geq 5$ and $d + 1$ are not powers of 2 and $\ell \geq 3$. We showed that there cannot be ℓ-skew embedding $\mathbb{R}^d \longrightarrow \mathbb{R}^N$ for $N \leq \frac{1}{2}(2^{\gamma(d)} - d - 1)(\ell - \epsilon(\ell)) + (d + 1)\ell - 2$. In order to see that Theorem 6.16(2) implies this result we show that the following difference is non-negative:

$$\begin{aligned}&\big((d-2e-1)(\ell-\alpha(\ell))+(d+1)\ell-2\big)\\&\quad-\big(\tfrac{1}{2}(2^{\gamma(d)}-d-1)(\ell-\epsilon(\ell))+(d+1)\ell-2\big)\\&\quad=(2^t-e-1)\Big(\ell-\alpha(\ell)-\frac{1}{2}(2^{t+1}-2^t-e-1)(\ell-\epsilon(\ell))\Big)\\&\quad=\frac{1}{2}(2^t-e-1)(\ell-\alpha(\ell)+\epsilon(\ell))\geq 0.\end{aligned}$$

(6) In the last case, Theorem 5.18(6), we have that $d\geq 5$ and $d+1$ are not powers of two, $2^{\gamma(d)}-d-1=2^{a_1}z$ where $a_1\geq 0$ and $z\geq 1$ is odd, and $\ell\geq 3$. We proved that in this case there is no ℓ-skew embedding $\mathbb{R}^d\longrightarrow\mathbb{R}^N$ for

$$N\leq\frac{1}{2}(2^{\gamma(d)}-d-1+2^{a_1})(\ell-\epsilon(\ell))-2^{a_1}\alpha(\ell)+(d+1)\ell-2.$$

In order to see that Theorem 6.16(2) implies this result we first note that in this case

$$d=2^t+e,\quad 1\leq e\leq 2^t-2,\quad 2^{a_1}z=2^{\gamma(d)}-d-1=2^t-e-1.$$

Now, as in the previous situations we consider the following difference and show that it is non-negative:

$$\begin{aligned}&\big((d-2e-1)(\ell-\alpha(\ell))+(d+1)\ell-2\big)\\&\quad-\Big(\frac{1}{2}(2^{\gamma(d)}-d-1+2^{a_1})(\ell-\epsilon(\ell))-2^{a_1}\alpha(\ell)+(d+1)\ell-2\Big)\\&\quad=(2^t-e-1)(\ell-\alpha(\ell))\\&\quad-\frac{1}{2}(2^t-e-1)(\ell-\epsilon(\ell))-2^{a_1-1}(\ell-\epsilon(\ell))+2^{a_1}\alpha(\ell)\\&\quad=2^{a_1-1}z(\ell-2\alpha(\ell)+\epsilon(\ell))-2^{a_1-1}(\ell-2\alpha(\ell)-\epsilon(\ell))\\&\quad=2^{a_1-1}\big((z-1)(\ell-2\alpha(\ell))+(z+1)\epsilon(\ell)\big)\geq 0.\end{aligned}$$

This concludes the proof of the corollary. □

In the same way as previous two corollaries we can show the following.

Corollary 6.19 *Let $d\geq 2$, $k\geq 1$ and $\ell\geq 1$ be integers, and let $d=2^t+e$ for some integers $t\geq 1$ and $0\leq e\leq 2^t-1$. Then*

$$\text{Theorem 6.16(3)}\implies\text{Theorem 5.22}.$$

6.4 Additional Bounds for the Existence of Complex Highly Regular Embeddings

In this section we derive additional consequence of Lemma 6.6 which yields further bounds for the existence of complex highly regular embeddings. In particular, we will improve bounds given in Theorems 5.26 and 5.28.

Let us denote by $\lambda_{d,m}^{\mathbb{C}} := \mathbb{C} \otimes \lambda_{d,m}$ and $\psi_{d,m}^{\mathbb{C}}[r_1^{\mathbb{C}}, \ldots, r_m^{\mathbb{C}}] := \mathbb{C} \otimes \psi_{d,m}$ complex versions of the real vector bundles $\lambda_{d,m}$ and $\psi_{d,m}[r_1, \ldots, r_m]$. Note that there is an isomorphism of real vector bundles $\lambda_{d,m}^{\mathbb{C}} \cong \lambda_{d,m}^{\oplus 2}$ and

$$\psi_{d,m}^{\mathbb{C}}[r_1^{\mathbb{C}}, \ldots, r_m^{\mathbb{C}}] \cong \psi_{d,m}[2r_1^{\mathbb{C}}, \ldots, 2r_m^{\mathbb{C}}].$$

The following lemma is a consequence of Lemma 6.6.

Lemma 6.20 *Let $m \geq 1$, $d^{\mathbb{C}} \geq 1$ and $\ell^{\mathbb{C}} \geq 1$ be integers, and let $r_1^{\mathbb{C}}, \ldots, r_m^{\mathbb{C}}$ be non-negative integers. If the binomial coefficient*

$$\binom{(r_1^{\mathbb{C}} 2^{m-2} + \cdots + r_{m-1}^{\mathbb{C}} 2^0) + r_m^{\mathbb{C}} - (d^{\mathbb{C}} - 1 + \ell^{\mathbb{C}})(2^{m-1} - 1) - \ell^{\mathbb{C}}}{d^{\mathbb{C}} - 1} \tag{6.18}$$

is odd, then

$$w_{2(d^{\mathbb{C}}-1)(2^m-1)}((\lambda_{2d^{\mathbb{C}}-1,m}^{\mathbb{C}})^{-\ell^{\mathbb{C}}} \oplus \psi_{2d^{\mathbb{C}}-1,m}^{\mathbb{C}}[r_1^{\mathbb{C}}, \ldots, r_m^{\mathbb{C}}]) \neq 0,$$

as an element of the cohomology group $H^{2(d^{\mathbb{C}}-1)(2^m-1)}(\mathrm{Pe}(\mathbb{R}^d, 2^m)/\mathcal{S}_{2^m}; \mathbb{F}_2)$. Here $(\lambda_{2d^{\mathbb{C}}-1,m}^{\mathbb{C}})^{-\ell^{\mathbb{C}}}$ denotes an inverse of the vector bundle $(\lambda_{2d^{\mathbb{C}}-1,m}^{\mathbb{C}})^{\oplus \ell}$.

Proof There is isomorphism of real vector bundles

$$(\lambda_{2d^{\mathbb{C}}-1,m}^{\mathbb{C}})^{-\ell^{\mathbb{C}}} \oplus \psi_{2d^{\mathbb{C}}-1,m}^{\mathbb{C}}[r_1^{\mathbb{C}}, \ldots, r_m^{\mathbb{C}}] \cong \lambda_{2d^{\mathbb{C}}-1,m}^{-2\ell^{\mathbb{C}}} \oplus \psi_{2d^{\mathbb{C}}-1,m}[2r_1^{\mathbb{C}}, \ldots, 2r_m^{\mathbb{C}}].$$

Thus, it suffices to prove that

$$w_{2(d^{\mathbb{C}}-1)(2^m-1)}(\lambda_{2d^{\mathbb{C}}-1,m}^{-2\ell^{\mathbb{C}}} \oplus \psi_{2d^{\mathbb{C}}-1,m}[2r_1^{\mathbb{C}}, \ldots, 2r_m^{\mathbb{C}}]) \neq 0.$$

If we set that $d = 2d^{\mathbb{C}} - 1$, $\ell = 2\ell^{\mathbb{C}}$, and $r_1 := 2r_1^{\mathbb{C}}, \ldots, r_1 := 2r_m^{\mathbb{C}}$, then by Lemma 6.6 we have that the Stiefel–Whitney class

$$\begin{aligned} &w_{2(d^{\mathbb{C}}-1)(2^m-1)}(\lambda_{2d^{\mathbb{C}}-1,m}^{-2\ell^{\mathbb{C}}} \oplus \psi_{2d^{\mathbb{C}}-1,m}[2r_1^{\mathbb{C}}, \ldots, 2r_m^{\mathbb{C}}]) \\ &\quad = w_{(d-1)(2^m-1)}(\lambda_{d,m}^{-\ell} \oplus \psi_{d,m}[r_1, \ldots, r_m]) \neq 0 \end{aligned}$$

does not vanish if and only if the binomial coefficient

$$\binom{(r_1 2^{m-2}+\cdots+r_{m-1}2^0)+r_m-(d-1+\ell)(2^{m-1}-1)-\ell}{d-1}$$
$$=\binom{2\big((r_1^{\mathbb{C}}2^{m-2}+\cdots+r_{m-1}^{\mathbb{C}}2^0)+r_m^{\mathbb{C}}-(d^{\mathbb{C}}-1+\ell^{\mathbb{C}})(2^{m-1}-1)-\ell^{\mathbb{C}}\big)}{2(d^{\mathbb{C}}-1)}$$

is odd. Since the binomial coefficient $\binom{2a}{2b}$ is odd if and only if the binomial coefficient $\binom{a}{b}$ is odd, then the assumption (6.18) implies that the binomial coefficient

$$\binom{2\big((r_1^{\mathbb{C}}2^{m-2}+\cdots+r_{m-1}^{\mathbb{C}}2^0)+r_m^{\mathbb{C}}-(d^{\mathbb{C}}-1+\ell^{\mathbb{C}})(2^{m-1}-1)-\ell^{\mathbb{C}}\big)}{2(d^{\mathbb{C}}-1)}$$

is odd. Hence, $w_{2(d^{\mathbb{C}}-1)(2^m-1)}((\lambda_{2d^{\mathbb{C}}-1,m}^{\mathbb{C}})^{-\ell^{\mathbb{C}}}\oplus\psi_{2d^{\mathbb{C}}-1,m}^{\mathbb{C}}[r_1^{\mathbb{C}},\ldots,r_m^{\mathbb{C}}])\neq 0$. □

Let us denote by $\mu_{d,m}^{\mathbb{C}}[j]:=\mathbb{C}\otimes\mu_{d,m}[j]$, for all $1\le j\le m$, the complex version of the real vector bundle $\mu_{d,m}[j]$. There is an isomorphism of real vector bundles $\mu_{d,m}^{\mathbb{C}}[j]\cong\mu_{d,m}[j]^{\oplus 2}$. Now, like in the case of Lemma 6.9 and Corollary 6.10, we deduce the following consequence of Lemma 6.20.

Corollary 6.21 *Let $d^{\mathbb{C}}\ge 2$ and $m\ge 1$ be integers, and let $d^{\mathbb{C}}=2^t+e$ for some integer $t\ge 1$ and $0\le e\le 2^t-1$. If $d=2d^{\mathbb{C}}-1$, then*

(1) $w_{2(d^{\mathbb{C}}-1)(2^m-1)}\big((\lambda_{d,m}^{\mathbb{C}})^{-1}\oplus(\mu_{d,m}^{\mathbb{C}}[2]\oplus\cdots\oplus\mu_{d,m}^{\mathbb{C}}[m])^{\oplus 2e}\big)\neq 0$,
(2) $w_{2(d^{\mathbb{C}}-1)(2^m-1)}\big((\lambda_{d,m}^{\mathbb{C}})^{-(d^{\mathbb{C}}+1)}\oplus(\mu_{d,m}^{\mathbb{C}}[1]\oplus\cdots\oplus\mu_{d,m}^{\mathbb{C}}[m])^{\oplus 2e}\big)\neq 0$.

Proof The proof is almost identical to the proof of Lemma 6.9. For the sake of completeness we present the proof in detail.

(1) Let $\ell^{\mathbb{C}}=1$, $r_1^{\mathbb{C}}=0$, and $r_2^{\mathbb{C}}=\cdots=r_m^{\mathbb{C}}=2e$. Then

$$\begin{aligned}
&(r_1^{\mathbb{C}}2^{m-2}+\cdots+r_{m-1}^{\mathbb{C}}2^0)+r_m^{\mathbb{C}}-(d^{\mathbb{C}}-1+\ell^{\mathbb{C}})(2^{m-1}-1)-\ell^{\mathbb{C}}\\
&=2e(2^{m-3}+\cdots+2^0)+2e-d^{\mathbb{C}}(2^{m-1}-1)-1\\
&=2^{m-1}(e-d^{\mathbb{C}})+2e+d^{\mathbb{C}}-1\\
&=2^{m-1}(e-d^{\mathbb{C}})+d^{\mathbb{C}}-1\\
&=\begin{cases} d^{\mathbb{C}}-1-2^{m-1+t}, & m\ge 2,\\ -1, & m=1.\end{cases}
\end{aligned}$$

Since $\binom{-1}{d^{\mathbb{C}}-1}=_{\mathbb{F}_2}1\in\mathbb{F}_2$ and $\binom{d^{\mathbb{C}}-1-2^{m-1+t}}{d^{\mathbb{C}}-1}=_{\mathbb{F}_2}\binom{d^{\mathbb{C}}-1}{d^{\mathbb{C}}-1}=_{\mathbb{F}_2}1\in\mathbb{F}_2$, the assumption of Lemma 6.20 is satisfied for the chosen parameters. Consequently,

$$w_{2(d^{\mathbb{C}}-1)(2^m-1)}\big((\lambda_{d,m}^{\mathbb{C}})^{-1}\oplus(\mu_{d,m}^{\mathbb{C}}[2]\oplus\cdots\oplus\mu_{d,m}^{\mathbb{C}}[m])^{\oplus 2e}\big)\neq 0.$$

(2) Let $\ell^{\mathbb{C}} = d + 1$, $r_1^{\mathbb{C}} = 0$, and $r_1^{\mathbb{C}} = \cdots = r_m^{\mathbb{C}} = 2e$. Hence

$$\begin{aligned}
&(r_1^{\mathbb{C}} 2^{m-2} + \cdots + r_{m-1}^{\mathbb{C}} 2^0) + r_m^{\mathbb{C}} - (d^{\mathbb{C}} - 1 + \ell^{\mathbb{C}})(2^{m-1} - 1) - \ell^{\mathbb{C}} \\
&= 2e(2^{m-2} + \cdots + 2^0) + 2e - 2d^{\mathbb{C}}(2^{m-1} - 1) - d^{\mathbb{C}} - 1 \\
&= 2^m(e - d^{\mathbb{C}}) + d^{\mathbb{C}} - 1 = d^{\mathbb{C}} - 1 - 2^{m+t}.
\end{aligned}$$

Thus, $\binom{d^{\mathbb{C}}-1-2^{m+t}}{d^{\mathbb{C}}-1} =_{\mathbb{F}_2} 1 \in \mathbb{F}_2$, because $d^{\mathbb{C}} - 1 < 2^{t+1} \leq 2^{t+m}$. Consequently, the assumption of Lemma 6.20 is satisfied for the chosen parameters, and so

$$w_{2(d^{\mathbb{C}}-1)(2^m-1)}((\lambda_{d,m}^{\mathbb{C}})^{-(d^{\mathbb{C}}+1)} \oplus (\mu_{d,m}^{\mathbb{C}}[1] \oplus \cdots \oplus \mu_{d,m}^{\mathbb{C}}[m])^{\oplus 2e}) \neq 0.$$

Just as in the proof of Corollary 6.14 we can derive the following estimate for non-vanishing of the relevant dual Stiefel–Whitney classes.

Corollary 6.22 *Let $d^{\mathbb{C}} \geq 2$ and $m \geq 1$ be integers, and let $d^{\mathbb{C}} = 2^t + e$ for some integer $t \geq 1$ and $0 \leq e \leq 2^t - 1$. Set $d = 2d^{\mathbb{C}} - 1$. Then there exist integers a and b with the property that*

$$(d^{\mathbb{C}} - 1)(2^m - 1) - 2(2^{m-1} - 1)e \leq a \leq (d^{\mathbb{C}} - 1)(2^m - 1),$$

$$(d^{\mathbb{C}} - 1)(2^m - 1) - 2(2^m - 1)e \leq b \leq (d^{\mathbb{C}} - 1)(2^m - 1),$$

and in addition, the dual Stiefel–Whitney classes

$$\overline{w}_{2a}(\lambda_{d,m}^{\mathbb{C}}) \neq 0 \qquad \text{and} \qquad \overline{w}_{2b}((\lambda_{d,m}^{\mathbb{C}})^{d^{\mathbb{C}}+1}) \neq 0,$$

do not vanish.

Now, using the criteria for the existence of complex k-regular embeddings and complex ℓ-skew embeddings given in Lemma 5.25, we can directly derive the following estimates—a complex analogue of Theorem 6.16.

Theorem 6.23 *Let $d^{\mathbb{C}} \geq 2$, $m \geq 1$, $k \geq 1$ and $\ell \geq 1$ be integers, and let $d^{\mathbb{C}} = 2^t + e$ for some integer $t \geq 1$ and $0 \leq e \leq 2^t - 1$. Set $d = 2d^{\mathbb{C}} - 1$. Then*

(i) *there is no complex k-regular embedding $\mathbb{R}^d \longrightarrow \mathbb{C}^N$ if*

$$N \leq (d^{\mathbb{C}} - 1 - e)(k - \alpha(k)) + e(\alpha_2(k) - \epsilon(k)) + k \quad 1,$$

(ii) *there is no complex ℓ-skew embedding $\mathbb{C}^{d^{\mathbb{C}}} \longrightarrow \mathbb{C}^N$ if*

$$N \leq (d^{\mathbb{C}} - 1 - 2e)(\ell - \alpha(\ell)) + (d^{\mathbb{C}} + 1)\ell - 2.$$

In a similar way as in Corollaries 6.17, 6.18 and 6.19 it can be verified that Theorem 6.23 implies Theorems 5.26 and 5.28.

Part III
Technical Tools

Chapter 7
Operads

In this chapter we recall basic notions from the theory of operads that we use. We follow the framework developed by May in [76], with some slight modifications in the notation.

Let Top be the category of compactly generated weak Hausdorff spaces with continuous maps as morphisms, and let $\mathrm{Top}_{\mathrm{pt}}$ denote the category of compactly generated weak Hausdorff spaces with non-degenerate base points where the morphisms are base point preserving continuous maps. Furthermore, assume that all the products as well as function spaces are endowed with compactly generated topology.

7.1 Definition and Basic Example

The notion of an operad and its first formal definition appeared in work of May in 1970s. For more details consult the original publication [76, Sec. 1].

Definition 7.1 An **operad** $\mathcal{O}$ is given by a family of topological spaces in Top

$$\mathcal{O} := \{\mathcal{O}(n) : n \geq 0\} \qquad \text{where} \qquad \mathcal{O}(0) := \{\mathrm{pt}\}$$

together with a family of continuous maps

$$\mu \colon (\mathcal{O}(n_1) \times \cdots \times \mathcal{O}(n_k)) \times \mathcal{O}(k) \longrightarrow \mathcal{O}(n)$$

P. V. M. Blagojević et al., *Equivariant Cohomology of Configuration Spaces Mod 2*, Lecture Notes in Mathematics 2282, https://doi.org/10.1007/978-3-030-84138-6_7

where $k, n_1, \ldots, n_k \geq 0$ and $n = n_1 + \cdots + n_k$, such that the following axioms are satisfied:

(1) For every $a \in O(k)$, $b_1 \in O(n_1), \ldots, b_k \in O(n_k)$, $c_1 \in O(m_1), \ldots, c_n \in O(m_n)$

$$\mu(c_1, \ldots, c_n; \mu(b_1, \ldots, b_k; a)) = \mu(d_1, \ldots, d_k; a)$$

where for $1 \leq i \leq k$

$$d_i := \begin{cases} \mu(c_{n_1+\cdots+n_{i-1}+1}, \ldots, c_{n_1+\cdots+n_i}; b_i), & n_i \neq 0 \\ \text{pt}, & n_i = 0. \end{cases}$$

(2) There exists an element $\mathbf{1} \in O(1)$ with the property that for every $a \in O(n)$ and every $b \in O(k)$

$$\mu(a; \mathbf{1}) = a, \qquad \mu(\mathbf{1}, \ldots, \mathbf{1}; b) = b.$$

(3) Every space $O(n)$ is endowed with a right action of the symmetric group $\mathfrak{S}_n$ that fulfills

$$\mu(b_1, \ldots, b_k; a \cdot \pi) = \mu(b_{\pi^{-1}(1)}, \ldots, b_{\pi^{-1}(k)}; a) \cdot \pi_{n_1,\ldots,n_k},$$

$$\mu(b_1 \cdot \pi_1, \ldots, b_k \cdot \pi_k; a) = \mu(b_1, \ldots, b_k; a) \cdot (\pi_1, \ldots, \pi_k),$$

where the permutation

- $\pi_{n_1,\ldots,n_k} \in \mathfrak{S}_n$ is given by permuting k blocks

$$(1, \ldots, n_1)(n_1 + 1, \ldots, n_1 + n_2) \cdots (n_1 + \cdots n_{k-1} + 1, \ldots, n_1 + \cdots n_k)$$

as the permutation $\pi \in \mathfrak{S}_k$ permutes $(1, \ldots, k)$, and
- $(\pi_1, \ldots, \pi_k) \in \mathfrak{S}_n$ denotes the image of $(\pi_1, \ldots, \pi_k) \in \mathfrak{S}_{n_1} \times \cdots \times \mathfrak{S}_{n_k}$ via the inclusion $\mathfrak{S}_{n_1} \times \cdots \times \mathfrak{S}_{n_k} \hookrightarrow \mathfrak{S}_n$.

If in addition each action of the symmetric group $\mathfrak{S}_n$ on $O(n)$ is free the operad O is called a **$\mathfrak{S}$-free operad**. The map $\mu_O := \mu$ is called the *structural map* of the operad O.

The defining requirement on the action of the symmetric group $\mathfrak{S}_n$ on $O(n)$ for $n \geq 0$, Definition 7.1(3), directly implies the following property of the structural map.

Lemma 7.2 *Let $k \geq 1$ be an integer, and let $n_1, \ldots, n_k$ be integers such that*

$$n_1 = \cdots = n_{i_1} < n_{i_1+1} = \cdots = n_{i_2} < \cdots < n_{i_{r-1}+1} = \cdots = n_{i_r}$$

where $1 \leq i_1 < i_2 < \cdots < i_{r-1} < i_r = k$. *Then*

(1) *The product* $(O(n_1) \times \cdots \times O(n_k)) \times O(k)$ *is endowed with a*

$$\mathcal{S}_{n_1,\ldots,n_k;k} =$$

$$(\mathfrak{S}_{n_{i_1}}^{i_1} \rtimes \mathfrak{S}_{i_1}) \times (\mathfrak{S}_{n_{i_2}}^{i_2-i_1+1} \rtimes \mathfrak{S}_{i_2-i_1+1}) \times \cdots \times (\mathfrak{S}_{n_{i_r}}^{i_r-i_{r-1}+1} \rtimes \mathfrak{S}_{i_r-i_{r-1}+1})$$

right action.

(2) *The structural map*

$$\mu\colon (O(n_1) \times \cdots \times O(n_k)) \times O(k) \longrightarrow O(n),$$

with $n := n_1 + \cdots + n_k$, *is an equivariant map with respect to the natural inclusion map of the groups* $\mathcal{S}_{n_1,\ldots,n_k;k} \hookrightarrow \mathfrak{S}_n$.

The first and the central example of an operad is the endomorphism operad associated with any topological space in Top_{pt}. The importance of the endomorphism operad will become apparent after the definition of an action of an operad on a topological space.

Example 7.3 Let X be a topological space in Top_{pt}. The **endomorphism operad** $\mathcal{E}nd_X$ associated to X is defined as follows:

- $\mathcal{E}nd_X(n) := \text{Mor}_{\text{Top}_{\text{pt}}}(X^n, X)$ where $n \geq 0$ and $X^0 := \text{pt}$,
- $\mu(g_1, \ldots, g_k; f) := f \circ (g_1 \times \cdots \times g_k)$ for $f \in \mathcal{E}nd_X(k)$, and $g_1 \in \mathcal{E}nd_X(n_1), \ldots, g_k \in \mathcal{E}nd_X(n_k)$, and
- $(f \cdot \pi)(x_1, \ldots, x_k) := f(x_{\pi^{-1}(1)}, \ldots, x_{\pi^{-1}(k)})$ where $f \in \mathcal{E}nd_X(k)$, $\pi \in \mathfrak{S}_k$, and $(x_1, \ldots, x_k) \in X^k$.

The endomorphism operad is not a $\mathfrak{S}$-free operad.

In order to have a category of operads we introduce a notion of a morphism between operads.

Definition 7.4 Let O and $\mathcal{D}$ be operads. A **morphism of operads** $\Phi\colon O \longrightarrow \mathcal{D}$ is a family of $\mathfrak{S}_n$-equivariant maps $\Phi_n\colon O(n) \longrightarrow \mathcal{D}(n)$ such that the following diagram commute:

$$\begin{array}{ccc} (O(n_1) \times \cdots \times O(n_k)) \times O(k) & \xrightarrow{\mu_O} & O(n) \\ \Big\downarrow{\scriptstyle (\Phi_{n_1} \times \cdots \times \Phi_{n_k}) \times \Phi_k} & & \Big\downarrow{\scriptstyle \Phi_n} \\ (\mathcal{D}(n_1) \times \cdots \times \mathcal{D}(n_k)) \times \mathcal{D}(k) & \xrightarrow{\mu_{\mathcal{D}}} & \mathcal{D}(n) \end{array}$$

for every collection of integers $k, n_1, \ldots, n_k \geq 0$ where $n := n_1 + \cdots + n_k$.

The **category of operads** Op consists of operads as objects and morphisms of operads as morphisms.

7.2 $\mathcal{O}$-Space

An action of an operad on a topological with a base point is defined as follows.

Definition 7.5 Let X be a space in $\mathrm{Top}_{\mathrm{pt}}$ and let $\mathcal{O}$ be an operad. An **action of the operad** $\mathcal{O}$ on the space X is a morphism of operads $\Theta\colon \mathcal{O} \longrightarrow \mathcal{E}nd_X$. The pairs (X, Θ) is called an **$\mathcal{O}$-space.**
A **morphism of $\mathcal{O}$-spaces** (X, Θ) and (X', Θ') is a continuous based map $f\colon X \longrightarrow X'$ in $\mathrm{Top}_{\mathrm{pt}}$ such that for every $n \geq 0$ and every $a \in \mathcal{O}(n)$ the following diagram commutes

$$\begin{array}{ccc} X^n & \xrightarrow{\Theta_n(a)} & X \\ \downarrow{\scriptstyle f^n} & & \downarrow{\scriptstyle f} \\ Y^n & \xrightarrow{\Theta'_n(a)} & Y. \end{array}$$

An equivalent definition of the action of an operad on a topological space is formulated in the following elementary lemma.

Lemma 7.6 *Let X be a space in* $\mathrm{Top}_{\mathrm{pt}}$*, and let $\mathcal{O}$ be an operad. An action $\Theta\colon \mathcal{O} \longrightarrow \mathcal{E}nd_X$ of the operad $\mathcal{O}$ on X determines and is determined by the family of continuous maps*

$$\Theta_n\colon X^n \times \mathcal{O}(n) \longrightarrow X,$$

where $\Theta_0\colon \mathrm{pt} \longrightarrow X$ is the inclusion of the base point, that satisfy the following properties:

(1) *For every $k, n_1, \ldots, n_k \geq 0$ and $n := n_1 + \cdots + n_k$ the following diagram commutes*

$$\begin{array}{ccc} X^n \times (\mathcal{O}(n_1) \times \cdots \times \mathcal{O}(n_k)) \times \mathcal{O}(k) & \xrightarrow{\mathrm{id} \times \mu} & X^n \times \mathcal{O}(n) \\ \downarrow{\scriptstyle s} & & \searrow{\scriptstyle \Theta_n} \\ & & X \\ & & \nearrow{\scriptstyle \Theta_k} \\ \big(X^{n_1} \times \mathcal{O}(n_1)\big) \times \cdots \times \big(X^{n_k} \times \mathcal{O}(n_k)\big) \times \mathcal{O}(k) & \xrightarrow{\Theta_{n_1} \times \cdots \times \Theta_{n_k} \times \mathrm{id}} & X^k \times \mathcal{O}(k) \end{array}$$

where

$$s\colon X^n \times (O(n_1) \times \cdots \times O(n_k)) \times O(k) \longrightarrow \\ \big(X^{n_1} \times O(n_1)\big) \times \cdots \times \big(X^{n_k} \times O(n_k)\big) \times O(k)$$

is the obvious shuffle homeomorphism.

(2) *For every $x \in X$*

$$\Theta_1(x; 1) = x.$$

(3) *For every $a \in O(n)$, $\pi \in \mathfrak{S}_n$, and every $(x_1, \dots, x_n) \in X^n$*

$$\Theta_n(x_1, \dots, x_n; a \cdot \pi) = \Theta_n(x_{\pi^{-1}(1)}, \dots, x_{\pi^{-1}(n)}; a).$$

Furthermore, if $f\colon (X, \Theta) \longrightarrow (X', \Theta')$ is a morphism of O-spaces, then for every $n \geq 0$ the following diagram commutes

$$\begin{array}{ccc} X^n \times O(n) & \xrightarrow{\Theta_n} & X \\ \Big\downarrow{\scriptstyle f^n \times \mathrm{id}} & & \Big\downarrow{\scriptstyle f} \\ (X')^n \times O(n) & \xrightarrow{\Theta'_n} & X'. \end{array}$$

7.3 Little Cubes Operad

In this section we give an example of an operad whose structural map is studied in the central part of this work.

Let $I := [0, 1] \subseteq \mathbb{R}$ be the unit interval, and then $I^d \subseteq \mathbb{R}^d$ is the associated d-cube. Denote by $p := (\frac{1}{2}, \dots, \frac{1}{2}) \in I^d$ the centre of the d-cube. A little d-cube is simply an embedding of a cube into I^d in such a way that then corresponding edges are parallel, see illustration in Fig. 7.1.

Definition 7.7 Let $d \geq 1$ be an integer. A **little d-cube** is an orientation preserving affine embedding $\mathbf{c}\colon I^d \longrightarrow I^d$ of a d-cube that can be presented as the product map $\mathbf{c} = c_1 \times \cdots \times c_d$ where each $c_i : I \longrightarrow I$, $1 \leq i \leq d$, is an orientation preserving affine embedding.

Now, the little cubes operad is defined as follows.

Definition 7.8 Let $d \geq 1$ be an integer. The **little d-cubes operad** C_d, for $n \geq 1$, is defined by the family of topological spaces of all n-tuples of little d-cubes whose interiors are pairwise disjoint, that is

$$C_d(n) := \{\alpha := (\mathbf{c}_1, \dots, \mathbf{c}_n) : \mathbf{c}_i(\operatorname{int} I^d) \cap \mathbf{c}_j(\operatorname{int} I^d) = \varnothing \text{ for } i \neq j\}.$$

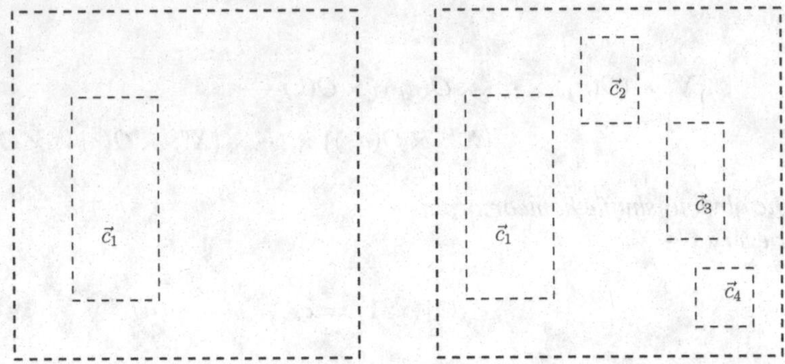

Fig. 7.1 Illustration of a 2-cube and an element of the little 2-cube operad space $C_2(4)$

The space $C_d(0)$ is the point, containing the unique "embedding" of the empty set into the cube I^d. The n-tuple $\alpha = (\mathbf{c}_1, \dots, \mathbf{c}_n)$ of little d-cubes can be seen also as a continuous map $\alpha\colon \coprod_{j=1}^n I^d \longrightarrow I^d$, where "$\coprod$" denotes the disjoint union. The space $C_d(0)$ is assumed to be a point interpreted as the unique embedding of the empty set into I^d. The space $C_d(n)$ is equipped with the subspace topology induced from the space of all continuous maps $\coprod_{j=1}^n I^d \longrightarrow I^d$.

The remaining necessary ingredients for the definition the little d-cube operad are given as follows.

(1) The structural map of the little d-cube operad

$$\mu\colon (C_d(n_1) \times \cdots \times C_d(n_k)) \times C_d(k) \longrightarrow C_d(n),$$

$n = n_1 + \cdots + n_k$, is defined as a composition

$$\alpha \circ (\beta_1 \uplus \cdots \uplus \beta_k)\colon \left(\coprod_{j=1}^{n_1} I^d\right) \amalg \cdots \amalg \left(\coprod_{j=1}^{n_k} I^d\right) \longrightarrow \left(\coprod_{j=1}^{k} I^d\right) \longrightarrow I^d$$

where $\alpha \in C(k)$ and $\beta_1 \in C_d(n_1), \dots, \beta_k \in C_d(n_k)$.

(2) The element $\mathbf{1} \in C_d(1)$ is the identity map id: $I^d \longrightarrow I^d$.

(3) The right action of the symmetric group $\mathfrak{S}_n$ on the space $C_d(n)$ is given by

$$(\mathbf{c}_1, \dots, \mathbf{c}_n) \cdot \pi := (\mathbf{c}_{\pi(1)}, \dots, \mathbf{c}_{\pi(n)}).$$

The little d-cubes operad is an $\mathfrak{S}$-free operad since for each n the symmetric group $\mathfrak{S}_n$ acts freely on $C_d(n)$.

There exists an $\mathfrak{S}_n$-equivariant map of little d-cube operad space $C_d(n)$ into the configuration space $\mathrm{F}(\mathbb{R}^d, n)$ given by evaluating each cube at the centre $p = (\frac{1}{2}, \ldots, \frac{1}{2}) \in I^d$:

$$\mathrm{ev}_{d,n} \colon C_d(n) \longrightarrow \mathrm{F}(\mathbb{R}^d, n), \qquad (\mathbf{c}_1, \ldots, \mathbf{c}_n) \longmapsto (\mathbf{c}_1(p), \ldots, \mathbf{c}_n(p)). \tag{7.1}$$

This map is an $\mathfrak{S}_n$-equivariant homotopy equivalence, see [76, Thm. 4.8].

Lemma 7.9 *For integers $d \geq 1$ and $n \geq 1$ the evaluation at centers of cubes map $\mathrm{ev}_{d,n} \colon C_d(n) \longrightarrow \mathrm{F}(\mathbb{R}^d, n)$ is an $\mathfrak{S}_n$-equivariant homotopy equivalence of the spaces $C_d(n)$ and $\mathrm{F}(\mathbb{R}^d, n)$.*

7.4 C_d-Spaces, An Example

In this section we give an example of a family of C_d-spaces. More precisely, for a pointed topological space X we define an action of the little cubes operad C_d on its d-fold loop space $\Omega^d X$. For more details consult for example [76, Thm. 5.1].

In order to define a C_d-action on the d-fold loop space $\Omega^d X$ we use Lemma 7.6 and give a family of functions

$$\Theta_{d,n} \colon (\Omega^d X)^n \times C_d(n) \longrightarrow \Omega^d X.$$

For $\omega_1, \ldots, \omega_n \in \Omega^d X$ and $(\mathbf{c}_1, \ldots, \mathbf{c}_n) \in C_d(n)$ the loop

$$\Theta_{d,n}(\omega_1, \ldots, \omega_n; \mathbf{c}_1, \ldots, \mathbf{c}_n) \colon I^d \longrightarrow X$$

is defined, when $x \in I^d$, by

$$\Theta_{d,n}(\omega_1, \ldots, \omega_n; \mathbf{c}_1, \ldots, \mathbf{c}_n)(x) := \begin{cases} \omega_i(\mathbf{c}_i^{-1}(x)), & \text{if } x \in \mathbf{c}_i(I^d) \text{ for some } 0 \leq i \leq n, \\ \mathrm{pt}, & \text{otherwise,} \end{cases}$$

where $\mathrm{pt} \in X$ is the base point. In the case $d = 1$ the elements $\omega_1, \ldots, \omega_n$ are pointed loops and the map $\Theta_{1,n}(\omega_1, \ldots, \omega_n; \mathbf{c}_1, \ldots, \mathbf{c}_n)$ is just a concatenation of loops $\omega_1, \ldots, \omega_n$ specified by the collection of, pairwise interior disjoint, intervals $(\mathbf{c}_1, \ldots, \mathbf{c}_n) \in C_1(n)$. For an illustration see Fig. 7.2. By direct inspection it can be verified that just defined family of functions satisfies all the necessary conditions of Lemma 7.6. Thus, for a space X in $\mathrm{Top}_{\mathrm{pt}}$ the d-fold loop space $\Omega^d X$ is a C_d-space.

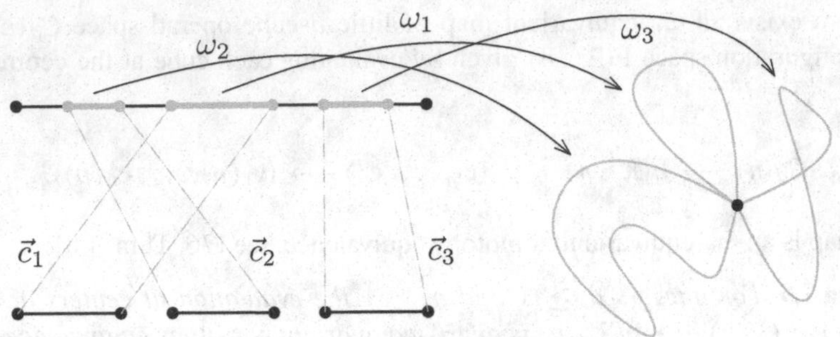

Fig. 7.2 An illustration of the loop $\Theta_{1,3}(\omega_1, \omega_2, \omega_1; \mathbf{c}_1, \mathbf{c}_2, \mathbf{c}_3)$

7.5 C_d-Spaces, a Free C_d-Space Over X

Let X be a pointed space with the base point $\mathrm{pt} \in X$, and let $d \geq 1$ be an integer. Then we define the C_d-space associated to X to be the quotient space

$$C_d(X) := \Big(\coprod_{m\geq 0} C_d(m) \times_{\mathfrak{S}_m} X^m\Big)/\approx$$

where for $(\mathbf{c}_1, \ldots, \mathbf{c}_m) \in C_d(m)$ and $(x_1, \ldots, x_{m-1}, \mathrm{pt}) \in X^m$ we define equivalence relation generated by

$$((\mathbf{c}_1, \ldots, \mathbf{c}_{m-1}, \mathbf{c}_m), (x_1, \ldots, x_{m-1}, \mathrm{pt})) \approx ((\mathbf{c}_1, \ldots, \mathbf{c}_{m-1}), (x_1, \ldots, x_{m-1})).$$

The C_d-action on $C_d(X)$ is induced from the structural maps of the little cubes operad C_d. For more details on the definition of $C_d(X)$, as a monad associated to the operad C_d, see [76, Constr. 2.4].

The C_d-space $C_d(X)$, associated to the pointed space X, can be called the **free C_d-space generated by** X, because for every C_d-space Y there exists a bijective correspondence

$$\begin{aligned}&(\text{morphisms of } C_d\text{-space from } C_d(X) \text{ to } Y) \longleftrightarrow \\ &\qquad\qquad (\text{morphisms of topological space from } X \text{ to } Y). \qquad (7.2)\end{aligned}$$

Consult [76, Prop 2.8 and Lem. 2.9].

Let $X \longrightarrow \Omega^d \Sigma^d X$ be the map associated to the identity map $\mathrm{id}\colon \Sigma^d X \longrightarrow \Sigma^d X$ along the adjunction relation

$$[A, \Omega^d B]_{\mathrm{pt}} \longleftrightarrow [\Sigma^d A, B]_{\mathrm{pt}}.$$

Here $[A, B]_{\text{pt}}$ denotes the set of all homotopy classes of pointed maps $A \longrightarrow B$ between the pointed spaces A and B. Next, let $\alpha_d \colon C_d(X) \longrightarrow \Omega^d \Sigma^d X$ denotes the morphism of C_d-space associated to the map $X \longrightarrow \Omega^d \Sigma^d X$ with respect to the correspondence (7.2). Now the Approximation theorem [76, Thm. 2.7 and Thm. 6.1] of May states the following.

Theorem 7.10 *Let $d \geq 1$ be an integer, or let $d = \infty$. If X is a path-connected space in* Top_{pt}*, then*

$$\alpha_d \colon C_d(X) \longrightarrow \Omega^d \Sigma^d X$$

is a weak homotopy equivalence.

7.6 Araki–Kudo–Dyer–Lashof Homology Operations

Following analogy with Dyer and Lashof [45, Def. 2.2] we review basic properties of the Araki–Kudo–Dyer–Lashof homology operations as defined by Cohen [33, Def. 5.6]. In the case $p = 2$ the homology operations were first introduced by Araki and Kudo [70].

Let $d \geq 1$ be an integer, or let $d = \infty$. Let Y be a compactly generated weak Hausdorff space with non-degenerate base point endowed with an action of the little cubes operad C_d. Then there exists a sequence of maps

$$Q_i \colon H_j(Y; \mathbb{F}_2) \longrightarrow H_{i+2j}(Y; \mathbb{F}_2), \qquad 0 \leq i \leq d-1,$$

called **Araki–Kudo–Dyer–Lashof homology operations.** Properties of these operations are listed in the next proposition, see also [45, Thm. 2.2, Cor. 1] and [33, Sec. 1]. In the following $\lambda_{d-1} \colon H_i(Y; \mathbb{F}_2) \otimes H_j(Y; \mathbb{F}_2) \longrightarrow H_{i+j+d-1}(Y; \mathbb{F}_2)$ denotes the Browder operation, consult [24] and [33, Thm. 1.2].

Proposition 7.11 *Let Y and Z be C_d-spaces. The following properties hold:*

(1) $Q_0(y) = y^2$ *for every* $x \in H_*(Y; \mathbb{F}_2)$.
(2) Q_i *is a homomorphism for every* $0 \leq i \leq d-2$.
(3) Q_{d-1} *is not homomorphism in general and*

$$Q_{d-1}(y_1 + y_2) = Q_{d-1}(y_1) + Q_{d-1}(y_2) + \lambda_{d-1}(y_1, y_2)$$

for $y_1, y_2 \in H_*(Y; \mathbb{F}_2)$ *with* $\deg(y_1) = \deg(y_2)$.
(4) *If Y is connected and* $1 \in H_0(Y; \mathbb{F}_2)$, *then* $Q_i(1) = 0$ *for* $0 < i \leq d-1$.
(5) *The operations Q_i are natural with respect to the morphisms of C_d-spaces.*

(6) *If* $y_r \in H_r(Y; \mathbb{F}_2)$, $Z_s \in H_s(Z; \mathbb{F}_2)$ *and* $0 \le i \le d-1$, *then in* $H_*(Y \times Z; \mathbb{F}_2)$ *holds*

$$Q_i(y_r \otimes z_s) = \sum_{j=0}^{i} Q_j(y_r) \otimes Q_{i-j}(z_s) + \varepsilon_n(y_r, z_s),$$

where $\varepsilon_n(y_r \otimes z_s)$ *is the "error term." For example, in the case when* $Y = Z = \Omega^d \Sigma^d S^L$, $d \ge 2$, *and* $L \ge 1$, *the "error term" vanishes.*

(7) *If* $y_r \in H_r(Y; \mathbb{F}_2)$, $y'_s \in H_s(Y; \mathbb{F}_2)$ *and* $0 \le i \le d-1$, *then in* $H_*(Y; \mathbb{F}_2)$ *holds*

$$Q_i(y_r \cdot y'_s) = \sum_{j=0}^{i} Q_j(y_r) \cdot Q_{i-j}(y'_s) + \varepsilon'_n(y_r, y'_s),$$

where "·" denotes the Pontryagin product and $\varepsilon'_n(y_r, y'_s)$ *stands for the error term. For example, in the case when* $Y = \Omega^d \Sigma^d S^L$, $d \ge 2$, *and* $L \ge 1$, *the error term vanishes.*

Assume that Y is still a C_d-space. Then the operations

$$Q_i \colon H_j(Y; \mathbb{F}_2) \longrightarrow H_{i+2j}(Y; \mathbb{F}_2)$$

are defined in [33, Def. 5.6] as follows. The C_d-action on Y yields a map

$$\Theta_2 \colon C_d(2) \times (Y \times Y) \longrightarrow Y$$

that behaves naturally with respect to the acton of $\mathfrak{S}_2 \cong \mathbb{Z}_2$. Consequently, it induces the quotient map (denoted in the same way)

$$\Theta_2 \colon C_d(2) \times_{\mathbb{Z}_2} (Y \times Y) \longrightarrow Y$$

where the action on the product $Y \times Y$ is given by interchanging factors. The space $C_d(2)$ is $\mathbb{Z}_2$ equivariantly homotopic to the sphere S^{d-1} equipped with the antipodal actions. Hence, the cohomology and homology of $C_d(2) \times_{\mathbb{Z}_2} (Y \times Y)$, with $\mathbb{F}_2$-coefficients, is completely described in Sect. 3.3. Using the notation from Theorem 3.7 we define

$$Q_i(y) := (\Theta_2)_*\big((y \otimes y) \otimes_{\mathbb{Z}_2} f_i\big)$$

for $y \in H_*(Y; \mathbb{F}_2)$ where $0 \le i \le d-1$.

Chapter 8
The Dickson Algebra

In this chapter we present all the facts about the Dickson algebra that are relevant for the calculations in this book. For this we rely on the sources [2, Sec. III.2], [27], and [101], where some of the results we present appeared already in the original paper of Dickson [43].

8.1 Rings of Invariants

Let $m \geq 1$ be an integer, let V be an m-dimensional vector space over the field $\mathbb{F}_2$, and let $\mathrm{Sym}(V)$ denotes the symmetric algebra of V over $\mathbb{F}_2$. Then for a choice of a basis $(x_1, \ldots, x_m)$ of V there is an isomorphism $\mathrm{Sym}(V) \cong \mathbb{F}_2[x_1, \ldots, x_m]$. The algebra $\mathrm{Sym}(V)$ is graded by setting $\deg(x_1) = \cdots = \deg(x_m) = 1$. The degree of a monomial is defined in the usual way by $\deg(x_1^{\alpha_1} \cdots x_m^{\alpha_m}) := \alpha_1 + \cdots + \alpha_m$. The set of all homogeneous polynomials of degree n is denoted by $\mathrm{Sym}^n(V)$ where $\mathrm{Sym}^0(V) \cong \mathbb{F}_2$. Consequently, $\mathrm{Sym}(V) \cong \bigoplus_{n \geq 0} \mathrm{Sym}^n(V)$.

The general linear group $\mathrm{GL}(V) \cong \mathrm{GL}_m(\mathbb{F}_2)$ of the vector space V acts from the left on $\mathrm{Sym}(V)$ and preserves the introduced grading. If G is a subgroup of $\mathrm{GL}_m(\mathbb{F}_2)$, then $\mathrm{Sym}(V)^G$ denotes the ring of G-invariants. In this section we will discuss the invariants only of the full general linear group $\mathrm{GL}_m(\mathbb{F}_2)$ and the subgroup $\mathrm{L}_m(\mathbb{F}_2) \subseteq \mathrm{GL}_m(\mathbb{F}_2)$ of all lower triangular matrices with 1's on the main diagonal. It is a known fact that $\mathrm{L}_m(\mathbb{F}_2)$ is a Sylow 2-subgroup of $\mathrm{GL}_m(\mathbb{F}_2)$. In fact, the order of the group $\mathrm{L}_m(\mathbb{F}_2)$ is $2^{\frac{m(m-1)}{2}}$ while its index in $\mathrm{GL}_m(\mathbb{F}_2)$ is $(2m-1)!! = 1 \cdot 3 \cdot 5 \cdots (2m-1)$.

P. V. M. Blagojević et al., *Equivariant Cohomology of Configuration Spaces Mod 2*, Lecture Notes in Mathematics 2282, https://doi.org/10.1007/978-3-030-84138-6_8

Let $(x_1, \dots, x_m)$ be a fixed basis of the vector space V, and let us introduce a complete flag of subspaces in V by

$$\{0\} = V_0 \subseteq V_1 \subseteq \cdots \subseteq V_{m-1} \subseteq V_m = V$$

where $V_i := \operatorname{span}\{x_m, \dots, x_{m-i+1}\}$, for $1 \leq i \leq m$. Thus the complete flag we consider is

$$\{0\} \subseteq \operatorname{span}\{x_m\} \subseteq \operatorname{span}\{x_m, x_{m-1}\} \subseteq \\ \cdots \subseteq \operatorname{span}\{x_m, x_{m-1}, \dots, x_2\} \subseteq \operatorname{span}\{x_m, x_{m-1}, \dots, x_2, x_1\}.$$

The rings of invariants of the polynomial ring $S(V) \cong \mathbb{F}_2[x_1, \dots, x_m]$ with respect to the groups $\mathrm{L}_m(\mathbb{F}_2)$ and $\mathrm{GL}_m(\mathbb{F}_2)$ are denoted as follows:

$$\mathfrak{H}_m := \operatorname{Sym}(V)^{\mathrm{L}_m(\mathbb{F}_2)} \cong \mathbb{F}_2[x_1, \dots, x_m]^{\mathrm{L}_m(\mathbb{F}_2)},$$

and

$$\mathfrak{D}_m := \operatorname{Sym}(V)^{\mathrm{GL}_m(\mathbb{F}_2)} \cong \mathbb{F}_2[x_1, \dots, x_m]^{\mathrm{GL}_m(\mathbb{F}_2)}.$$

Assuming the previously introduced notation a result of Mùi [82, Thm. 6.4] gives the following described the ring of invariants $\mathfrak{H}_m$, see also [27, Thm. 3.1]. The ring of invariants $\mathfrak{H}_m$ with the natural algebra structure is call the **Mùi algebra**.

Theorem 8.1 *Let* $h_i := \prod_{v \in V_{i-1}} (x_{m-i+1} + v)$ *for* $1 \leq i \leq m$, *a polynomial of degree* 2^{i-1} *in* $\operatorname{Sym}(V) \cong \mathbb{F}_2[x_1, \dots, x_m]$. *Then*

$$\mathfrak{H}_m = \mathbb{F}_2[h_1, \dots, h_m].$$

The polynomials $h_1, \dots, h_m$ are called the **Mùi invariants**. For example, if $m = 3$ then

$$h_1 = x_3, \qquad h_2 = x_2(x_2 + x_3), \qquad h_3 = x_1(x_1 + x_2)(x_1 + x_3)(x_1 + x_2 + x_3).$$

For the ring of invariants $\mathfrak{D}_m$ we rely on a result of Dickson [43] and have the following description of the ring, see also [101, Thm. 1.2].

Theorem 8.2 *Let* $d_{m,0}, \dots, d_{m,m-1}, d_{m,m}$ *be the polynomials in* $\mathbb{F}_2[x_1, \dots, x_m]$ *defined as the coefficients of the polynomial*

$$f_m(T) := \prod_{v \in V} (T + v) = \sum_{i=0}^{m} d_{m,i}\, T^{2^i}$$

in $\mathbb{F}_2[x_1, \ldots, x_m][T]$. *Then* $\deg(d_{m,i}) = 2^m - 2^i$ *for* $0 \leq i \leq m$, $d_{m,m} = 1$, *and*

$$\mathfrak{D}_m = \mathbb{F}_2[d_{m,0}, \ldots, d_{m,m-1}].$$

The polynomials $d_{m,0}, \ldots, d_{m,m-1}$ are called the **Dickson invariants** of $\mathrm{Sym}(V)$ or of the polynomial ring $\mathbb{F}_2[x_1, \ldots, x_m]$. The ring of invariants $\mathfrak{D}_m$ with the natural algebra structure is call the **Dickson algebra**.

For example, if $m = 2$ then

$$f_2(T) = T(T+x_1)(T+x_2)(T+x_1+x_2) = x_1x_2(x_1+x_2)T + (x_1x_2+x_1^2+x_2^2)T^2.$$

Hence,

$$d_{2,0} = x_1x_2(x_1 + x_2) \qquad \text{and} \qquad d_{2,1} = x_1x_2 + x_1^2 + x_2^2.$$

On the other hand $h_1 = x_2$ and $h_2 = x_1(x_1 + x_2)$, and consequently

$$d_{2,0} = h_1h_2 \qquad \text{and} \qquad d_{2,1} = 1 \cdot h_2 + \chi_2 h_1^2,$$

where $\chi_2 \in \mathrm{GL}_2(F_2)$ is the variable substitution given by the matrix $\begin{pmatrix} 0 & 1 \\ 1 & 0 \end{pmatrix}$.

The formula connecting generators of the rings of invariants $\mathfrak{H}_m$ and $\mathfrak{D}_m$ is as follows, see for more details [101, Prop. 1.3 (b)].

Proposition 8.3 *For* $0 \leq i \leq m-1$ *and setting* $d_{m-1,-1} = 0$ *we have that*

$$d_{m,i} = (\chi_m d_{m-1,i})h_m + (\chi_m d_{m-1,i-1})^2, \tag{8.1}$$

where $\chi_m \in \mathrm{GL}_m(\mathbb{F}_2)$ *is the change of variables given by the matrix*

$$\begin{pmatrix} 0\,0 \cdots 0\,1 \\ 0\,0 \cdots 1\,0 \\ \cdots \\ 1\,0 \cdots 0\,0 \end{pmatrix}$$

If we set $k_j := \chi_m h_j$ for $1 \leq j \leq m$ then the relation (8.1) becomes

$$d_{m,i} = d_{m-1,i}k_m + d_{m-1,i-1}^2, \tag{8.2}$$

where $0 \leq i \leq m-1$. It is not hard to see that $k_1, \ldots, k_m$ are generators of the ring of invariants with respect to the subgroup $\mathrm{U}_m(\mathbb{F}_2)$ of $\mathrm{GL}_m(\mathbb{F}_2)$ that consists of all upper triangular matrices with 1's on the main diagonal. More precisely,

$$\mathrm{Sym}(V)^{\mathrm{U}_m(\mathbb{F}_2)} \cong \mathbb{F}_2[x_1, \ldots, x_m]^{\mathrm{U}_m(\mathbb{F}_2)} = \mathbb{F}_2[k_1, \ldots, k_m].$$

From the relation (8.2) we can derive a presentation of each Dickson invariant $d_{m,i}$ in term of $k_1, \dots, k_m$. Indeed, let $0 \leq r \leq m-1$ be an integer, and let $J = (j_1, \dots, j_r) \in [m]^r$ with $1 \leq j_1 < \cdots < j_r \leq m$. We denote particular monomials in $k_1, \dots, k_n$ as follows:

$$k[J] := (k_1 \cdots k_{j_1-1})^{2^r} (k_{j_1+1} \cdots k_{j_2-1})^{2^{r-1}} \cdots (k_{j_r+1} \cdots k_m)^{2^0}.$$

Then as in [27, Thm. 4.3] we have that

$$\begin{aligned} d_{m,r} &= \sum_{J\in[m]^r \,:\, j_1<\cdots<j_r} k[J] \qquad (8.3)\\ &= \sum_{J\in[m]^r \,:\, j_1<\cdots<j_r} (k_1 \cdots k_{j_1-1})^{2^r} (k_{j_1+1} \cdots k_{j_2-1})^{2^{r-1}} \cdots (k_{j_r+1} \cdots k_m)^{2^0}. \end{aligned}$$

In particular,

$$d_{m,0} = k_1 \cdots k_m \qquad \text{and} \qquad d_{m,m-1} = k_1^{2^{m-1}} + k_2^{2^{m-2}} + \cdots + k_m^{2^0}. \qquad (8.4)$$

8.2 The Dickson Invariants as Characteristic Classes

Let $m \geq 1$ be an integer. Consider the sequence of group embeddings

$$\mathcal{E}_m := \mathbb{Z}_2^{\oplus m} \xrightarrow{\text{(reg)}} \mathfrak{S}_{2^m} \xrightarrow{\iota_{2^m}} \mathrm{O}(2^m),$$

where

- (reg): $\mathbb{Z}_2^{\oplus m} \longrightarrow \mathfrak{S}_{2^m}$ is the regular embedding that is given by the left translation action of $\mathbb{Z}_2^{\oplus m}$ on itself, consult [2, Ex. III.2.7], and
- $\iota_{2^m} : \mathfrak{S}_{2^m} \longrightarrow \mathrm{O}(2^m)$ is the embedding given by the permutation representation.

This sequence of embeddings induces a sequence of maps of the corresponding classifying spaces

$$\mathrm{B}\mathcal{E}_m \xrightarrow{\mathrm{B(reg)}} \mathrm{B}\mathfrak{S}_{2^m} \xrightarrow{\mathrm{B}(\iota_{2^m})} \mathrm{BO}(2^m).$$

Let γ_{2^m} denotes the tautological vector bundle over $\mathrm{BO}(2^m)$, and let us denote the pullbacks as follows:

$$\xi_{2^m} := \mathrm{B}(\iota_{2^m})^* \gamma_{2^m} \qquad \text{and} \qquad \nu_{2^m} := (\mathrm{B}(\iota_{2^m}) \circ \mathrm{B(reg)})^* \gamma_{2^m}.$$

Then the bundles γ_{2^m}, ξ_{2^m} and ν_{2^m} induce the following commutative diagram of vector bundle morphisms:

$$\begin{array}{ccccc}
\mathrm{E}\mathcal{E}_m \times_{\mathcal{E}_m} \mathbb{R}^{2^m} & \longrightarrow & \mathrm{E}\mathfrak{S}_{2^m} \times_{\mathfrak{S}_{2^m}} \mathbb{R}^{2^m} & \longrightarrow & \mathrm{EO}(2^m) \times_{\mathrm{O}(2^m)} \mathbb{R}^{2^m} \\
\downarrow \nu_{2^m} & & \downarrow \xi_{2^m} & & \downarrow \gamma_{2^m} \\
\mathrm{B}\mathcal{E}_m & \xrightarrow{\mathrm{B(reg)}} & \mathrm{B}\mathfrak{S}_{2^m} & \xrightarrow{\mathrm{B}(\iota_{2^m})} & \mathrm{BO}(2^m).
\end{array}$$

Recall that the cohomology ring of the elementary abelian group $\mathcal{E}_m$ with coefficients in the field $\mathbb{F}_2$ is a polynomial ring on m generators in degree 1, that is $H^*(\mathcal{E}_m; \mathbb{F}_2) \cong \mathbb{F}_2[y_1, \dots, y_m]$ with $\deg(y_1) = \cdots = \deg(y_m) = 1$.

The relationship of the introduced bundles and the Dickson invariants is given by the following theorem, see [72, Lem. 3.26].

Theorem 8.4 *The total Stiefel–Whitney class of the vector bundle ν_{2^m} is*

$$w(\nu_{2^m}) = 1 + d_{m,m-1} + d_{m,m-2} + \cdots + d_{m,0},$$

where $d_{m,m-1}, \dots, d_{m,0}$ are the Dickson invariants of $\mathbb{F}_2[y_1, \dots, y_m] \cong H^*(\mathcal{E}_m; \mathbb{F}_2)$. This means that

$$w_i(\nu_{2^m}) = \begin{cases} d_{m,j}, & i = 2^m - 2^j,\ 0 \le j \le m-1, \\ 1, & i = 0, \\ 0, & \text{otherwise.} \end{cases}$$

Chapter 9
The Stiefel–Whitney Classes of the Wreath Square of a Vector Bundle

In this chapter we introduce notions called the wreath square of a vector bundle [11, Sec. 3] and the $(d-1)$-partial wreath square of a vector bundle. Furthermore, we collect all necessary facts we use in Chap. 6. Our presentation partially follows [11, Sec. 3] and is given in the generality necessary for our computations.

9.1 The Wreath Square and the $(d-1)$-Partial Wreath Square of a Vector Bundle

Let X be a CW-complex which a priori is not finite. The product $X \times X$ has a natural action of the group $\mathbb{Z}_2$ given by interchanging the copies, that is $\omega \cdot (x_1, x_2) := (x_2, x_1)$, where ω generates $\mathbb{Z}_2$ and $(x_1, x_2) \in X \times X$. The spaces $(X \times X) \times \mathrm{E}\mathbb{Z}_2$ and $(X \times X) \times S^{d-1}$ with the diagonal $\mathbb{Z}_2$-actions are free $\mathbb{Z}_2$-spaces. The action on the sphere S^{d-1} is assumed to be antipodal. Thus, the projection maps

$$p_1 \colon (X \times X) \times \mathrm{E}\mathbb{Z}_2 \longrightarrow X \times X \qquad \text{and} \qquad p_1 \colon (X \times X) \times S^{d-1} \longrightarrow X \times X$$

given by $(x_1, x_1, e) \longmapsto (x_1, x_2)$ are $\mathbb{Z}_2$-maps. Here the model for $\mathrm{E}\mathbb{Z}_2$ is assumed to be the infinite sphere $S^\infty := \mathrm{colim}_{d\to\infty} S^{d-1}$ equipped with the antipodal action inherited from the action on S^{d-1}.

Now, we define the functors

$$S^2 \colon \mathrm{Top}_{\mathrm{cw}} \longrightarrow \mathrm{Top}_{\mathrm{cw}} \qquad \text{and} \qquad S^{2,d} \colon \mathrm{Top}_{\mathrm{cw}} \longrightarrow \mathrm{Top}_{\mathrm{cw}}$$

that are on the objects given by

$$S^2 X := ((X \times X) \times \mathrm{E}\mathbb{Z}_2)/\mathbb{Z}_2 = (X \times X) \times_{\mathbb{Z}_2} \mathrm{E}\mathbb{Z}_2,$$

P. V. M. Blagojević et al., *Equivariant Cohomology of Configuration Spaces Mod 2*, Lecture Notes in Mathematics 2282, https://doi.org/10.1007/978-3-030-84138-6_9

and

$$S^{2,d}X := ((X \times X) \times S^{d-1})/\mathbb{Z}_2 = (X \times X) \times_{\mathbb{Z}_2} S^{d-1},$$

where X is a CW-complex and $d \geq 1$ integer. For a morphism $h\colon X \longrightarrow Y$ of CW-complexes, a continuous map, we set

$$S^2 h := (h \times h) \times_{\mathbb{Z}_2} \mathrm{id}\colon (X \times X) \times_{\mathbb{Z}_2} \mathrm{E}\mathbb{Z}_2 \longrightarrow (Y \times Y) \times_{\mathbb{Z}_2} \mathrm{E}\mathbb{Z}_2,$$

and

$$S^{2,d} h := (h \times h) \times_{\mathbb{Z}_2} \mathrm{id}\colon (X \times X) \times_{\mathbb{Z}_2} S^{d-1} \longrightarrow (Y \times Y) \times_{\mathbb{Z}_2} S^{d-1},$$

to be the maps induced by the product maps

$$(h \times h) \times \mathrm{id}\colon (X \times X) \times \mathrm{E}\mathbb{Z}_2 \longrightarrow (Y \times Y) \times \mathrm{E}\mathbb{Z}_2$$

and

$$(h \times h) \times \mathrm{id}\colon (X \times X) \times S^{d-1} \longrightarrow (Y \times Y) \times S^{d-1}$$

by passing to the $\mathbb{Z}_2$-orbits.

Consider now a real n-dimensional vector bundle $\xi := (E(\xi) \xrightarrow{p_\xi} B(\xi))$ over the CW-complex $B(\xi)$ whose fiber is $F(\xi)$. Here we denote by $E(\xi)$ the total space of ξ, and by p_ξ the corresponding projection map. The pull-back vector bundle $p_1^*(\xi \times \xi)$ of the product vector bundle $\xi \times \xi$ along the p_1:

$$\begin{array}{ccc} E(p_1^*(\xi \times \xi)) & \longrightarrow & E(\xi \times \xi) \cong E(\xi) \times E(\xi) \\ \downarrow & & \downarrow \\ (B(\xi) \times B(\xi)) \times \mathrm{E}\mathbb{Z}_2 & \xrightarrow{p_1} & B(\xi \times \xi) \cong B(\xi) \times B(\xi), \end{array}$$

is equipped with a free $\mathbb{Z}_2$-action. Moreover, the projection map of the pull-back bundle $E(p_1^*(\xi \times \xi)) \longrightarrow (B(\xi) \times B(\xi)) \times \mathrm{E}\mathbb{Z}_2$ is a $\mathbb{Z}_2$-map. Hence, after passing to $\mathbb{Z}_2$-orbits we get the $2n$-dimensional vector bundle $S^2\xi$ over $S^2 B(\xi)$:

$$E(p_1^*(\xi \times \xi))/\mathbb{Z}_2 \longrightarrow (B(\xi) \times B(\xi)) \times_{\mathbb{Z}_2} \mathrm{E}\mathbb{Z}_2.$$

The bundle $S^2\xi$ is called the **wreath square** of the vector bundle ξ.

Next we consider the following pull-back diagram induced by the $\mathbb{Z}_2$-inclusion $i\colon S^{d-1} \longrightarrow S^{\infty}$:

$$\begin{array}{ccc} E(((\mathrm{id}\times\mathrm{id})\times_{\mathbb{Z}_2} i)^*S^2\xi) & \longrightarrow & E(p_1^*(\xi\times\xi))/\mathbb{Z}_2 = E(S^2\xi) \\ \downarrow & & \downarrow \\ (B(\xi)\times B(\xi))\times_{\mathbb{Z}_2} S^{d-1} & \xrightarrow{(\mathrm{id}\times\mathrm{id})\times_{\mathbb{Z}_2} i} & (B(\xi)\times B(\xi))\times_{\mathbb{Z}_2} \mathrm{E}\mathbb{Z}_2 = B(S^2\xi). \end{array}$$

The pull-back vector bundle $((\mathrm{id}\times\mathrm{id})\times_{\mathbb{Z}_2} i)^*S^2\xi$ is called the **$(d-1)$-partial wreath square** of the vector bundle ξ and is denoted by $S^{2,d}\xi$. In other words,

$$S^{2,d}\xi := ((\mathrm{id}\times\mathrm{id})\times_{\mathbb{Z}_2} i)^*S^2\xi,$$

with the base space $S^{2,d}B(\xi)$.

The wreath square, as well as the $(d-1)$-partial wreath square, of a vector bundle is natural with respect to morphisms of vector bundles. Indeed, a morphism between vector bundles $\xi \longrightarrow \eta$ induces morphisms between associated wreath squares $S^2\xi \longrightarrow S^2\eta$ and $(d-1)$-partial wreath squares $S^{2,d}\xi \longrightarrow S^{2,d}\eta$. These morphisms satisfy all expected properties with respect to the composition of morphisms. Furthermore, the wreath square and the $(d-1)$-partial wreath square of a vector bundle behave naturally with respect to the Whitney sum of vector bundles. This means that for arbitrary vector bundles ξ and η there are isomorphisms of vector bundles

$$S^2(\xi\oplus\eta)\cong S^2(\xi)\oplus S^2(\eta) \qquad \text{and} \qquad S^{2,d}(\xi\oplus\eta)\cong S^{2,d}(\xi)\oplus S^{2,d}(\eta). \tag{9.1}$$

9.2 Cohomology of $B(S^2\xi) = S^2B(\xi)$

In this section, based on the material presented in Sect. 3.3, we describe the cohomology of a typical base space of the wreath square of a vector bundle.

Let X be the base space of the vector bundle ξ, that is $X = B(\xi)$. Assume that X in addition is a CW-complex. Then the base space of the vector bundle $S^2\xi$ is S^2X, and also the total space of the fiber bundle

$$X\times X \longrightarrow (X\times X)\times_{\mathbb{Z}_2} \mathrm{E}\mathbb{Z}_2 \longrightarrow \mathrm{B}\mathbb{Z}_2. \tag{9.2}$$

Note that similarly the base space of the vector bundle $S^{2,d}\xi$ is the total space of the fiber bundle

$$X\times X \longrightarrow (X\times X)\times_{\mathbb{Z}_2} S^{d-1} \longrightarrow \mathbb{R}\mathrm{P}^{d-1}. \tag{9.3}$$

The Serre spectral sequence associated to the fiber bundle (9.2) has E_2-term given by

$$E_2^{i,j}(X) = H^i(\mathrm{B}\mathbb{Z}_2; \mathcal{H}^j(X \times X; \mathbb{F}_2)) \cong H^i(\mathbb{Z}_2; H^j(X \times X; \mathbb{F}_2)). \tag{9.4}$$

As discussed in Sect. 3.3 this spectral sequence collapses at the E_2-term, that means $E_2^{i,j}(X) \cong E_\infty^{i,j}(X)$ for all $i, j \in \mathbb{Z}$. For more details consult Theorem 3.4 or [2, Thm. IV.1.7].

In the description of the total Stiefel–Whitney classes of the wreath square of a vector bundle the following maps turn out to be very useful. At first, consider the map (not a homomorphism)

$$P\colon H^j(X; \mathbb{F}_2) \longrightarrow H^{2j}(X \times X; \mathbb{F}_2)^{\mathbb{Z}_2} \cong E_2^{0,2j} \cong E_\infty^{0,2j}$$

given by

$$P(u) := u \otimes u,$$

for $u \in H^j(X; \mathbb{F}_2)$ and $j \geq 0$ an integer. By a direct inspection we see that the map P is not an additive map, but it is a multiplicative map. The second map we consider is

$$T\colon H^j(X \times X; \mathbb{F}_2) \longrightarrow H^j(X \times X; \mathbb{F}_2)^{\mathbb{Z}_2}$$

defined by

$$T(u \otimes u') := u \otimes u' + u' \otimes u,$$

where $u \otimes u' \in H^j(X \times X; \mathbb{F}_2)$ and $j \geq 0$ is an integer. The map T is an additive map, but not a multiplicative map.

With the help of just introduced maps P and T, based on Lemma 3.2 and Theorem 3.4, we can describe the E_∞-term of the Serre spectral sequence (5.5) as follows:

- $E_2^{*,0} \cong E_\infty^{*,0} \cong H^*(\mathbb{Z}_2; \mathbb{F}_2) \cong \mathbb{F}_2[f]$, $\deg(f) = 1$,
- $E_2^{0,*} \cong E_\infty^{0,*} \cong H^*(X \times X; \mathbb{F}_2)^{\mathbb{Z}_2}$,
- $E_2^{0,j} \cong E_\infty^{0,j} \cong P(H^{j/2}(X; \mathbb{F}_2)) \oplus T(H^j(X \times X; \mathbb{F}_2))$ for $j \geq 2$ even,
- $E_2^{i,j} \cong E_\infty^{i,j} \cong P(H^{j/2}(X; \mathbb{F}_2)) \otimes H^*(\mathbb{Z}_2; \mathbb{F}_2)$ for $j \geq 2$ even, and $i \geq 1$.

Furthermore, we have that the generator f of $H^*(\mathbb{Z}_2; \mathbb{F}_2)$ annihilates the image of the map T, that is

$$T(H^*(X \times X; \mathbb{F}_2)) \cdot f = 0. \tag{9.5}$$

Note that the Serre spectral sequence associated to the fiber bundle (9.3) can be describe in a similar way—as discussed in Sect. 3.3.

9.3 The Total Stiefel–Whitney Class of the Wreath Square of a Vector Bundle

Let ξ be a real n-dimensional vector bundle over the CW-complex $B(\xi)$. To the vector bundle ξ we associate characteristic classes

$$s^2(\xi) := w(S^2\xi) \in H^*(B(S^2\xi); \mathbb{F}_2),$$

and

$$s^{2,d}(\xi) := w(S^{2,d}\xi) \in H^*(B(S^{2,d}\xi); \mathbb{F}_2).$$

These are the total Stiefel–Whitney classes of the real $2n$-dimensional vector bundles $S^2\xi$ and $S^{2,d}\xi$. The assignments

$$\xi \longmapsto s^2(\xi) \qquad \text{and} \qquad \xi \longmapsto s^{2,d}(\xi)$$

are natural with respect to the continuous maps. This means that for a continuous map $h\colon K \longrightarrow B(\xi)$ from a CW-complex K into the base space $B(\xi)$ of ξ the following equalities hold

$$(S^2h)^*(s^2(\xi)) = s^2(h^*\xi) \qquad \text{and} \qquad (S^{2,d}h)^*(s^{2,d}(\xi)) = s^{2,d}(h^*\xi)$$

Here $h^*\xi$ denotes the pull-back vector bundle of ξ along the map h.

The characteristic class $s^2(\xi)$ was calculated, in term of Stiefel–Whitney classes of ξ, in [11, Thm. 3.4] and is given in the following theorem.

Theorem 9.1 *Let ξ be a real n-dimensional vector bundle over a CW-complex. Then*

$$s^2(\xi) = w(S^2\xi) = \sum_{0\leq r<s\leq n} T(w_r(\xi)\otimes w_s(\xi)) + \sum_{0\leq r\leq n} P(w_r(\xi))\cdot(1+f)^{n-r}$$

In particular, if ξ is 1 dimensional, then

$$\begin{aligned} s^2(\xi) = w(S^2\xi) &= T(w_0(\xi)\otimes w_1(\xi)) + P(w_0(\xi))\cdot(1+f) + P(w_1(\xi)) \\ &= 1 + \big(f + 1\otimes w_1(\xi) + w_1(\xi)\otimes 1\big) + w_1(\xi)\otimes w_1(\xi). \end{aligned}$$

From the computations in Sect. 3.3 and the fact that by definition $S^{2,d}\xi = ((\mathrm{id}\times\mathrm{id})\times_{\mathbb{Z}_2} i)^* S^2\xi$ we get the following consequence of Theorem 9.1.

Corollary 9.2 *Let ξ be a real n-dimensional vector bundle over a CW-complex, and let $d \geq 2$ be an integer. Then*

$$\begin{aligned} s^{2,d}(\xi) &= w(S^{2,d}\xi) \\ &= \sum_{0\leq r<s\leq n} T(w_r(\xi)\otimes w_s(\xi)) \\ &\quad + \sum_{0\leq r\leq n}\ \sum_{0\leq j\leq \min\{n-r,d-1\}} \binom{n-r}{j} P(w_r(\xi)) f^j. \end{aligned}$$

In particular, Theorem 9.1 and Corollary 9.2 give formulas for the evaluation of mod 2 Euler classes (top Stiefel–Whitney class) of the vector bundles $S^2\xi$ and $S^{2,d}\xi$ in term of the mod 2 Euler class of the vector bundle ξ. In the following we present these formulas and show them with a direct proof which does not rely on the previous claims.

Corollary 9.3 *Let ξ be a real n-dimensional vector bundle over a CW-complex. Then*

$$\mathfrak{e}(S^2\xi) = P(\mathfrak{e}(\xi)) \qquad \textit{or} \qquad w_{2n}(S^2\xi) = P(w_n(\xi)),$$

and

$$\mathfrak{e}(S^{2,d}\xi) = P(\mathfrak{e}(\xi)) \qquad \textit{or} \qquad w_{2n}(S^{2,d}\xi) = P(w_n(\xi)).$$

In particular, if $\mathfrak{e}(\xi)\neq 0$, then $\mathfrak{e}(S^2\xi)\neq 0$ and $\mathfrak{e}(S^{2,d}\xi)\neq 0$.

Here $\mathfrak{e}(S^2\xi)$, $\mathfrak{e}(S^{2,d}\xi)$ and $\mathfrak{e}(\xi)$ denote mod 2 Euler classes of the vector bundles $S^2\xi$, $S^{2,d}\xi$ and ξ, respectively. Note the abuse of notation: the map P is not the same in the formulas for $\mathfrak{e}(S^2\xi)$ and $\mathfrak{e}(S^{2,d}\xi)$, since it operates on different spectral sequences.

Proof Let $u \in H^n(D(\xi), S(\xi); \mathbb{F}_2)$ be the Thom class of the n-dimensional vector bundle ξ. Here $D(\xi)$ and $S(\xi)$ denote respectively the disk and sphere bundles associate to the vector bundle ξ. The mod 2 Euler class of the vector bundle ξ, by definition, equals to $\mathfrak{e}(\xi) = i_\xi^*(u)$ where $i_\xi : (E(\xi), \varnothing) \longrightarrow (D(\xi), S(\xi))$ is the zero section.

Consider the following commutative diagram

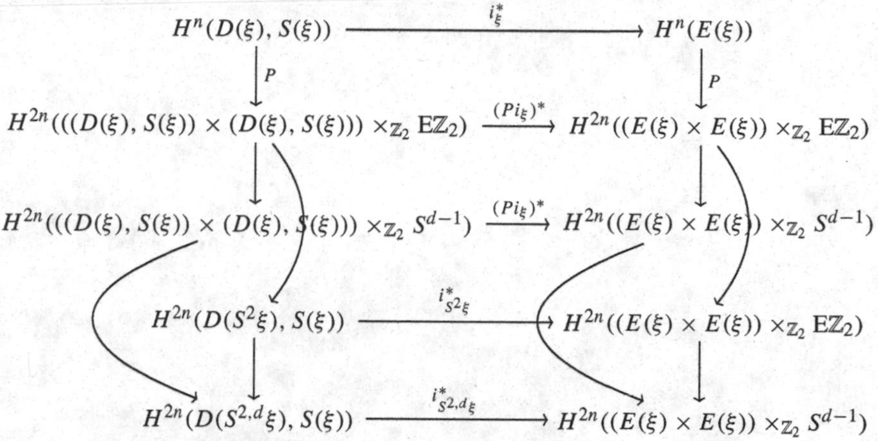

$$
\begin{array}{ccc}
H^n(D(\xi), S(\xi)) & \xrightarrow{\quad i_\xi^* \quad} & H^n(E(\xi)) \\
\downarrow P & & \downarrow P \\
H^{2n}(((D(\xi), S(\xi)) \times (D(\xi), S(\xi))) \times_{\mathbb{Z}_2} \mathrm{E}\mathbb{Z}_2) & \xrightarrow{(Pi_\xi)^*} & H^{2n}((E(\xi) \times E(\xi)) \times_{\mathbb{Z}_2} \mathrm{E}\mathbb{Z}_2) \\
\downarrow & & \downarrow \\
H^{2n}(((D(\xi), S(\xi)) \times (D(\xi), S(\xi))) \times_{\mathbb{Z}_2} S^{d-1}) & \xrightarrow{(Pi_\xi)^*} & H^{2n}((E(\xi) \times E(\xi)) \times_{\mathbb{Z}_2} S^{d-1}) \\
 & & \\
H^{2n}(D(S^2\xi), S(\xi)) & \xrightarrow{\quad i_{S^2\xi}^* \quad} & H^{2n}((E(\xi) \times E(\xi)) \times_{\mathbb{Z}_2} \mathrm{E}\mathbb{Z}_2) \\
\downarrow & & \downarrow \\
H^{2n}(D(S^{2,d}\xi), S(\xi)) & \xrightarrow{\quad i_{S^{2,d}\xi}^* \quad} & H^{2n}((E(\xi) \times E(\xi)) \times_{\mathbb{Z}_2} S^{d-1})
\end{array}
$$

where the coefficients are assumed to be in the field $\mathbb{F}_2$. Note that the curly left arrows are isomorphism while the right ones are equalities.

We claim that $P(u)$, appropriately interpreted, is the Thom class of the vector bundle $S^2\xi$, respectively $S^{2,d}\xi$. To check this it is enough to show that it restricts to the generator in each fibre. Thus we can reduce to the case in which $E(\xi)$ and $\mathrm{B}\mathbb{Z}_2$, respectively $\mathbb{RP}^{d-1}$, are just points. Then it is an elementary fact that the square map P:

$$\big(H^n(D(\mathbb{R}^n), S(\mathbb{R}^n); \mathbb{F}_2) \cong \mathbb{F}_2\big) \longrightarrow \big(H^{2n}(D(\mathbb{R}^{2n}), S(\mathbb{R}^{2n}); \mathbb{F}_2) \cong \mathbb{F}_2\big)$$

maps the generator to the generator.

The assertion now follows from the commutativity of the corresponding diagram:

$$\mathfrak{e}(S^2\xi) = i_{S^2\xi}^*(P(u)) = P(i_\xi^*(u)) = P(\mathfrak{e}(\xi)),$$

and similarly

$$\mathfrak{e}(S^{2,d}\xi) = i_{S^{2,d}\xi}^*(P(u)) = P(i_\xi^*(u)) = P(\mathfrak{e}(\xi)).$$

□

Chapter 10
Miscellaneous Calculations

In order give a smoother presentation of the main computational components in the main body of the book in this chapter we present details of some auxiliary computations.

10.1 Detecting Group Cohomology

In this section we review some basic facts from classical work of Quillen [90] in generality we need in this book. For more details consult for example [2, Sec. IV.4 and VI.1].

Let G be a finite group, and let p be a prime. The family of subgroups $\{H_i : i \in I\}$ of the group G **detects the cohomology** of G modulo $\mathbb{F}_p$, or is a **detecting family of subgroups**, if the homomorphism

$$H^*(G;\mathbb{F}_p) \longrightarrow \prod_{i\in I} H^*(H_i;\mathbb{F}_p)$$

induced by the restrictions $\operatorname{res}^G_{H_i}$ is a monomorphism. If $G^{(p)}$ is Sylow p-subgroup of G, then the restriction $\operatorname{res}^G_{G^{(p)}}$ is a monomorphism, and consequently $G^{(p)}$ detects the cohomology of G modulo $\mathbb{F}_p$.

First we recall an auxiliary lemma that is particularly useful for us, see [72, Lem. 3.22].

Lemma 10.1 *Let G be a finite group, and let p be a prime. Then $G \wr \mathbb{Z}_p$ is detected by $G \times \mathbb{Z}_p$ and $\underbrace{G \times \cdots \times G}_{p\ times}$.*

Next, let p be a prime, and let $n = p^m$ for an integer $m \geq 1$. Consider symmetric group $\mathfrak{S}_n = \mathfrak{S}_{p^m}$ as a group of permutation of the set $\mathbb{Z}_p^{\oplus m}$. Then the elementary

P. V. M. Blagojević et al., *Equivariant Cohomology of Configuration Spaces Mod 2*, Lecture Notes in Mathematics 2282, https://doi.org/10.1007/978-3-030-84138-6_10

abelian group $\mathcal{E}_m$ of all translations (seen as permutations) of the vector space $\mathbb{F}_p^{\oplus m}$, the so called regular embedded subgroup [2, Ex. III.2.7], is isomorphic to the elementary abelian group $\mathbb{Z}_p^{\oplus m}$. Let

$$S_{p^m} = E_{m,1} \wr \cdots \wr E_{m,m} \cong \underbrace{\mathbb{Z}_p \wr \cdots \wr \mathbb{Z}_p}_{m \text{ times}}$$

denotes the Sylow p-subgroup of $\mathfrak{S}_{p^m}$ containing $\mathcal{E}_m$. Here $E_{m,i} \cong \mathbb{Z}_p$ is the subgroup of $\mathcal{E}_m$ generated by the ith basis element $(0, \ldots, 0, 1, 0, \ldots, 0)$ of the vector space $\mathbb{Z}_p^{\oplus m}$.

Now we proceed with the following detection property of the cohomology of the symmetric group $\mathfrak{S}_{p^m}$ and its Sylow p-subgroup S_{p^m}, consult [2, Cor. VI.1.4].

Theorem 10.2 *Let p be a prime, and let $n = p^m$ for an integer $m \geq 1$.*

(1) *The cohomology $H^*(\mathfrak{S}_{p^m}; \mathbb{F}_p)$ of the symmetric group $\mathfrak{S}_{p^m}$ is detected modulo $\mathbb{F}_p$ by the elementary abelian subgroup $\mathcal{E}_m$ and the product subgroup*

$$\underbrace{\mathfrak{S}_{p^{m-1}} \times \cdots \times \mathfrak{S}_{p^{m-1}}}_{p \text{ times}}.$$

(2) *The cohomology $H^*(S_{p^m}; \mathbb{F}_p)$ of the Sylow p-subgroup S_{p^m} of the symmetric group $\mathfrak{S}_{p^m}$ is detected modulo $\mathbb{F}_p$ by the elementary abelian subgroup $\mathcal{E}_m$ and the product subgroup*

$$\underbrace{S_{p^{m-1}} \times \cdots \times S_{p^{m-1}}}_{p \text{ times}}.$$

Finally we state the classical result of Quillen about detection with the family of elementary abelian subgroups, see [2, Thm. VI.1.2].

Theorem 10.3 *Let $n \geq 1$ be an integer, and let p be a prime. The cohomology $H^*(\mathfrak{S}_n; \mathbb{F}_p)$ of the symmetric group $\mathfrak{S}_n$ with $\mathbb{F}_p$ coefficients is detected by the family of its elementary abelian p-subgroups.*

10.2 The Image of a Restriction Homomorphism

Let G be a finite group and let H a subgroup of G. In this section we want to give a description of the image of the restriction homomorphism

$$\operatorname{im}\big(\operatorname{res}_H^G \colon H^*(G; M) \longrightarrow H^*(H; M)\big)$$

where M is a trivial G-module.

Let G be a (finite) group. Any contractible free G-CW complex equipped with the right G cellular action is called a model for an EG space. The Milnor model is given by $\mathrm{E}G = \mathrm{colim}_{n\in\mathbb{N}} G^{*n}$ where G stands for a 0-dimensional free G-simplicial complex whose vertices are indexed by the elements of the group G and the action on G is given by the right translation, and G^{*n} is an n-fold join of the 0-dimensional simplicial complex with induced diagonal (right) action. A typical point in EG can be presented as follows

$$\sum_{i\geq 1} \lambda_i g_i \equiv (\lambda_1 g_1, \lambda_2 g_2, \lambda_3 g_3, \ldots),$$

where $g_i \in G$ and $\lambda_i \geq 0$ for all $i \geq 1$, the set $I := \{\lambda_i \neq 0 : i \geq 1\}$ is finite, and $\sum_{i\in I} \lambda_i = 1$. The quotient space $\mathrm{B}G = \mathrm{E}G/G$ is called a classifying space a the group G. For a trivial G-module M the group cohomology $H^*(G; M)$ can be defined as a singular cohomology $H^*(\mathrm{B}G; M)$.

Let H and G be (finite) groups, and let $f: H \longrightarrow G$ be a homomorphism. Then f induces the following G-equivariant map $\mathrm{E}(f): \mathrm{E}H \longrightarrow \mathrm{E}G$ by

$$\sum_{i\geq 1} \lambda_i h_i \equiv (\lambda_1 h_1, \lambda_2 h_2, \lambda_3 h_3, \ldots)$$
$$\longmapsto \sum_{i\geq 1} \lambda_i g(h_i) \equiv (\lambda_1 f(h_1), \lambda_2 f(h_2), \lambda_3 f(h_3), \ldots)$$

where the H-action on G is induced by the homomorphism f. Since the map $\mathrm{E}(f)$ is H-equivariant map it induces a map between quotient spaces $\mathrm{B}(f): \mathrm{B}H \longrightarrow \mathrm{B}G$. In particular, if H is a subgroup of G, $i: H \longrightarrow G$ the inclusion map, then the induced homomorphism in cohomology $\mathrm{B}(i)^*$ by definition is the restriction homomorphism res^G_H, that is $\mathrm{B}(i)^* = \mathrm{res}^G_H$.

We prove an auxiliary lemma following [44, Prop. I.6.14] and [2, Thm. II.1.9].

Lemma 10.4 *Let G be a finite group.*

(i) *Any two continuous G-equivariant maps $f, g: \mathrm{E}G \longrightarrow \mathrm{E}G$ are G-homotopic.*
(ii) *Let $a \in G$, and let $k_a: G \longrightarrow G$ be the conjugation homomorphism $k_a(g) := aga^{-1}$. Then the induced map $\mathrm{B}(k): \mathrm{B}G \longrightarrow \mathrm{B}G$ is homotopic to the identity $\mathrm{id}: \mathrm{B}G \longrightarrow \mathrm{B}G$.*

Proof

(i) This is a presentation of the proof of [44, Prop. I.6.14]. For $e \in \mathrm{E}G$ let us denote images $f(e)$ and $g(e)$ as follows

$$f(e) = (\lambda_1(e) f_1(e), \lambda_2(e) f_2(e), \ldots)$$

and

$$g(e) = (\mu_1(e)g_1(e), \mu_2(e)g_2(e), \ldots).$$

We define two additional maps $\bar{f}, \bar{g}\colon \mathrm{E}G \longrightarrow \mathrm{E}G$ by

$$\bar{f}(e) = (\lambda_1(e)f_1(e), 0, \lambda_2(e)f_2(e), 0, \ldots)$$

and

$$\bar{g}(e) = (0, \mu_1(e)g_1(e), 0, \mu_2(e)g_2(e), 0, \ldots).$$

The pairs of maps f, $\bar{f}$ and g, $\bar{g}$ are G-homotopic, that is $f \simeq_G \bar{f}$ and $g \simeq_G \bar{g}$. In order to construct a G-homotopy, for example, between f and $\bar{f}$ we proceed as follows. Let $j \geq 1$ be an integer and let the G-homotopy $H_j\colon \mathrm{E}G \times I \longrightarrow \mathrm{E}G$ be defined by

$$\begin{aligned} H_j(e,t) := \big(&\lambda_1(e)f_1(e), \ldots, \lambda_j(e)f_j(e), \\ &t\lambda_{j+1}(e)f_{j+1}(e), (1-t)\lambda_{j+1}(e)f_{j+1}(e), \\ &t\lambda_{j+2}(e)f_{j+2}(e), (1-t)\lambda_{j+2}(e)f_{j+2}(e), \ldots\big). \end{aligned}$$

Starting with $\bar{f}(e) = H_1(e, 0)$ and applying consecutively $H_1, H_2, \ldots$ we reach $f(e)$. Hence we have define a G-homotopy between $\bar{f}$ and f. Since in the presentation of every point $e = (\lambda_1 g_1, \lambda_2 g_2, \ldots) \in \mathrm{E}G$ there are only finitely many non-zero λ's the definition we just gave is correct and the map is moreover continuous. Therefore, $f \simeq_G \bar{f}$ and $g \simeq_G \bar{g}$.
It suffices to prove that $\bar{f} \simeq_G \bar{g}$. For this we give the G-homotopy $H : \mathrm{E}G \times I \longrightarrow \mathrm{E}G$ by

$$\begin{aligned} H(e,t) := \big((1-t)\lambda_1(e)f_1(e),\ &t\mu_1(e)g_1(e), \\ &(1-t)\lambda_2(e)f_2(e),\ t\mu_2(e)g_2(e), \ldots\big). \end{aligned}$$

This concludes the proof of part (i).

(ii) The homomorphisms $\mathrm{id}\colon G \longrightarrow G$ and $k_a\colon G \longrightarrow G$ induce G-equivariant maps

$$\mathrm{E}(\mathrm{id}) = \mathrm{id}\colon \mathrm{E}G \longrightarrow \mathrm{E}G \qquad \text{and} \qquad \mathrm{E}(k_a)\colon \mathrm{E}G \longrightarrow \mathrm{E}G.$$

From the part (i) of this lemma we have that $\mathrm{E}(\mathrm{id}) = \mathrm{id}$ and $\mathrm{E}(k_a)$ are G-homotopic. Consequently, $\mathrm{B}(\mathrm{id}) = \mathrm{id}$ and $\mathrm{B}(k_a)$ are homotopic. □

Now we are ready to give a description of the image a restriction that we use in the central part of the book. Consult [2, Lem. II.3.1].

Lemma 10.5 *Let G be a finite group, H its subgroup, $N_G(H)$ the normalizer of H in G, $W_G(H) := N_G(H)/H$ the corresponding Weyl group, and M a trivial G-module. There is an action of $W_G(H)$ on $H^*(H; M)$ such that*

$$\operatorname{im}\left(\operatorname{res}^G_H\colon H^*(G; M) \longmapsto H^*(H; M)\right) \subseteq H^*(H; M)^{W_G(H)}.$$

Proof First we introduce an action of the normalizer $N_G(H)$ on $\mathrm{B}H$. Let $a \in N_G(H)$, and let $k_a\colon H \longrightarrow H$ be a homomorphism defined by $k_a(h) := aha^{-1}$ for $h \in H$. For $a, b \in N_G(H)$ the following relation obviously holds $k_{ab} = k_a \circ k_b$. Applying the functor B on the maps k_a, for every $a \in N_G(H)$, we get an action of $N_G(H)$ on $\mathrm{B}H$. This action naturally extends to an action on the cohomology

$$N_G(H) \times H^*(H; M) \longrightarrow H^*(H; M).$$

In the case when $a \in H \subseteq N_G(H)$ from Lemma 10.4 we get that the homomorphism $\mathrm{B}(k_a)^*\colon H^*(H; M) \longrightarrow H^*(H; M)$ is the identity. Consequently the action of $N_G(H)$ factors through an action of the Weyl group $W_G(H)$, that gives us a commutative diagram

$$\begin{array}{ccc} N_G(H) \times H^*(H; M) & \longrightarrow & H^*(H; M) \\ & \searrow \quad \nearrow & \\ & W_G(H) \times H^*(H; M). & \end{array}$$

The group $N_G(H)$ acts on $\mathrm{B}G$ in the identical way. Now for $a \in N_G(H)$ we set $k_a\colon G \longrightarrow G$ to be again defined by $k_a(g) := aga^{-1}$ for $g \in G$. Applying the functor B we now get an action of $N_G(H)$ on $\mathrm{B}G$ and consequently on $H^*(G; M)$. In this case Lemma 10.4 implies that each map $\mathrm{B}(k_a)$ is homotopic to the identity. Hence, the induced action of $N_G(H)$ on $H^*(G; M)$ is a trivial action.
Since the action of $N_G(H)$ on $\mathrm{B}G$ is an extension of the action of $N_G(H)$ on $\mathrm{B}H$ we have the following commutative diagram

$$\begin{array}{ccc} \mathrm{B}H & \xrightarrow{\mathrm{B}(i)} & \mathrm{B}G \\ {\scriptstyle \mathrm{B}(k_a)}\downarrow & & \downarrow{\scriptstyle \mathrm{B}(k_a)} \\ \mathrm{B}H & \xrightarrow{\mathrm{B}(i)} & \mathrm{B}G \end{array} \tag{10.1}$$

for every $a \in N_G(H)$ where $i\colon H \longrightarrow G$ is the inclusion. Applying the cohomology functor on the commutative diagram (10.1) we get that

$$\mathrm{B}(i)^* \circ \mathrm{B}(k_a)^* = \mathrm{B}(k_a)^* \circ \mathrm{B}(i)^*.$$

Since $\mathrm{B}(i)^* = \mathrm{res}^G_H$, and $\mathrm{B}(k_a)^*\colon H^*(G; M) \longrightarrow H^*(G; M)$ is the identity for every $a \in N_G(H)$, we get that $\mathrm{res}^G_H = \mathrm{B}(k_a)^* \circ \mathrm{res}^G_H$, and consequently

$$\mathrm{im}\left(\mathrm{res}^G_H\colon H^*(G; M) \longmapsto H^*(H; M)\right) \subseteq H^*(H; M)^{N_G(H)} = H^*(H; M)^{W_G(H)}.$$

10.3 Weyl Groups of an Elementary Abelian Group

In this section we identify Weyl groups $W_{\mathcal{S}_{2^m}}(\mathcal{E}_m)$ and $W_{\mathfrak{S}_{2^m}}(\mathcal{E}_m)$ for every $m \geq 1$. First, following Mùi [82, Proof of Lem. II.5.1] we prove the following fact.

Lemma 10.6 *Let $m \geq 0$ be an integer. Then*

$$W_{\mathcal{S}_{2^m}}(\mathcal{E}_m) \cong \mathrm{L}_m(\mathbb{F}_2).$$

Proof Let $\kappa\colon N_{\mathcal{S}_{2^m}}(\mathcal{E}_m) \longrightarrow \mathrm{Aut}(\mathcal{E}_m)$ be the homomorphism defined by $\kappa(a) := k_a$, where as before $k_a\colon \mathcal{E}_m \longrightarrow \mathcal{E}_m$ is the conjugation automorphism $k_a(e) = aea^{-1}$, $e \in \mathcal{E}_m$. The kernel of the homomorphism κ is the centralizer of $\mathcal{E}_m$ in $\mathcal{S}_{2^m}$, which is $\ker(\kappa) = C_{\mathcal{S}_{2^m}}(\mathcal{E}_m)$. Furthermore, in our situation, $C_{\mathcal{S}_{2^m}}(\mathcal{E}_m) = \mathcal{E}_m$. Thus, there is an exact sequence of groups

$$1 \longrightarrow \mathcal{E}_m \longrightarrow \mathrm{N}_{\mathcal{S}_{2^m}}(\mathcal{E}_m) \longrightarrow \mathrm{im}(\kappa) \longrightarrow 1. \tag{10.2}$$

In other words $N_{\mathcal{S}_{2^m}}(\mathcal{E}_m)$ is a semi-direct product $\mathcal{E}_m \rtimes \mathrm{im}(\kappa)$. Consequently,

$$W_{\mathcal{S}_{2^m}}(\mathcal{E}_m) = N_{\mathcal{S}_{2^m}}(\mathcal{E}_m)/\mathcal{E}_m \cong \mathrm{im}(\kappa).$$

Since $\mathrm{Aut}(\mathcal{E}_m) \cong \mathrm{GL}_m(\mathbb{F}_2)$ we can say that $\mathrm{im}(\kappa) \subseteq \mathrm{GL}_m(\mathbb{F}_2)$ and therefore the Weyl group $W_{\mathcal{S}_{2^m}}(\mathcal{E}_m)$ can be seen as a subgroup of $\mathrm{GL}_m(\mathbb{F}_2)$. Thus it remains to identify $\mathrm{im}(\kappa)$, and this will be done in two steps.

First we prove that $\mathrm{L}_m(\mathbb{F}_2) \subseteq N_{\mathcal{S}_{2^m}}(\mathcal{E}_m)$ using the induction on $m \geq 1$. Notice that obviously $\mathrm{L}_m(\mathbb{F}_2) \subseteq N_{\mathfrak{S}_{2^m}}(\mathcal{E}_m)$. For $m = 1$ the group $\mathrm{L}_1(\mathbb{F}_2)$ is trivial group while $\mathcal{E}_1 = \mathcal{S}_2 = N_{\mathcal{S}_2}(\mathcal{E}_1) \cong \mathbb{Z}_2$. Let us assume that for $m \geq 2$ the following inclusions $\mathrm{L}_{m-1}(\mathbb{F}_2) \subseteq N_{\mathcal{S}_{2^{m-1}}}(\mathcal{E}_{m-1}) \subseteq \mathcal{S}_{2^{m-1}}$ holds. Consider the embedding $\mathrm{L}_{m-1}(\mathbb{F}_2) \longrightarrow \mathrm{L}_m(\mathbb{F}_2)$ of group given by

$$A \longmapsto \begin{pmatrix} 1 & 0 \\ 0 & A \end{pmatrix}$$

for $A \in \mathrm{L}_{m-1}(\mathbb{F}_2)$. Note the "difference" between zeroes in the upper matrix. Since for every $i \in \mathbb{F}_2$ and $e \in \mathbb{F}_2^{m-1}$

$$\begin{pmatrix} 1 & 0 \\ 0 & A \end{pmatrix}\begin{pmatrix} i \\ e \end{pmatrix} = \begin{pmatrix} i \\ Ae \end{pmatrix}$$

we have that $\mathrm{L}_{m-1}(\mathbb{F}_2)$ is a subgroup of $\delta(\mathcal{S}_{2^{m-1}}) \subseteq \mathcal{S}_{2^{m-1}} \times \mathcal{S}_{2^{m-1}}$ where δ denotes the diagonal embedding. An arbitrary element of the group $\mathrm{L}_m(\mathbb{F}_2)$ can be presented in the form

$$\begin{pmatrix} 1 & 0 \\ a & A \end{pmatrix}$$

where $A \in \mathrm{L}_{m-1}(\mathbb{F}_2)$ and $a \in \mathbb{F}_2^{m-1}$. Then for $i \in \mathbb{F}_2$ and $e \in \mathbb{F}_2^{m-1}$ we have

$$\begin{pmatrix} 1 & 0 \\ a & A \end{pmatrix}\begin{pmatrix} i \\ e \end{pmatrix} = \begin{pmatrix} i \\ Ae + ai \end{pmatrix}.$$

Consequently, $\mathrm{L}_m(\mathbb{F}_2) \subseteq (\mathrm{L}_{m-1}(\mathbb{F}_2) \cdot \mathcal{E}_{m-1}) \times (\mathrm{L}_{m-1}(\mathbb{F}_2) \cdot \mathcal{E}_{m-1})$. From the induction hypothesis we have more

$$\mathrm{L}_m(\mathbb{F}_2) \subseteq (\mathrm{L}_{m-1}(\mathbb{F}_2) \cdot \mathcal{E}_{m-1}) \times (\mathrm{L}_{m-1}(\mathbb{F}_2) \cdot \mathcal{E}_{m-1}) \subseteq \mathcal{S}_{2^{m-1}} \times \mathcal{S}_{2^{m-1}} \subseteq \mathcal{S}_{2^m}.$$

As we have seen $\mathrm{L}_m(\mathbb{F}_2) \subseteq N_{\mathfrak{S}_{2^m}}(\mathcal{E}_m)$ and thus

$$\mathrm{L}_m(\mathbb{F}_2) \subseteq N_{\mathcal{S}_{2^m}}(\mathcal{E}_m),$$

which concludes the induction.

Now we continue identification of $\mathrm{im}(\kappa)$. Since $\mathrm{L}_m(\mathbb{F}_2) \cap \mathcal{E}_m = \{1\}$ we have that the exact sequence (10.2) gives us an embedding of $\mathrm{L}_m(\mathbb{F}_2)$ into $\mathrm{im}(\kappa)$. Hence we have the following inclusions of 2-groups (with the obvious abuse of notation)

$$\mathrm{L}_m(\mathbb{F}_2) \subseteq \mathrm{im}(\kappa) \subseteq \mathrm{GL}_m(\mathbb{F}_2).$$

The image $\mathrm{im}(\kappa)$ is a 2-group as an image of the 2-group $N_{\mathcal{S}_{2^m}}(\mathcal{E}_m)$. Because $\mathrm{L}_m(\mathbb{F}_2)$ is a Sylow 2-subgroup of $\mathrm{GL}_m(\mathbb{F}_2)$ we have that $\mathrm{L}_m(\mathbb{F}_2) = \mathrm{im}(\kappa)$ and consequently

$$W_{\mathcal{S}_{2^m}}(\mathcal{E}_m) \cong \mathrm{im}(\kappa) \cong \mathrm{L}_m(\mathbb{F}_2).$$

Next, adapting the proof of the previous lemma and following [2, Ex. III.2.7] we determine the Weyl group of $\mathcal{E}_m$ now inside the symmetric group $\mathfrak{S}_{2^m}$.

Lemma 10.7 *Let $m \geq 0$ be an integer. Then*

$$W_{\mathfrak{S}_{2^m}}(\mathcal{E}_m) \cong \mathrm{GL}_m(\mathbb{F}_2).$$

Proof Let $\kappa\colon N_{\mathfrak{S}_{2^m}}(\mathcal{E}_m) \longrightarrow \mathrm{Aut}(\mathcal{E}_m)$ be the homomorphism defined by $\kappa(a) := k_a$, where as before $k_a\colon \mathcal{E}_m \longrightarrow \mathcal{E}_m$ is the conjugation automorphism $k_a(e) = aea^{-1}, e \in \mathcal{E}_m$. As in the proof of the previous lemma $\ker(\kappa) = C_{\mathfrak{S}_{2^m}}(\mathcal{E}_m) = \mathcal{E}_m$.

Hence, we get an exact sequence of groups

$$1 \longrightarrow \mathcal{E}_m \longrightarrow N_{\mathfrak{S}_{2^m}}(\mathcal{E}_m) \longrightarrow \operatorname{im}(\kappa) \longrightarrow 1.$$

Consequently, $N_{\mathfrak{S}_{2^m}}(\mathcal{E}_m)$ is a semi-direct product $\mathcal{E}_m \rtimes \operatorname{im}(\kappa)$ implying that

$$W_{\mathfrak{S}_{2^m}}(\mathcal{E}_m) = N_{\mathfrak{S}_{2^m}}(\mathcal{E}_m)/\mathcal{E}_m \cong \operatorname{im}(\kappa).$$

Since $\operatorname{Aut}(\mathcal{E}_m) \cong \mathrm{GL}_m(\mathbb{F}_2)$, in order to complete the proof of the lemma, it suffices to prove that $\operatorname{im}(\kappa) = \operatorname{Aut}(\mathcal{E}_m)$. For that fix $\alpha \in \operatorname{Aut}(\mathcal{E}_m)$, and denote by $\bar{\alpha} \in \mathfrak{S}_{2^m}$ the permutation given with $e \longmapsto \alpha(e)$ for $e \in \mathcal{E}_m$. Here we use the fact that $\mathfrak{S}_{2^m}$ is group of permutations of the set $\mathcal{E}_m$. Then $\bar{\alpha} \in N_{\mathfrak{S}_{2^m}}(\mathcal{E}_m)$ and $\kappa(\bar{\alpha}) = \alpha$. Indeed, for $e, h \in \mathcal{E}_m$ holds:

$$\bar{\alpha}e\bar{\alpha}^{-1}(h) = (\bar{\alpha}e)(\alpha^{-1}(h)) = \bar{\alpha}(e + \alpha^{-1}(h)) = \alpha(e + \alpha^{-1}(h)) = \alpha(e) + h.$$

consequently we have an equality of permutations $\bar{\alpha}e\bar{\alpha}^{-1} = \alpha(e) \in \mathcal{E}_m$ that implies $\bar{\alpha} \in N_{\mathfrak{S}_{2^m}}(\mathcal{E}_m)$ and $\kappa(\bar{\alpha}) = \alpha$. This concludes the proof of surjectivity of κ and of the lemma. □

10.4 Cohomology of the Real Projective Space with Local Coefficients

For $d \geq 2$ we compute the cohomology of the projective space $H^*(\mathbb{RP}^{d-1}; \mathcal{M})$ where the local coefficient system $\mathcal{M}$ is additively $\mathbb{F}_2 \oplus \mathbb{F}_2$, and the action of $\pi_1(\mathbb{RP}^{d-1}) = \langle t \rangle$ on $\mathcal{M}$ is given by $t \cdot (a_1, a_2) = (a_2, a_1)$ where $(a_1, a_2) \in \mathbb{F}_2 \oplus \mathbb{F}_2$. We consider two separate cases: $d = 2$ when $\pi_1(\mathbb{RP}^{d-1}) \cong \mathbb{Z}$, and $d \geq 3$ when $\pi_1(\mathbb{RP}^{d-1}) \cong \mathbb{Z}_2$. First we recall the definition of the cohomology with local coefficients, consult for example [60, Sec. 3.H].

Let X be a path-connected CW-complex, $\widetilde{X}$ its universal cover, and let $\pi := \pi_1(X)$ be its fundamental group. Denote by $\mathcal{L}$ a local coefficient system on X, that is a $\mathbb{Z}[\pi]$-module. Assume that we are given a structure of $\mathbb{Z}[\pi]$-CW-complex on $\widetilde{X}$ with associated cellular chain complex

$$\cdots \longrightarrow C_{n+1}(\widetilde{X}) \xrightarrow{d_{n+1}} C_n(\widetilde{X}) \xrightarrow{d_n} C_{n-1}(\widetilde{X}) \xrightarrow{d_{n-1}} \cdots$$

$$\cdots \xrightarrow{d_2} C_1(\widetilde{X}) \xrightarrow{d_1} C_0(\widetilde{X}) \longrightarrow 0,$$

where $C_n(\widetilde{X})$ is a $\mathbb{Z}[\pi]$-module, and $d_n \colon C_n(\widetilde{X}) \longrightarrow C_{n-1}(\widetilde{X})$ is a $\mathbb{Z}[\pi]$-module homomorphism, for every $n \in \mathbb{Z}$. The cohomology $H^*(X; \mathcal{L})$ of X with local coefficients in $\mathcal{L}$ is cohomology of the cochain complex

$$\cdots \longleftarrow \hom_{\mathbb{Z}[\pi]}(C_{n+1}(\widetilde{X});\mathcal{L}) \longleftarrow \hom_{\mathbb{Z}[\pi]}(C_n(\widetilde{X});\mathcal{L}) \longleftarrow \hom_{\mathbb{Z}[\pi]}(C_{n-1}(\widetilde{X});\mathcal{L})$$

$$\cdots \longleftarrow \hom_{\mathbb{Z}[\pi]}(C_1(\widetilde{X});\mathcal{L}) \longleftarrow \hom_{\mathbb{Z}[\pi]}(C_0(\widetilde{X});\mathcal{L}) \longleftarrow 0\,.$$

This means that

$$H^n(X;\mathcal{M}) := \frac{\ker\big(\hom_{\mathbb{Z}[\pi]}(C_n(\widetilde{X});\mathcal{L})) \longrightarrow \hom_{\mathbb{Z}[\pi]}(C_{n+1}(\widetilde{X});\mathcal{L})\big)}{\operatorname{im}\big(\hom_{\mathbb{Z}[\pi]}(C_{n-1}(\widetilde{X});\mathcal{L}) \longrightarrow \hom_{\mathbb{Z}[\pi]}(C_n(\widetilde{X});\mathcal{L})\big)}.$$

For $d = 2$ the universal cover of the projective space $\mathbb{RP}^1$ is the real line $\mathbb{R}^1$, and the fundamental group $\pi = \pi_1(\mathbb{RP}^1) = \langle t\rangle$ is the infinite cyclic group. An associated $\mathbb{Z}[\pi]$-CW-complex of the universal cover $\widetilde{\mathbb{RP}^1} = \mathbb{R}^1$ is defined as follows: The 0-cells are all integers $\{x_0^i := \{i\} : i \in \mathbb{Z}\}$, while the 1-cells are intervals $\{x_1^i := [i, i+1] : i \in \mathbb{Z}\}$. The action of the generator t of the fundamental group on the cells of the $\mathbb{Z}[\pi]$-CW-complex is given by $t \cdot x_j^i = x_j^{i+1}$ for $j \in \{0, 1\}$. Hence, the induced cellular chain complex of $\mathbb{Z}[\pi]$-modules is given by

$$C_0(\widetilde{\mathbb{RP}^1}) = \mathbb{Z}[\mathbb{Z}] =_{\mathbb{Z}[\mathbb{Z}]\text{-module}} \langle x_0^1\rangle \qquad \text{and} \qquad C_1(\widetilde{\mathbb{RP}^1}) = \mathbb{Z}[\mathbb{Z}] =_{\mathbb{Z}[\mathbb{Z}]\text{-module}} \langle x_1^1\rangle$$

where the only non-trivial boundary homomorphism d_1 is defined on the generator of $C_1(\widetilde{\mathbb{RP}^1})$ by $d_1(x_1^1) := x_0^1 - x_0^0 = (t-1)\cdot x_0^0$. Thus, the cellular chain complex we consider in this case is

$$0 \longrightarrow \mathbb{Z}[\mathbb{Z}] \xrightarrow{(t-1)\cdot} \mathbb{Z}[\mathbb{Z}] \longrightarrow 0.$$

After applying the functor $\hom_{\mathbb{Z}[\mathbb{Z}]}(\cdot, \mathcal{M})$ we get the cochain complex

$$0 \longleftarrow \mathcal{M} \xleftarrow{(t-1)\cdot} \mathcal{M} \longleftarrow 0.$$

If we recall that additively $\mathcal{M} = \mathbb{F}_2 \oplus \mathbb{F}_2$ by direct inspection we see that

$$H^r(\mathbb{RP}^1;\mathcal{M}) \cong \begin{cases} \ker\big(\mathcal{M} \xrightarrow{(t-1)\cdot} \mathcal{M}\big) \cong \mathbb{F}_2, & r = 0,\\ \mathcal{M}/\operatorname{im}\big(\mathcal{M} \xrightarrow{(t-1)\cdot} \mathcal{M}\big) \cong \mathbb{F}_2, & r = 1,\\ 0, & \text{otherwise}. \end{cases}$$

For further use we denote the generator of the group $H^1(\mathbb{RP}^{d-1};\mathcal{M})$ by z_1.

Now, let $d \geq 3$ be an integer. Then the universal cover of the projective space $\mathbb{RP}^{d-1}$ is the sphere S^{d-1}, and the fundamental group is $\pi := \pi_1(\mathbb{RP}^{d-1}) \cong \mathbb{Z}_2 =$

$\langle t\rangle$. We associate an $\mathbb{Z}[\pi]$-CW-complex to the universal cover $\widetilde{\mathbb{R}\mathrm{P}^{d-1}} = S^{d-1}$ as follows: In each dimension i, where $0 \le i \le d-1$, there are two cells x_i^0 and x_i^1. The generator t of the fundamental group π acts on the cells by $t \cdot x_j^0 = x_j^1$ and $t \cdot x_j^1 = x_j^0$ for $0 \le j \le d-1$. Thus, the induced cellular chain complex of $\mathbb{Z}[\pi]$-modules is given by

$$C_n(\widetilde{\mathbb{R}\mathrm{P}^{d-1}}) = \mathbb{Z}[\mathbb{Z}_2] =_{\mathbb{Z}[\mathbb{Z}_2]\text{-module}} \langle x_n^0\rangle, \qquad \text{for } 0 \le n \le d-1,$$

where the boundary homomorphism $d_n\colon C_n(\widetilde{\mathbb{R}\mathrm{P}^{d-1}}) \longrightarrow C_{n-1}(\widetilde{\mathbb{R}\mathrm{P}^{d-1}})$ on the generator x_n^0 is

$$\begin{aligned} d_n(x_n^0) &:= x_{n-1}^1 - x_{n-1}^0 = (t-1)\cdot x_{n-1}^0, && \text{for } 1 \le n \le d-1 \text{ odd},\\ d_n(x_n^0) &:= x_{n-1}^1 + x_{n-1}^0 = (t+1)\cdot x_{n-1}^0, && \text{for } 1 \le n \le d-1 \text{ even,} \end{aligned}$$

and otherwise zero. The cellular chain complex we obtained in this case is

$$0 \longrightarrow \mathbb{Z}[\mathbb{Z}_2] \xrightarrow{(t+1)\cdot} \mathbb{Z}[\mathbb{Z}_2] \xrightarrow{(t-1)\cdot} \cdots \xrightarrow{(t+1)\cdot} \mathbb{Z}[\mathbb{Z}_2] \xrightarrow{(t-1)\cdot} \mathbb{Z}[\mathbb{Z}_2] \longrightarrow 0,$$

when d is odd, and

$$0 \longrightarrow \mathbb{Z}[\mathbb{Z}_2] \xrightarrow{(t-1)\cdot} \mathbb{Z}[\mathbb{Z}_2] \xrightarrow{(t+1)\cdot} \cdots \xrightarrow{(t+1)\cdot} \mathbb{Z}[\mathbb{Z}_2] \xrightarrow{(t-1)\cdot} \mathbb{Z}[\mathbb{Z}_2] \longrightarrow 0,$$

when d is even.

Applying the functor $\hom_{\mathbb{Z}[\mathbb{Z}_2]}(\cdot, \mathcal{M})$ we get the cochain complex which is isomorphic to the following cochain complex

$$0 \longleftarrow \mathcal{M} \xleftarrow{(t+1)\cdot} \mathcal{M} \xleftarrow{(t-1)\cdot} \cdots \xleftarrow{(t+1)\cdot} \mathcal{M} \xleftarrow{(t-1)\cdot} \mathcal{M} \longleftarrow 0,$$

when d is odd, and

$$0 \longleftarrow \mathcal{M} \xleftarrow{(t-1)\cdot} \mathcal{M} \xleftarrow{(t+1)\cdot} \cdots \xleftarrow{(t+1)\cdot} \mathcal{M} \xleftarrow{(t-1)\cdot} \mathcal{M} \longleftarrow 0,$$

when d is even. Since $\mathcal{M} = \mathbb{F}_2 \oplus \mathbb{F}_2$, and the homomorphisms $\mathcal{M} \xrightarrow{(t-1)\cdot} \mathcal{M}$ and $\mathcal{M} \xrightarrow{(t+1)\cdot} \mathcal{M}$ coincide on $\mathcal{M}$, we conclude that

$$H^r(\mathbb{R}\mathrm{P}^{d-1}; \mathcal{M}) \cong \begin{cases} \ker\left(\mathcal{M} \xrightarrow{(t-1)\cdot} \mathcal{M}\right) \cong \mathbb{F}_2, & r = 0,\\ \mathcal{M}/\operatorname{im}\left(\mathcal{M} \xrightarrow{(t-1)\cdot} \mathcal{M}\right) \cong \mathbb{F}_2, & r = d-1,\\ 0, & \text{otherwise.} \end{cases}$$

We denote the generator of the group $H^{d-1}(\mathbb{R}\mathrm{P}^{d-1}; \mathcal{M})$ by z_{d-1}.

10.5 Homology of the Real Projective Space with Local Coefficients

For $d \geq 2$, along the lines of the previous section, we compute the homology of the projective space $H_*(\mathbb{RP}^{d-1}; \mathcal{M})$ where the local coefficient system $\mathcal{M}$ is additively $\mathbb{F}_2 \oplus \mathbb{F}_2$, and the action of $\pi_1(\mathbb{RP}^{d-1}) = \langle t \rangle$ on $\mathcal{M}$ is given by $t \cdot (a_1, a_2) = (a_2, a_1)$ where $(a_1, a_2) \in \mathbb{F}_2 \oplus \mathbb{F}_2$.

Like in the case of cohomology we consider two separate cases: $d = 2$ when $\pi_1(\mathbb{RP}^{d-1}) \cong \mathbb{Z}$, and $d \geq 3$ when $\pi_1(\mathbb{RP}^{d-1}) \cong \mathbb{Z}_2$. First we recall the definition of the homology with local coefficients, consult for example [60, Sec. 3.H].

For a path-connected CW-complex X let $\widetilde{X}$ denote its universal cover, and let $\pi := \pi_1(X)$ be its fundamental group. Denote by $\mathcal{L}$ a local coefficient system on X, that is a $\mathbb{Z}[\pi]$-module. Assume that we are given a structure of $\mathbb{Z}[\pi]$-CW-complex on $\widetilde{X}$ with associated cellular chain complex

$$\cdots \longrightarrow C_{n+1}(\widetilde{X}) \xrightarrow{d_{n+1}} C_n(\widetilde{X}) \xrightarrow{d_n} C_{n-1}(\widetilde{X}) \xrightarrow{d_{n-1}} \cdots$$

$$\cdots \xrightarrow{d_2} C_1(\widetilde{X}) \xrightarrow{d_1} C_0(\widetilde{X}) \longrightarrow 0\,,$$

where $C_n(\widetilde{X})$ is a $\mathbb{Z}[\pi]$-module, and $d_n\colon C_n(\widetilde{X}) \longrightarrow C_{n-1}(\widetilde{X})$ is a $\mathbb{Z}[\pi]$-module homomorphism, for every $n \in \mathbb{N}$.

The homology $H_*(X; \mathcal{L})$ of X with local coefficients in $\mathcal{L}$ is homology of the chain complex

$$\cdots \longrightarrow C_{n+1}(\widetilde{X}) \otimes_{\mathbb{Z}[\pi]} \mathcal{L} \longrightarrow C_n(\widetilde{X}) \otimes_{\mathbb{Z}[\pi]} \mathcal{L} \longrightarrow C_{n-1}(\widetilde{X}) \otimes_{\mathbb{Z}[\pi]} \mathcal{L} \longrightarrow \cdots$$

$$\cdots \longrightarrow C_1(\widetilde{X}) \otimes_{\mathbb{Z}[\pi]} \mathcal{L} \longrightarrow C_0(\widetilde{X}) \otimes_{\mathbb{Z}[\pi]} \mathcal{L} \longrightarrow 0\,.$$

This means that

$$H_n(X; \mathcal{M}) := \frac{\ker\big(C_n(\widetilde{X}) \otimes_{\mathbb{Z}[\pi]} \mathcal{L} \longrightarrow C_{n-1}(\widetilde{X}) \otimes_{\mathbb{Z}[\pi]} \mathcal{L}\big)}{\operatorname{im}\big(C_{n+1}(\widetilde{X}) \otimes_{\mathbb{Z}[\pi]} \mathcal{L} \longrightarrow C_n(\widetilde{X}) \otimes_{\mathbb{Z}[\pi]} \mathcal{L}\big)}.$$

First we consider the case $d = 2$. The universal cover of the projective space $\mathbb{RP}^1$ is the real line $\mathbb{R}^1$, and the fundamental group $\pi = \pi_1(\mathbb{RP}^1) = \langle t \rangle$ is the infinite cyclic group. An associated $\mathbb{Z}[\pi]$-CW-complex of the universal cover $\widetilde{\mathbb{RP}^1} = \mathbb{R}^1$ is defined as follows: The 0-cells are all integers $\{x_0^i := \{i\} : i \in \mathbb{Z}\}$, while the 1-cells are intervals $\{x_1^i := [i, i+1] : i \in \mathbb{Z}\}$. The action of the generator t of the fundamental group on the cells of the $\mathbb{Z}[\pi]$-CW-complex is given by $t \cdot x_j^i = x_j^{i+1}$ for $j \in \{0, 1\}$. Hence, the induced cellular chain complex of $\mathbb{Z}[\pi]$-modules is given by

$$C_0(\widetilde{\mathbb{RP}^1}) = \mathbb{Z}[\mathbb{Z}] =_{\mathbb{Z}[\mathbb{Z}]\text{-module}} \langle x_0^1 \rangle \qquad \text{and} \qquad C_1(\widetilde{\mathbb{RP}^1}) = \mathbb{Z}[\mathbb{Z}] =_{\mathbb{Z}[\mathbb{Z}]\text{-module}} \langle x_1^1 \rangle$$

where the only non-trivial boundary homomorphism d_1 is defined on the generator of $C_1(\widetilde{\mathbb{RP}^1})$ by $d_1(x_1^1) := x_0^1 - x_0^0 = (t-1) \cdot x_0^0$. Thus, the cellular chain complex we consider in this case is

$$0 \longrightarrow \mathbb{Z}[\mathbb{Z}] \xrightarrow{(t-1)\cdot} \mathbb{Z}[\mathbb{Z}] \longrightarrow 0.$$

After applying the functor $\cdot \otimes_{\mathbb{Z}[\mathbb{Z}]} \mathcal{M}$ we get the chain complex

$$0 \longrightarrow \mathcal{M} \xrightarrow{(t-1)\cdot} \mathcal{M} \longrightarrow 0.$$

Since additively $\mathcal{M} = \mathbb{F}_2 \oplus \mathbb{F}_2$ by direct inspection we see that

$$H_r(\mathbb{RP}^1; \mathcal{M}) \cong \begin{cases} \mathcal{M}/\operatorname{im}\left(\mathcal{M} \xrightarrow{(t-1)\cdot} \mathcal{M}\right) \cong \mathbb{F}_2, & r = 0, \\ \ker\left(\mathcal{M} \xrightarrow{(t-1)\cdot} \mathcal{M}\right) \cong \mathbb{F}_2, & r = 1, \\ 0, & \text{otherwise.} \end{cases}$$

For further use we denote the generator of the group $H_1(\mathbb{RP}^{d-1}; \mathcal{M})$ by h_1.

When $d \geq 3$ the universal cover of the projective space $\mathbb{RP}^{d-1}$ is the sphere S^{d-1}, and the fundamental group is $\pi := \pi_1(\mathbb{RP}^{d-1}) \cong \mathbb{Z}_2 = \langle t \rangle$. An $\mathbb{Z}[\pi]$-CW-complex is associated to the universal cover $\widetilde{\mathbb{RP}^{d-1}} = S^{d-1}$ as follows: In each dimension i, where $0 \leq i \leq d-1$, there are two cells x_i^0 and x_i^1. The generator t of the fundamental group π acts on the cells by $t \cdot x_j^0 = x_j^1$ and $t \cdot x_j^1 = x_j^0$ for $0 \leq j \leq d-1$. Thus, the induced cellular chain complex of $\mathbb{Z}[\pi]$-modules is given by

$$C_n(\widetilde{\mathbb{RP}^{d-1}}) = \mathbb{Z}[\mathbb{Z}_2] =_{\mathbb{Z}[\mathbb{Z}_2]\text{-module}} \langle x_n^0 \rangle, \qquad \text{for } 0 \leq n \leq d-1,$$

where the boundary homomorphism $d_n \colon C_n(\widetilde{\mathbb{RP}^{d-1}}) \longrightarrow C_{n-1}(\widetilde{\mathbb{RP}^{d-1}})$ on the generator x_n^0 is

$$d_n(x_n^0) := x_{n-1}^1 - x_{n-1}^0 = (t-1) \cdot x_{n-1}^0, \qquad \text{for } 1 \leq n \leq d-1 \text{ odd},$$
$$d_n(x_n^0) := x_{n-1}^1 + x_{n-1}^0 = (t+1) \cdot x_{n-1}^0, \qquad \text{for } 1 \leq n \leq d-1 \text{ even},$$

and otherwise zero. The cellular chain complex we obtained in this case is

$$0 \longrightarrow \mathbb{Z}[\mathbb{Z}_2] \xrightarrow{(t+1)\cdot} \mathbb{Z}[\mathbb{Z}_2] \xrightarrow{(t-1)\cdot} \cdots \xrightarrow{(t+1)\cdot} \mathbb{Z}[\mathbb{Z}_2] \xrightarrow{(t-1)\cdot} \mathbb{Z}[\mathbb{Z}_2] \longrightarrow 0,$$

when d is odd, and

$$0 \longrightarrow \mathbb{Z}[\mathbb{Z}_2] \xrightarrow{(t-1)\cdot} \mathbb{Z}[\mathbb{Z}_2] \xrightarrow{(t+1)\cdot} \cdots \xrightarrow{(t+1)\cdot} \mathbb{Z}[\mathbb{Z}_2] \xrightarrow{(t-1)\cdot} \mathbb{Z}[\mathbb{Z}_2] \longrightarrow 0,$$

when d is even.

Applying the functor $\cdot \otimes_{\mathbb{Z}[\mathbb{Z}]} \mathcal{M}$ yields a chain complex isomorphic to the chain complex:

$$0 \longrightarrow \mathcal{M} \xrightarrow{(t+1)\cdot} \mathcal{M} \xrightarrow{(t-1)\cdot} \cdots \xrightarrow{(t+1)\cdot} \mathcal{M} \xrightarrow{(t-1)\cdot} \mathcal{M} \longrightarrow 0,$$

when d is odd, and

$$0 \longrightarrow \mathcal{M} \xrightarrow{(t-1)\cdot} \mathcal{M} \xrightarrow{(t+1)\cdot} \cdots \xrightarrow{(t+1)\cdot} \mathcal{M} \xrightarrow{(t-1)\cdot} \mathcal{M} \longrightarrow 0,$$

when d is even. Since $\mathcal{M} = \mathbb{F}_2 \oplus \mathbb{F}_2$, and the homomorphisms $\mathcal{M} \stackrel{(t-1)\cdot}{\longrightarrow} \mathcal{M}$ and $\mathcal{M} \stackrel{(t+1)\cdot}{\longrightarrow} \mathcal{M}$ coincide on $\mathcal{M}$, we have that

$$H_r(\mathbb{R}\mathrm{P}^{d-1}; \mathcal{M}) \cong \begin{cases} \mathcal{M}/\operatorname{im}\left(\mathcal{M} \stackrel{(t-1)\cdot}{\longrightarrow} \mathcal{M}\right) \cong \mathbb{F}_2, & r = 0, \\ \ker\left(\mathcal{M} \stackrel{(t-1)\cdot}{\longrightarrow} \mathcal{M}\right) \cong \mathbb{F}_2, & r = d-1, \\ 0, & \text{otherwise.} \end{cases}$$

We denote the generator of the group $H_{d-1}(\mathbb{R}\mathrm{P}^{d-1}; \mathcal{M})$ by h_{d-1}, and the generator of the group $H_0(\mathbb{R}\mathrm{P}^{d-1}; \mathcal{M})$ by h_0.

[illegible] exact [illegible]

$$0 \longrightarrow [illegible] \longrightarrow 0$$

[illegible]

Applying the functor [illegible] yields a chain complex [illegible]

$$0 \longrightarrow [illegible] \longrightarrow M \longrightarrow [illegible]$$

[illegible]

$$[illegible] \longrightarrow M \longrightarrow [illegible] \longrightarrow 0,$$

[illegible] and the homomorphisms $M \longrightarrow M$ and [illegible]

$$[illegible]$$

otherwise.

[illegible]

References

1. Alejandro Adem, John Maginnis, and R. James Milgram, *Symmetric invariants and cohomology of groups*, Math. Ann. **287** (1990), no. 3, 391–411.
2. Alejandro Adem and James R. Milgram, *Cohomology of Finite Groups*, 2nd ed., Grundlehren der Mathematischen Wissenschaften, vol. 309, Springer-Verlag, Berlin, 2004.
3. Vladimir Igorevich Arnol'd, *A remark on the branching of hyperelliptic integrals as functions of the parameters*, Funkcional. Anal. i Priložen. **2** (1968), no. 3, 1–3.
4. ———, *Singularities of smooth mappings*, Uspehi Mat. Nauk **23** (1968), no. 1, 3–44.
5. ———, *The cohomology ring of the group of dyed braids*, Mat. Zametki **5** (1969), 227–231.
6. ———, *Certain topological invariants of algebrac functions*, Trudy Moskov. Mat. Obšč. **21** (1970), 27–46.
7. Emil Artin, *Theorie der Zöpfe*, Abh. Math. Sem. Univ. Hamburg **4** (1925), no. 1, 47–72.
8. ———, *Braids and permutations*, Ann. of Math. (2) **48** (1947), 643–649.
9. ———, *Theory of braids*, Ann. of Math. (2) **48** (1947), 101–126.
10. Michael F. Atiyah, *Power operations in K-theory*, Quart. J. Math. Oxford Ser. (2) **17** (1966), 165–193.
11. Djordje Baralić, Pavle V. M. Blagojević, Roman Karasev, and Aleksandar Vučić, *Index of Grassmann manifolds and orthogonal shadows*, Forum Math. **30** (2018), no. 6, 1539–1572.
12. Joan S. Birman, *Braids, links, and mapping class groups*, Princeton University Press, Princeton, N.J.; University of Tokyo Press, Tokyo, 1974, Annals of Mathematics Studies, No. 82.
13. Anders Björner and Günter M. Ziegler, *Combinatorial stratification of complex arrangements*, J. American Math. Soc. **5** (1992), no. 1, 105–149.
14. Pavle V. M. Blagojević, Frederick R. Cohen, Wolfgang Lück, and Günter M. Ziegler, *On complex highly regular embeddings and the extended Vassiliev conjecture*, Int. Math. Res. Not. IMRN (2016), no. 20, 6151–6199.
15. Pavle V. M. Blagojević, Wolfgang Lück, and Günter M. Ziegler, *On highly regular embeddings*, Trans. Amer. Math. Soc. **368** (2016), no. 4, 2891–2912.
16. Pavle V. M. Blagojević and Günter M. Ziegler, *Convex equipartitions via equivariant obstruction theory*, Israel J. Math. **200** (2014), no 1, 49–77.
17. Michael J. Boardman and Rainer M. Vogt, *Homotopy Invariant Algebraic Structures on Topological Spaces*, Lecture Notes in Mathematics, vol. 347, Springer-Verlag, Berlin-New York, 1973.

P. V. M. Blagojević et al., *Equivariant Cohomology of Configuration Spaces Mod 2*,
Lecture Notes in Mathematics 2282, https://doi.org/10.1007/978-3-030-84138-6

18. Carl-Friedrich Bödigheimer, *Stable splittings of mapping spaces*, Algebraic Topology (Seattle, Wash., 1985), Lecture Notes in Mathematics, vol. 1286, Springer, Berlin, 1987, pp. 174–187.
19. Carl-Friedrich Bödigheimer and Frederick R. Cohen, *Rational cohomology of configuration spaces of surfaces*, Algebraic topology and transformation groups (Göttingen, 1987), Lecture Notes in Math., vol. 1361, Springer, Berlin, 1988, pp. 7–13.
20. Carl-Friedrich Bödigheimer, Frederick R. Cohen, and Laurence R. Taylor, *On the homology of configuration spaces*, Topology **28** (1989), no. 1, 111–123.
21. Vladimir Grigorévich Boltjanskiĭ, Sergeĭ Sergeevich Ryškov, and Yu. A. Šaškin, *On k-regular imbeddings and their application to the theory of approximation of functions*, Amer. Math. Soc. Transl. (2) **28** (1963), 211–219.
22. Karol Borsuk, *On the k-independent subsets of the Euclidean space and of the Hilbert space*, Bull. Acad. Polon. Sci. Cl. III. **5** (1957), 351–356, XXIX.
23. Glen E. Bredon, *Topology and Geometry*, Graduate Texts in Math., vol. 139, Springer, New York, 1993.
24. William Browder, *Homology operations and loop spaces*, Illinois J. Math. **4** (1960), 347–357.
25. Kenneth S. Brown, *Cohomology of Groups*, Graduate Texts in Math., vol. 87, Springer-Verlag, New York, 1994.
26. Jarosław Buczyński, Tadeusz Januszkiewicz, Joachim Jelisiejew, and Mateusz Michałek, *Constructions of k-regular maps using finite local schemes*, J. Eur. Math. Soc. (JEMS) **21** (2019), no. 6, 1775–1808.
27. Eddy H. E. A. Campbell and Ian P. Hughes, *The ring of upper triangular invariants as a module over the Dickson invariants*, Math. Annalen **306** (1996), no. 3, 429–443.
28. Jeffrey Caruso, Frederick R. Cohen, J. Peter May, and Laurence R. Taylor, *James maps, Segal maps, and the Kahn-Priddy theorem*, Trans. Amer. Math. Soc. **281** (1984), no. 1, 243–283.
29. Michael E. Chisholm, *k-regular mappings of 2^n-dimensional Euclidean space*, Proc. Amer. Math. Soc. **74** (1979), no. 1, 187–190.
30. Wei-Liang Chow, *On the algebraical braid group*, Ann. of Math. (2) **49** (1948), 654–658.
31. Frederick R. Cohen, *Cohomology of braid spaces*, Bull. Amer. Math. Soc. **79** (1973), 763–766.
32. ______, *Homology of $\Omega^{(n+1)}\Sigma^{(n+1)}X$ and $C_{(n+1)}X$, $n > 0$*, Bull. Amer. Math. Soc. **79** (1973), 1236–1241 (1974).
33. ______, *The homology of C_{n+1}-spaces, $n \geq 0$*, in: "The Homology of Iterated Loop Spaces", Lecture Notes in Mathematics, vol. 533, Springer-Verlag, Heidelberg, 1976, pp. 207–351.
34. ______, *The unstable decomposition of $\Omega^2\Sigma^2X$ and its applications*, Math. Z. **182** (1983), no. 4, 553–568.
35. Frederick R. Cohen, Ralph L. Cohen, Nicholas J. Kuhn, and Joseph A. Neisendorfer, *Bundles over configuration spaces*, Pacific J. Math. **104** (1983), no. 1, 47–54.
36. Frederick R. Cohen and David Handel, *k-regular embeddings of the plane*, Proc. Amer. Math. Soc. **72** (1978), no. 1, 201–204.
37. Frederick R. Cohen, Thomas J. Lada, and J. Peter May, *The homology of iterated loop spaces*, Lecture Notes in Mathematics, Vol. 533, Springer-Verlag, Berlin-New York, 1976.
38. Frederick R. Cohen, Mark E. Mahowald, and James R. Milgram, *The stable decomposition for the double loop space of a sphere*, Algebraic and Geometric Topology (Proc. Sympos. Pure Math., Stanford Univ., Stanford, Calif., 1976), Part 2, Proc. Sympos. Pure Math., XXXII, Amer. Math. Soc., Providence, R.I., 1978, pp. 225–228.
39. Frederick R. Cohen, J. Peter May, and Laurence R. Taylor, *Splitting of certain spaces CX*, Math. Proc. Camb. Philos. Soc. **84** (1978), 465–496.
40. ______, *Splitting of some more spaces*, Math. Proc. Camb. Philos. Soc. **86** (1979), 227–236.
41. ______, *James maps and E_n ring spaces*, Trans. Amer. Math. Soc. **281** (1984), no. 1, 285–295.
42. Michael C. Crabb, *The topological Tverberg theorem and related topics*, J. Fixed Point Theory Appl. **12** (2012), no. 1-2, 1–25.

43. Leonard Eugene Dickson, *A fundamental system of invariants of the general modular linear group with a solution of the form problem*, Trans. Amer. Math. Soc. **12** (1911), no. 1, 75–98.
44. Tammo tom Dieck, *Transformation Groups*, Studies in Mathematics, vol. 8, Walter de Gruyter, Berlin, 1987.
45. Eldon Dyer and Richard K. Lashof, *Homology of iterated loop spaces*, Amer. J. Math. **84** (1962), 35–88.
46. Edward Fadell, *Homotopy groups of configuration spaces and the string problem of Dirac*, Duke Math. J. **29** (1962), 231–242.
47. Edward Fadell and Lee Neuwirth, *Configuration spaces*, Math. Scand. **10** (1962), 111–118.
48. Edward Fadell and James Van Buskirk, *On the braid groups of E^2 and S^2*, Bull. Amer. Math. Soc. **67** (1961), 211–213.
49. Edward R. Fadell and Sufian Y. Husseini, *Geometry and topology of configuration spaces*, Springer Monographs in Mathematics, Springer-Verlag, Berlin, 2001.
50. Yves Félix and Jean-Claude Thomas, *Rational Betti numbers of configuration spaces*, Topology Appl. **102** (2000), no. 2, 139–149.
51. Mark Feshbach, *The mod 2 cohomology rings of the symmetric groups and invariants*, Topology **41** (2002), no. 1, 57–84.
52. Ralph Hartzler Fox and Lee Neuwirth, *The braid groups*, Math. Scand. **10** (1962), 119–126.
53. Dmitry Borisovich Fuks, *Cohomology of the braid group* mod 2, Funkcional. Anal. i Priložen. **4** (1970), no. 2, 62–73.
54. Mohammad Ghomi and Sergei Tabachnikov, *Totally skew embeddings of manifolds*, Math. Z. **258** (2008), no. 3, 499–512.
55. Chad Giusti, Paolo Salvatore, and Dev Sinha, *The mod-2 cohomology rings of symmetric groups*, J. Topol. **5** (2012), no. 1, 169–198.
56. Mikhail L. Gromov, *Isoperimetry of waists and concentration of maps*, Geom. Funct. Anal. **13** (2003), no. 1, 178–215.
57. David Handel, *Obstructions to 3-regular embeddings*, Houston J. Math. **5** (1979), no. 3, 339–343.
58. ______, *2k-regular maps on smooth manifolds*, Proc. Amer. Math. Soc. **124** (1996), no. 5, 1609–1613.
59. David Handel and Jack Segal, *On k-regular embeddings of spaces in Euclidean space*, Fund. Math. **106** (1980), no. 3, 231–237.
60. Allen Hatcher, *Algebraic Topology*, Cambridge University Press, Cambridge, England, 2002.
61. Nguyên Hữu Viêt Hưng, *The mod 2 cohomology algebras of symmetric groups*, Acta Math. Vietnam. **6** (1981), no. 2, 41–48 (1982).
62. ______, *The mod 2 equivariant cohomology algebras of configuration spaces*, Acta Math. Vietnam. **7** (1982), no. 1, 95–100.
63. ______, *The modulo 2 cohomology algebras of symmetric groups*, Japan. J. Math. (N.S.) **13** (1987), no. 1, 169–208.
64. ______, *The mod 2 equivariant cohomology algebras of configuration spaces*, Pac. J. Math. **143** (1990), no. 2, 251–286.
65. Adolf Hurwitz, *Ueber Riemann'sche Flächen mit gegebenen Verzweigungspunkten*, Math. Ann. **39** (1891), no. 1, 1–60.
66. Daniel S. Kahn and Stewart B. Priddy, *The transfer and stable homotopy theory*, Math. Proc. Cambridge Philos. Soc. **83** (1978), no. 1, 103–111.
67. Roman Karasev, Alfredo Hubard, and Boris Aronov, *Convex equipartitions: the spicy chicken theorem*, Geom. Dedicata **170** (2014), 263–279.
68. Roman Karasev and Peter Landweber, *Estimating the higher symmetric topological complexity of spheres*, Algebr. Geom. Topol. **12** (2012), no. 1, 75–94.
69. Roman Karasev and Alexey Volovikov, *Waist of the sphere for maps to manifolds*, Topology Appl. **160** (2013), no. 13, 1592–1602.
70. Tatsuji Kudo and Shôrô Araki, *Topology of H_n-spaces and H-squaring operations*, Mem. Fac. Sci. Kyūsyū Univ. Ser. A. **10** (1956), 85–120.

71. Ian J. Leary, *On the integral cohomology of wreath products*, J. Algebra **198** (1997), no. 1, 184–239.
72. Ib Madsen and James R. Milgram, *The Classifying Spaces for Surgery and Cobordism of Manifolds*, Annals of Mathematics Studies, vol. 92, Princeton University Press, Princeton, NJ, 1979.
73. Benjamin M. Mann, *The cohomology of the symmetric groups*, Trans. Amer. Math. Soc. **242** (1978), 157–184.
74. Benjamin Matschke, *A parametrized version of the Borsuk-Ulam-Bourgin-Yang-Volovikov theorem*, J. Topol. Anal. **6** (2014), no. 2, 263–280.
75. J. Peter May, *A general algebraic approach to Steenrod operations*, The Steenrod Algebra and its Applications (Proc. Conf. to Celebrate N. E. Steenrod's Sixtieth Birthday, Battelle Memorial Inst., Columbus, Ohio, 1970), Lecture Notes in Mathematics, vol. 168, Springer, Berlin, 1970, pp. 153–231.
76. ______, *The geometry of iterated loop spaces*, Lectures Notes in Mathematics, vol. 271, Springer-Verlag, Berlin, New York, 1972.
77. ______, *The homology of E_∞ spaces*, in: "The Homology of Iterated Loop Spaces" (Heidelberg), Lecture Notes in Mathematics, vol. 533, Springer-Verlag, 1976, pp. 1–68.
78. Dusa McDuff, *Configuration spaces of positive and negative particles*, Topology **14** (1975), 91–107.
79. Dusa McDuff [printed: MacDuff], *Configuration spaces, K*-theory and operator algebras (Proc. Conf., Univ. Georgia, Athens, Ga., 1975), 1977, pp. 88–95. Lecture Notes in Math., Vol. 575.
80. John W. Milnor and James D. Stasheff, *Characteristic Classes*, Annals of Mathematics Studies, vol. 76, Princeton University Press, Princeton, NJ, 1974.
81. Robert E. Mosher and Martin C. Tangora, *Cohomology Operations and Applications in Homotopy Theory*, Harper & Row, Publishers, New York-London, 1968.
82. Huỳnh Mùi, *Modular invariant theory and cohomology algebras of symmetric groups*, J. Fac. Sci. Univ. Tokyo Sect. IA Math. **22** (1975), no. 3, 319–369.
83. ______, *Duality in the infinite symmetric products*, Acta Math. Vietnam. **5** (1980), no. 1, 100–149 (1981).
84. James R. Munkres, *Elements of algebraic topology*, Addison-Wesley Publishing Company, Menlo Park, CA, 1984.
85. Tokusi Nakamura, *On cohomology operations*, Japan. J. Math. **33** (1963), 93–145.
86. Minoru Nakaoka, *Decomposition theorem for homology groups of symmetric groups*, Ann. of Math. (2) **71** (1960), 16–42.
87. ______, *Homology of the infinite symmetric group*, Ann. of Math. (2) **73** (1961), 229–257.
88. ______, *Note on cohomology algebras of symmetric groups*, J. Math. Osaka City Univ. **13** (1962), 45–55.
89. Daniel Quillen, *The Adams conjecture*, Topology **10** (1971), 67–80.
90. ______, *The spectrum of an equivariant cohomology ring. I, II*, Ann. of Math. (2) **94** (1971), 549–572; ibid. (2) 94 (1971), 573–602.
91. Daniel Quillen and Boris B. Venkov, *Cohomology of finite groups and elementary abelian subgroups*, Topology **11** (1972), 317–318.
92. Graeme Segal, *Configuration-spaces and iterated loop-spaces*, Invent. Math. **21** (1973), 213–221.
93. Paul A. Smith, *Fixed-point theorems for periodic transformations*, Amer. J. Math. **63** (1941), 1–8.
94. Victor P. Snaith, *A stable decomposition of $\Omega^n S^n X$*, J. London Math. Soc. (2) **7** (1974), 577–583.
95. Norman E. Steenrod, *Cohomology Operations (Lectures by N. E. Steenrod written and revised by D. B. A. Epstein)*, Annals of Mathematics Studies, vol. 50, Princeton University Press, Princeton NJ, 1962.
96. Gordana Stojanović, *Embeddings with multiple regularity*, Geom. Dedicata **123** (2006), 1–10.

97. Aleksander Nikolaevič Varčenko, *The branching of multiple integrals which depend on parameters*, Funkcional. Anal. i Priložen **3** (1969), no. 4, 79–80.
98. Victor A. Vassiliev, *Braid group cohomologies and algorithm complexity*, Funct. Anal. Appl. **22** (1988), no. 3, 182–190.
99. ———, *Complements of Discriminants of Smooth Maps: Topology and Applications*, Translations of Mathematical Monographs, vol. 98, American Mathematical Society, Providence, RI, 1992, Translated from the Russian by B. Goldfarb.
100. ———, *On r-neighborly submanifolds in* $\mathbb{R}^N$, Topological Methods in Nonlinear Analysis. Journal of the Juliusz Schauder Center **11** (1998), 273–281.
101. Clarence Wilkerson, *A primer on the Dickson invariants*, Proceedings of the Northwestern Homotopy Theory Conference (Evanston, Ill., 1982), Contemp. Math., vol. 19, Amer. Math. Soc., Providence, RI, 1983, pp. 421–434.
102. Oscar Zariski, *The topological discriminant group of a Riemann surface of genus p*, Amer. J. Math. **59** (1937), no. 2, 335–358.

Index

ℓ-skew embedding, 95, 114, 127, 128, 153, 155, 159
C_d-space, 77, 84, 169–172
$\mathcal{O}$-space, 166
$\mathfrak{S}$-free operad, 164
k-regular embedding, ix, 95–97, 113, 153, 154, 159
k-regular-ℓ-skew embedding, 95, 128, 129, 131, 132, 153

action of operad, 166
additive basis, 7, 34, 38, 39, 42–44, 47, 48, 50
Approximation theorem, 79, 171
Araki–Kudo–Dyer–Lashof homology operations, x, 18, 19, 79, 80, 86, 171
Artin's presentation of the braid group, 3

Bockstein homomorphism, 14
braid, 2
braid group, xiii, 2, 5, 6, 9, 10

category of compactly generated weak Hausdorff spaces, xiv, 163
category of compactly generated weak Hausdorff spaces with non-degenerate base points, xiv, 77, 163
category of CW-complexes, xiv
category of operads, 166
cell complex model, 7, 11
centralizer of a subgroup, xiii
classifying space, 10, 52
cohomology ring, 14, 36, 51, 52
complex k-regular embedding, 133, 134
complex ℓ-skew embedding, 133, 134
cross-section, 4

De Rham cohomology, 8
detecting family of subgroups, 187
Dickson algebra, 173
Dickson invariants, 54, 57, 61, 72, 105, 119, 123, 175–177
dihedral group, 35

Eilenberg–Mac Lane space, 4, 6, 46
Eilenberg–Zilber isomorphism, 61
elementary abelian group, xiii, 31, 64, 188, 192
endomorphism operad, 165, 166
equivariant cochain, 13
equivariant cohomology, 33
Euler class, xi, 146, 184

fibration, 4
filtration, 17, 78, 80, 81
filtration refinement, 17
Finiteness, Repetition and Stability theorem, 9
free action of the symmetric group, vii
free C_d-space, 170

P. V. M. Blagojević et al., *Equivariant Cohomology of Configuration Spaces Mod 2*,
Lecture Notes in Mathematics 2282, https://doi.org/10.1007/978-3-030-84138-6

geometric braid, 2

height of an algebra, 98
height of an element, 98
homology of the unordered configuration space, vii, 19
homotopy groups of ordered configuration space, 4
Hopf algebra, 88, 92
Hopf ring, 88

integral cohomology ring, vii, 7
iterated loop space, 79

key lemma, 140
Künneth formula, 102, 111, 121, 126, 130, 152

lexicographic ordering on the coordinates, 5
lexicographic stratification, 11
little cubes, vii
little cubes epicycle embedding, xiv
little cubes epicycles space, xiv, 27
little cubes operad, 26, 76, 84, 87, 102, 111, 150, 167, 168, 171
little d-cube, 167

Mùi algebra, 174
Mùi invariants, 53, 57, 174
morphism between spectral sequences, 36, 41, 47, 49
morphism of operads, 165
morphism of $\mathcal{O}$-spaces, 166

normalizer of a subgroup, xiii, 191

operad, 163, 164, 166, 167
ordered configuration space, vii, xiv, 1, 3, 13, 23, 64
orthogonal group, xiii

partial wreath square, 142, 179, 181
Poincaré duality, 11
Poincaré polynomial, 7
Pontryagin ring, 79
presentation of the braid group, 3
product filtration, 18
Ptolemaic epicycles embedding, xiv, 24, 29
Ptolemaic epicycles space, xiv, 64
pure braid group, xiii, 3, 4, 7

restriction homomorphism, 52, 54, 57, 59, 64, 72, 188, 189

Serre spectral sequence, 7, 15, 36, 38, 39, 41, 43–46, 67, 182, 183
singular cochain complex, 13
space of ordered n-tuples of interior disjoint little d-cubes, xiv
Stiefel manifold, 5, 96
Stiefel–Whitney classes, xi, 54, 58–61, 64, 71, 97, 100–103, 106, 107, 110–114, 116, 120, 125, 128, 130, 139, 141, 143, 145, 146, 148–150, 152, 155, 157–159, 177, 182–184
strings of the braid, 2
Sylow p-subgroup, 15, 187, 188
Sylow 2-subgroup, xiii, 64, 173, 193
symmetric group, xiii, 23, 31, 59, 64, 84, 96, 164, 187, 188

Thom class, 184, 185
Thom isomorphism theorem, 19
Thom space, 19, 78
transfer homomorphism, 64

unitary group, xiii
Universal Coefficient theorem, 48
unordered configuration space, vii, viii, xi, xiv, 1, 9, 11, 13, 16, 82

Vanishing Theorem, 15
vector bundle, xv, 54, 58–60, 78, 96, 100–103, 106, 110, 114, 116, 120, 125, 130, 133, 137–139, 141, 143, 146, 149, 152, 157, 158, 176, 179–181, 183

waist of a sphere, xi
Weyl group, xiii, 53, 191–193
wreath square, 179–181

LECTURE NOTES IN MATHEMATICS

Editorial Policy

1. Lecture Notes aim to report new developments in all areas of mathematics and their applications – quickly, informally and at a high level. Mathematical texts analysing new developments in modelling and numerical simulation are welcome.
   Manuscripts should be reasonably self-contained and rounded off. Thus they may, and often will, present not only results of the author but also related work by other people. They may be based on specialised lecture courses. Furthermore, the manuscripts should provide sufficient motivation, examples and applications. This clearly distinguishes Lecture Notes from journal articles or technical reports which normally are very concise. Articles intended for a journal but too long to be accepted by most journals, usually do not have this "lecture notes" character. For similar reasons it is unusual for doctoral theses to be accepted for the Lecture Notes series, though habilitation theses may be appropriate.
2. Besides monographs, multi-author manuscripts resulting from SUMMER SCHOOLS or similar INTENSIVE COURSES are welcome, provided their objective was held to present an active mathematical topic to an audience at the beginning or intermediate graduate level (a list of participants should be provided).

   The resulting manuscript should not be just a collection of course notes, but should require advance planning and coordination among the main lecturers. The subject matter should dictate the structure of the book. This structure should be motivated and explained in a scientific introduction, and the notation, references, index and formulation of results should be, if possible, unified by the editors. Each contribution should have an abstract and an introduction referring to the other contributions. In other words, more preparatory work must go into a multi-authored volume than simply assembling a disparate collection of papers, communicated at the event.
3. Manuscripts should be submitted either online at www.editorialmanager.com/lnm to Springer's mathematics editorial in Heidelberg, or electronically to one of the series editors. Authors should be aware that incomplete or insufficiently close-to-final manuscripts almost always result in longer refereeing times and nevertheless unclear referees' recommendations, making further refereeing of a final draft necessary. The strict minimum amount of material that will be considered should include a detailed outline describing the planned contents of each chapter, a bibliography and several sample chapters. Parallel submission of a manuscript to another publisher while under consideration for LNM is not acceptable and can lead to rejection.
4. In general, **monographs** will be sent out to at least 2 external referees for evaluation.

   A final decision to publish can be made only on the basis of the complete manuscript, however a refereeing process leading to a preliminary decision can be based on a pre-final or incomplete manuscript.

   Volume Editors of **multi-author works** are expected to arrange for the refereeing, to the usual scientific standards, of the individual contributions. If the resulting reports can be forwarded to the LNM Editorial Board, this is very helpful. If no reports are forwarded or if other questions remain unclear in respect of homogeneity etc, the series editors may wish to consult external referees for an overall evaluation of the volume.

5. Manuscripts should in general be submitted in English. Final manuscripts should contain at least 100 pages of mathematical text and should always include

   – a table of contents;
   – an informative introduction, with adequate motivation and perhaps some historical remarks: it should be accessible to a reader not intimately familiar with the topic treated;
   – a subject index: as a rule this is genuinely helpful for the reader.
   – For evaluation purposes, manuscripts should be submitted as pdf files.

6. Careful preparation of the manuscripts will help keep production time short besides ensuring satisfactory appearance of the finished book in print and online. After acceptance of the manuscript authors will be asked to prepare the final LaTeX source files (see LaTeX templates online: https://www.springer.com/gb/authors-editors/book-authors-editors/manuscriptpreparation/5636) plus the corresponding pdf- or zipped ps-file. The LaTeX source files are essential for producing the full-text online version of the book, see http://link.springer.com/bookseries/304 for the existing online volumes of LNM). The technical production of a Lecture Notes volume takes approximately 12 weeks. Additional instructions, if necessary, are available on request from lnm@springer.com.
7. Authors receive a total of 30 free copies of their volume and free access to their book on SpringerLink, but no royalties. They are entitled to a discount of 33.3 % on the price of Springer books purchased for their personal use, if ordering directly from Springer.
8. Commitment to publish is made by a *Publishing Agreement*; contributing authors of multiauthor books are requested to sign a *Consent to Publish form*. Springer-Verlag registers the copyright for each volume. Authors are free to reuse material contained in their LNM volumes in later publications: a brief written (or e-mail) request for formal permission is sufficient.

Addresses:
Professor Jean-Michel Morel, CMLA, École Normale Supérieure de Cachan, France
E-mail: moreljeanmichel@gmail.com

Professor Bernard Teissier, Equipe Géométrie et Dynamique,
Institut de Mathématiques de Jussieu – Paris Rive Gauche, Paris, France
E-mail: bernard.teissier@imj-prg.fr

Springer: Ute McCrory, Mathematics, Heidelberg, Germany,
E-mail: lnm@springer.com

Printed in the United States
by Baker & Taylor Publisher Services

Printed in the United States
by Baker & Taylor Publisher Services